S0-BSI-097

Real-
Life
Math

Real-Life Math

Volume 2: M–Z

Field of Application Index
General Index

K. Lee Lerner & Brenda Wilmoth Lerner,
Editors

THOMSON
GALE™

Detroit • New York • San Francisco • San Diego • New Haven, Conn. • Waterville, Maine • London • Munich

Real-Life Math

K. Lee Lerner and Brenda Wilmoth Lerner, Editors

Project Editor
Kimberley A. McGrath

Editorial
Luann Brennan, Meggin M. Condino, Madeline Harris, Paul Lewon, Elizabeth Manar

Editorial Support Services
Andrea Lopeman

Indexing
Factiva, a Dow Jones & Reuters Company

Rights and Acquisitions
Margaret Abendroth, Timothy Sisler

Imaging and Multimedia
Lezlie Light, Denay Wilding

Product Design
Pamela Galbreath, Tracey Rowens

Composition
Evi Seoud, Mary Beth Trimper

Manufacturing
Wendy Blurton, Dorothy Maki

For more information, contact:
Thomson Gale
27500 Drake Rd.
Farmington Hills, MI 48331-3535
Or you can visit our Internet site at
http://www.gale.com

LIBRARY OF CONGRESS CATALOGING-IN-PUBLICATION DATA

Real-life math / K. Lee Lerner and Brenda Wilmoth Lerner, editors.
p. cm.
Includes bibliographical references and index.
ISBN 0-7876-9422-3 (set : hardcover: alk. paper)—
ISBN 0-7876-9423-1 (v. 1)—ISBN 0-7876-9424-X (v. 2)
1. Mathematics—Encyclopedias.
I. Lerner, K. Lee. II. Lerner, Brenda Wilmoth.

QA5.R36 2006
510'.3—dc22 2005013141

This title is also available as an e-book, ISBN 1414404999 (e-book set).
ISBN: 0-7876-9422-3 (set); 0-7876-9423-1 (v1); 0-7876-9424-X (v2)
Contact your Gale sales representative for ordering information.

Printed in the United States of America
10 9 8 7 6 5 4 3 2 1

Table of Contents

Volume 1: A–L

Volume 2: M–Z

Entries (With Areas of Discussion)

Addition

Algebra

Algorithms

Architectural Math

Area

Average

Base

Business Math

Calculator Math

Calculus

Calendars

Cartography

Charts

Computers and Mathematics

Conversions

Coordinate Systems

Decimals

Demographics

Discrete Mathematics

Division

Domain and Range

Elliptic Functions

Estimation

Exponents

Factoring

Financial Calculations, Personal

Fractals

Fractions

Functions

Game Math

Game Theory

Geometry

Graphing

Imaging

Information Theory

Inverse

Iteration

Linear Mathematics

Logarithms

Logic

Matrices and Determinants

Measurement

Medical Mathematics

Modeling

Multiplication

Music and Mathematics

Nature and Numbers

Negative Numbers

Number Theory

Odds

Percentages

Perimeter

Perspective

Photography Math

Plots and Diagrams

Powers

Prime Numbers

Probability

Proportion

Quadratic, Cubic, and Quartic Equations

Ratio

Rounding

Rubric

Sampling

Scale

Scientific Math

Scientific Notation

Sequences, Sets, and Series

Sports Math

Square and Cube Roots

Statistics

Subtraction

Symmetry

Tables

Topology

Trigonometry

Vectors

Volume

Word Problems

Zero-Sum Games

Introduction

Real-Life Math takes an international perspective in exploring the role of mathematics in everyday life and is intended for high school age readers. As *Real-Life Math* (*RLM*) is intended for a younger and less mathematically experienced audience, the authors and editors faced unique challenges in selecting and preparing entries.

The articles in the book are meant to be understandable by anyone with a curiosity about mathematical topics. *Real-Life Math* is intended to serve all students of math such that an 8th- or 9th-grade student just beginning their study of higher maths can at least partially comprehend and appreciate the value of courses to be taken in future years. Accordingly, articles were constructed to contain material that might serve all students. For example, the article, "Calculus" is intended to be able to serve students taking calculus, students finished with prerequisites and about to undertake their study of calculus, and students in basic math or algebra who might have an interest in the practical utility of a far-off study of calculus. Readers should anticipate that they might be able to read and reread articles several times over the course of their studies in maths. *Real-Life Math* challenges students on multiple levels and is designed to facilitate critical thinking and reading-in-context skills. The beginning student is not expected to understand more mathematically complex text dealing, for example, with the techniques for calculus, and so should be content to skim through these sections as they read about the practical applications. As students progress through math studies, they will naturally appreciate greater portions of more advanced sections designed to serve more advanced students.

To be of maximum utility to students and teachers, most of the 80 topics found herein—arranged alphabetically by theory or principle—were predesigned to correspond to commonly studied fundamental mathematical concepts as stated in high school level curriculum objectives. However, as high school level maths generally teach concepts designed to develop skills toward higher maths of greater utility, this format sometimes presented a challenge with regard to articulating understandable or direct practical applications for fundamental skills without introducing additional concepts to be studied in more advanced math classes. It was sometimes difficult to isolate practical applications for fundamental concepts because it often required more complex mathematical concepts to most accurately convey the true relationship of mathematics to our advancing technology. Both the authors and editors of the project made exceptional efforts to smoothly and seamlessly incorporate the concepts necessary (and at an accessible level) within the text.

Although the authors of *Real-Life Math* include math teachers and professors, the bulk of the writers are practicing engineers and scientists who use math on a daily basis. However, *RLM* is not intended to be a book about real-life applications as used by mathematicians and scientists but rather, wherever possible, to illustrate and discuss applications within the experience—and that are understandable and interesting—to younger readers.

RLM is intended to maximize readability and accessibility by minimizing the use of equations, example problems, proofs, etc. Accordingly, *RLM* is not a math textbook, nor is it designed to fully explain the mathematics involved in each concept. Rather, *RLM* is intended to compliment the mathematics curriculum by serving a general reader for maths by remaining focused on fundamental math concepts as opposed to the history of math, biographies of mathematicians, or simply interesting applications. To be sure, there are inherent difficulties in presenting mathematical concepts without the use of mathematical notation, but the authors and editors of *RLM* sought to use descriptions and concepts instead of mathematical notation, problems, and proofs whenever possible.

To the extent that *RLM* meets these challenges it becomes a valuable resource to students and teachers of mathematics.

The editors modestly hope that *Real-Life Math* serves to help students appreciate the scope of the importance and influence of math on everyday life. *RLM* will achieve its highest purposes if it intrigues and inspires students to continue their studies in maths and so advance their understanding of the both the utility and elegance of mathematics.

"[The universe] cannot be read until we have learnt the language and become familiar with the characters in which it is written. It is written in mathematical language, and the letters are triangles, circles, and other geometrical figures, without which means it is humanly impossible to comprehend a single word." Galilei, Galileo (1564–1642)

K. Lee Lerner and Brenda Wilmoth Lerner, Editors

List of Advisors and Contributors

In compiling this edition, we have been fortunate in being able to rely upon the expertise and contributions of the following scholars who served as contributing advisors or authors for *Real-Life Math*, and to them we would like to express our sincere appreciation for their efforts:

William Arthur Atkins

Mr. Atkins holds a BS in physics and mathematics as well as an MBA. He lives at writes in Perkin, Illinois.

Juli M. Berwald, PhD

In addition to her graduate degree in ocean sciences, Dr. Berwald holds a BA in mathematics from Amherst College, Amherst, Massachusetts. She currently lives and writes in Chicago, Illinois.

Bennett Brooks

Mr. Brooks is a PhD graduate student in mathematics. He holds a BS in mathematics, with departmental honors, from University of Redlands, Redlands, California, and currently works as a writer based in Beaumont, California.

Rory Clarke, PhD

Dr. Clark is a British physicist conducting research in the area of high-energy physics at the University of Bucharest, Romania. He holds a PhD in high energy particle physics from the University of Birmingham, an MSc in theoretical physics from Imperial College, and a BSc degree in physics from the University of London.

Raymond C. Cole

Mr. Cole is an investment banking financial analyst who lives in New York. He holds an MBA from the Baruch Zicklin School of Business and a BS in business administration from Fordham University.

Bryan Thomas Davies

Mr. Davies holds a Bachelor of Laws (LLB) from the University of Western Ontario and has served as a criminal prosecutor in the Ontario Ministry of the Attorney General. In addition to his legal experience, Mr. Davies is a nationally certified basketball coach.

John F. Engle

Mr. Engle is a medical student at Tulane University Medical School in New Orleans, Louisiana.

William J. Engle
Mr. Engle is a retired petroleum engineer who lives in Slidell, Louisiana.

Paul Fellows
Dr. Fellows is a physicist and mathematician who lives in London, England.

Renata A. Ficek
Ms. Ficek is a graduate mathematics student at the University of Queensland, Australia.

Larry Gilman, PhD
Dr. Gilman holds a PhD in electrical engineering from Dartmouth College and an MA in English literature from Northwestern University. He lives in Sharon, Vermont.

Amit Gupta
Mr. Gupta holds an MS in information systems and is managing director of Agarwal Management Consultants P. Ltd., in Ahmedabad, India.

William C. Haneberg, PhD
Dr. Haneberg is a professional geologist and writer based in Seattle, Washington.

Bryan D. Hoyle, PhD
Dr. Hoyle is a microbiologist and science writer who lives in Halifax, Nova Scotia, Canada.

Kenneth T. LaPensee, PhD
In addition to professional research in epidemiology, Dr. LaPensee directs Skylands Healthcare Consulting located in Hampton, New Jersey.

Holly F. McBain
Ms. McBain is a science and math writer who lives near New Braunfels, Texas.

Mark H. Phillips, PhD
Dr. Phillips serves as an assistant professor of management at Abilene Christian University, located in Abilene, Texas.

Nephele Tempest
Ms. Tempest is a writer based in Los Angeles, California.

David Tulloch
Mr. Tulloch holds a BSc in physics and an MS in the history of science. In addition to research and writing he serves as a radio broadcaster in Ngaio, Wellington, New Zealand.

James A. Yates
Mr. Yates holds a MMath degree from Oxford University and is a teacher of maths in Skegnes, England.

ACKNOWLEDGMENTS

The editors would like to extend special thanks to Connie Clyde for her assistance in copyediting. The editors also wish to especially acknowledge Dr. Larry Gilman for his articles on calculus and exponents as well as his skilled corrections of the entire text. The editors are profoundly grateful to their assistant editors and proofreaders, including Lynn Nettles and Bill Engle, who read and corrected articles under the additional pressures created by evacuations mandated by Hurricane Katrina. The final editing of this book was interrupted as Katrina damaged the Gulf Coast homes and offices of several authors, assistant editors, and the editors of *RLM* just as the book was being prepared for press. Quite literally, many pages were read and corrected by light produced by emergency generators—and in some cases, pages were corrected from evacuation shelters. The editors are forever grateful for the patience and kind assistance of many fellow scholars and colleagues during this time.

The editors gratefully acknowledge the assistance of many at Thompson Gale for their help in preparing *Real-Life Math.* The editors wish to specifically thank Ms. Meggin Condino for her help and keen insights while launching this project. The deepest thanks are also offered to Gale Senior Editor Kim McGrath for her tireless, skilled, good-natured, and intelligent guidance.

Matrices and Determinants

Overview

In mathematics, a matrix is a group of numbers that have been arranged in a rectangle. The word for more than one matrix is matrices. The mathematics of handling matrices is called matrix algebra or linear algebra. Matrices are one of the most widely applied of all mathematical tools. They are used to solve problems in the design of machines, the layout by oil and trucking companies of efficient shipping routes, the playing of competitive "games" in war and business, mapmaking, earthquake prediction, imaging the inside of the body, prediction of both short-term weather and global climate change, and thousands of other purposes.

Fundamental Mathematical Concepts and Terms

Matrices are usually printed with square brackets around them. The matrix depicted in Figure 1 contains four numbers or "elements."

A column in a matrix is a vertical stack of numbers: in this matrix, 3 and 7 form the first column and 5 and 4 form the second. A row in a matrix is a horizontal line of numbers: in this matrix, 3 and 5 form the first row and 7 and 4 form the second. Matrices are named by how tall and wide they are. In this example the matrix is two elements tall and two elements wide, so it is a 2×2 matrix. The matrix in Figure 2 is 3 elements tall and five elements wide, so it is a 3×5 matrix.

A flat matrix that could be written on the squares of a chessboard, like these two examples, is called "two dimensional" because we need two numbers to say where each element of the matrix is. For instance, the number "10" in the 3×5 matrix would be indicated "row 2, column 4." A matrix can also be three-dimensional: in this case, numbers are arranged as if on the squares of a stack of chessboards, and to point to a particular number you have to name its row, its column, and which board in the stack it is on. There is no limit to the number of dimensions that a matrix can have. We cannot form mental pictures of matrices with four, five, or more dimensions, but they are just as mathematically real.

The numbers in a matrix can stand for anything. They might stand for the brightnesses of the dots in an image, or for the percentages of spotted owls in various age groups that survive to older ages. In one of the most important practical uses of matrices, the numbers in the matrix stand for the coefficients of linear equations. A linear equation is an equation that consists of a sum of variables (unknown numbers), each multiplied by a coefficient (a known number). Here are two linear equations: $2x + 3y = 11$ and $7x + 9y = 0$. Where the variables are x and y and the coefficients are the numbers that multiply them (namely 2, 3, 7, and 9). Together, these two equations form a "system." This system of equations can also be written as a 2×2 matrix times a 2×1 matrix (or "vector"), set equal to a second 2×1 matrix, as depicted in Figure 3.

The information that is in the matrix equation is also in the original system of equations, and is in almost the same arrangement on the paper. The only thing that has changed is the way the information is written down. For very large systems of equations (with tens or hundreds of variables, not just x and y), matrix equations are much more efficient.

Say that we wish to solve this matrix equation. This means we want to find a value of x and a value of y for which the equation is true. In this case, the only solution is

$$x = -33, y = \frac{77}{3}$$

(If you try these values for x and y in the equations $2x + 3y = 11$ and $7x + 9y = 0$, you'll find that both equations work out as true. No other values of x and y will work.) Finding solutions to matrix equations is one of the most important uses of computers in science, engineering, and business today, because thousands of practical problems can be described using systems of linear equations (sometimes very large systems, with matrices of many dimensions containing thousands or millions of numbers). Computers are solving larger matrix equations faster and faster, making many new products and scientific discoveries possible.

The rules for doing math with matrices, including solving matrix equations, are described by the field of mathematics called "matrix algebra." Matrices of the same size can be added, subtracted, or multiplied. One number that can be calculated from any square matrix—that is, any matrix that has the same number of rows as it has columns—is the determinant. Every square matrix has a determinant. The determinant is calculated by multiplying the elements of the matrix by each other and then adding the products according to a certain rule. For example, the rule for the determinant of a 2×2 matrix is as follows:

$$\begin{bmatrix} a_1 & b_1 \\ a_2 & b_2 \end{bmatrix} = a_1b_2 - a_2b_1$$

$$\begin{bmatrix} 3 & 5 \\ 7 & 4 \end{bmatrix}$$

Figure 1: A matrix with four numbers or "elements."

$$\begin{bmatrix} 5 & 9 & 4 & 2 & 5 \\ 3 & 1 & 0 & 10 & 2 \\ 9 & 9 & 5 & 2 & 7 \end{bmatrix}$$

Figure 2: A 3×5 matrix.

$$\begin{bmatrix} 2 & 3 \\ 7 & 9 \end{bmatrix} \times \begin{bmatrix} x \\ y \end{bmatrix} = \begin{bmatrix} 11 \\ 0 \end{bmatrix}$$

Figure 3.

(Here the small numbers attached to the variables are just labels to help us tell them apart). For a 3×3 matrix, the rule is more complicated:

$$\begin{bmatrix} a_1 & b_1 & c_1 \\ a_2 & b_2 & c_2 \\ a_3 & b_3 & c_3 \end{bmatrix} = a_1b_2c_3 + a_3b_1c_2 + a_2b_3c_1 - a_1b_3c_2 - a_2b_1c_3 - a_3b_2c_1$$

—and the rules get more and more complicated for larger matrices and higher dimensions. But that isn't a problem, because computers are good at calculating determinants.

Determinants are always studied by students learning matrix algebra, where they have many technical uses in matrix algebra. However, they are less important today in matrix theory than they were before the invention of computers. About a hundred years ago, a major mathematical reference work was published that merely summarized the properties of determinants that had been discovered up to that time: it filled four entire volumes. Today, mathematicians are less concerned with determinants than they once were. As one widely used textbook says, "After all, a single number can tell only so much about a matrix."

Real-life Applications

DIGITAL IMAGES

A digital camera produces a matrix of numbers when it takes a picture. The lens of the camera focuses an image

Key Terms

Matrix: A rectangular array of variables or numbers, often shown with square brackets enclosing the array. Here "rectangular" means composed of columns of equal length, not two-dimensional. A matrix equation can represent a system of linear equations.

Vector: A quantity consisting of magnitude and direction, usually represented by an arrow whose length represents the magnitude and whose orientation in space represents the direction.

on a flat rectangular surface covered with tiny light-sensitive electronic devices. The devices detect the color and brightness of the image in focus, and this information is saved as a matrix of numbers in the camera's memory. When a picture is downloaded from a camera to a computer and altered using image-editing software, it is subjected to mathematical manipulations described by matrix algebra. The picture may also be "compressed" so as to take up less computer memory or transmit over the Internet more quickly. When an image is compressed, the similarities between some of the numbers in its original matrix are used to generate a smaller matrix that takes up less memory but describes the same image with as little of image sharpness lost as possible.

FLYING THE SPACE SHUTTLE

In the early days of flight, pilots pushed and pulled on a joystick connected to wires. The wires ran over pulleys to the wings and rudder, which steered the plane. It would not be possible to fly as complex a craft as the Space Shuttle, which is steered not only by movable pieces of wing, but by 44 thruster jets, by directly mechanical means like these. Steering must be done by computer, in response to measurements of astronaut hand pressure on controls. In this method the flight computer combines measurements from sensors that detect how the ship is moving with measurements from the controls. These measurements are fed through the flight computers of the Shuttle as vectors, that is, as $n \times 1$ matrices, where the measurements from ship and pilot are the numbers in the vectors. The ship's computer performs calculations on these vectors using matrix algebra in order to decide how to move the control surfaces (moveable parts of the wing and tail) and how to fire the 44 steering jets.

POPULATION BIOLOGY

One of the things that biologists try to do is predict how populations of animals change in the wild. This is known as the study of population dynamics because in science or math, anything that is changing or moving is in a "dynamic" state. In population biology, a matrix equation describes how many members of a population shift from one stage of their reproductive life to the next, year to year. Such a matrix equation has appeared in the debate over whether the spotted owl of the Pacific Northwest (United States) is endangered or not. If the numbers of juvenile, subadult, and adult owls in year k are written as J_k, S_k, and A_k, respectively (where the small letters are labels to mark the year), and if the populations for the next year, year $k + 1$, are written as J_{k+1}, S_{k+1}, and A_{k+1}, then biologists have found that the following matrix equation relates one years' population to the next:

$$\begin{bmatrix} 0 & 0 & .33 \\ .18 & 0 & 0 \\ 0 & .71 & .94 \end{bmatrix} \begin{bmatrix} J_k \\ S_k \\ A_k \end{bmatrix} = \begin{bmatrix} J_{k+1} \\ S_{k+1} \\ A_{k+1} \end{bmatrix}$$

By analyzing this equation using advanced tools of matrix algebra such as eigenvalues, biologists have shown that if recent rates of decline of habitat loss (caused by clearcutting) continue, the spotted owl may be doomed to extinction. Owls, like all predators, need large areas of land in which to hunt—for spotted owls, about 4 square miles per breeding pair.

DESIGNING CARS

Before the 1970s, car makers designed new cars by making first drawings, then physical models, then the actual cars. Since the 1970s, they have also used a tool called computer-aided design (CAD). CAD is now taught in many high schools using software far more sophisticated than was available to the big auto makers in the beginning, but the principles are the same. In automotive CAD, the first step is still a drawing by an artist using their imagination—a design for how the car will look, often scrawled on paper. When a new image has been agreed on, the next step is the creation of a "wireframe" model. The wireframe model is a mass of lines, defined by numbers stored in matrices, that outline the shape of every major part of the car. The numbers specify the three-dimensional coordinates of enough points on the surface of the car to define its shape. The wireframe model may be created directly or by using lasers to scan a clay model in three dimensions. The wireframe car model is stored as a collection of many matrices, each describing one part. This model can be displayed, rotated, and adjusted for good looks. More importantly, by using matrix-based mathematical techniques called finite element methods, the car company can use the wireframe model to predict how the design will behave in crashes and how smoothly air will flow over it when it is in motion (which affects how much gas the car uses). These features can be experimentally improved by changing numbers in matrices rather than by building expensive test models.

Where to Learn More

Books

Lay, David C. *Linear Algebra and its Applications*, 2nd ed. New York: Addison-Wesley, 1999.

Strang, Gilbert. *Linear Algebra and its Applications*, 3rd ed. New York: Harcourt Brace Jovanovich College Publishers, 1988.

Measurement

Overview

Measurement is the quantifying of an exact physical value. The thing being measured is normally called a variable, because it can take different values in different circumstances.

The most common type of measurement is that of distance and time. As science and mathematics have developed, accuracy has increased dramatically in both microscopic and macroscopic scenarios. Without the ability to measure, many of the things that societies take for granted simply would not exist. Scientists can now accurately measure quantities ranging from the height of the tallest mountains, to the force of gravity on a distant moon through to the distance between stars light years away. The list is practically endless. Yet, the most remarkable aspect is that much of this can be achieved without a direct physical measurement. It is this theoretical aspect, along with the obvious practical consequences, that makes the area of measurement so fascinating.

Fundamental Mathematical Concepts and Terms

Distance is nearly always defined to be the shortest length between two points. If it were not so defined then there would be an infinite number of distances. To realize this, just plot two points and attempt to sketch all the different-length curves that exist between them. It would take literally forever.

Time is a much harder item to quantify. There are many philosophical discussions about the nature of time and its existence. In physics, time is the regular passing of the present to the future. This is perfectly adequate, until one starts to travel extremely fast. Believe it or not, time starts to slow down the faster one travels. This theory leads into a whole branch of science known as relativity.

The majority of science is also concerned with the measurement of forces, namely the interactions that exist between different objects. The most common forces are gravity and magnetism. The ability to measure these has immediate consequences in terms of space travel and electricity, respectively, as well as the interaction and motion of objects.

Finally, the measurements of speed, velocity, and acceleration have profound implications on travel. Velocity is defined to be the speed of an object with the direction of travel specified. Acceleration the rate at which velocity is changing. A large acceleration indicates that speed is increasing quickly.

Measurement can also be made of the geometrical aspects of objects. This has led to the development of angle-measuring techniques. Much of the development in navigation and exploration has come from the consequences of measuring geometry accurately.

A Brief History of Discovery and Development

The human race has been assigning measurements for thousands of years. An early unit was the cubit. This was defined to be the distance from a man's elbow to the end of his outstretched middle finger. However, as trade began to develop, the need for a standardized system became more important. It made little sense to have two workers producing planks of two cubits, when the two people would differently define a cubit. Inconsistencies would cause considerable problems in engineering projects.

It is acknowledged that the Babylonians were the first to standardize weight. This was particularly important in trade. They had specified stones of fixed weight and these were used to weigh and, hence, value precious gems and jewels. This led to the common terminology of stone, though even this had a variance; the stone used by a fisherman being half that used by a wool merchant. Yet, this wouldn't be a problem because the two trades were distinct. Definitions of mass and its measurement were especially important in the trade of gems.

It was King Edward I who first standardized the yard, a measurement used predominantly in Britain. There was an iron bar named the iron ulna from which all yardsticks were derived. This allowed for a standardized measurement. However, the metric system used today was first developed during the 1790s, Napoleon's time, by the French government. The meter was initially defined to be one ten-millionth of the distance from the North Pole to the equator passing through Paris.

It was not until 1832 that any legal standardized lengths existed in the United States. Indeed, it was a bill in 1866 that finally accepted the metric system. Finally, in 1975, Congress passed the Metric Conversion Act and so the metric system became the predominant measuring system in the United States.

Shortest Distance Between Two Cities

When looking for the distance between two cities in geography the temptation is to pick up a two-dimensional map, measure how far apart the cities are with a ruler, and use a suitable scale to evaluate this route. However, this method is completely erroneous. The best way to understand the following idea is to get a globe and actually test the following for your self. Pick two places of your choice. Get a piece of string and lay it on your flat map between the two cities marking where the cities are. Cut the string to this length. Apart from very few cases, you will be able to place the string on the globe so as to find a route that involves less string.

This is a simple example of spherical geometry. An analogous way to think about the difficulties is to imagine you are walking along when confronted by a hill. It is often the case that the shortest route passes around the hill, and not straight over the summit. The globe is a perfect form of a hill. There is a simple was to find the shortest route between two points. The globe can be cut into Great Circles. These are circles that cut the globe exactly in half. The Great Circle that passes through both cities provides the shortest route.

Real-life Applications

MEASURING DISTANCE

It is extremely easy to measure the distance between two objects if they are of close proximity and, perhaps more importantly, if travel can be done easily between these two points. Travel plans are based on such distances, and such factors to consider include buying gas or deciding whether to walk or take public transport.

All companies involved in travel need to know the distance between start and destination. Customers like to know how far they are going so that they can plan for their arrival. On a smaller scale, production companies need to be able to produce items of a specified length. It is impossible to build a building and yet be unable to specify the lengths of materials required. Or, in the production of parts in either an automobile of a plane, it would be rather difficult to manufacture such products if the pieces didn't fit together. It would be as useless as a jigsaw with incorrect pieces.

DIMENSIONS

Travel and objects do not exist in only one dimension. Instead, there are three aspects to most items: length, depth, and height. These are called the three dimensions. It is

Activity: Running Tracks

Most people would state that a running track is defined to be 400 meters. After all, in a 400-meter race they run once around the track. Yet, a confusing fact is that at the start of the race the runners all start spread out. Is it the case that runners on the outside lanes (of which there are normally eight) are given an unfair advantage?

The solution to this comes simply from careful mathematical thought. Indeed mathematics is required by designers to ensure that any running track meets defined standards. A multi-billion dollar Olympic industry depends on such accuracy. Consider the running track to be two straights and two half circles. It is quite clear that on the outer lane, the runner is passing along a greater circle. We can simplify the problem by joining the two half circles together to create one whole circle. If every lane is a fixed distance d apart, then as you progressively move to outer lanes, each runner would have to run an extra $2 \times \pi \times d$. This equates to an approximate distance of six meters per lane; hence the reason for the staggered starts.

therefore easy to specify the size of a brick required in construction by just stating the three measurements required.

Of course it is not always the most efficient way to specify an object. Circular laminas, thin plates, are often used in construction. In this case, the radius is a much better way of defining the shape.

Mathematicians often define different situations by using different systems of dimensions. This is done to make both the mathematics and its application in real life as simple as possible. Dimensions can also be distorted by the real world. For instance, if a person travels from New York to Los Angeles, can they actually travel in a straight line, even ignoring heights above sea level? The simple answer is no, because the actual world itself is curved to form a global shape. This raises a whole new area of mathematics called curvature, which explores how curved surfaces affect distances.

ACCURACY IN MEASUREMENT

Given a specified measurement in a specified dimension, the next task often required is to be able to do such a measurement accurately, often within a distance imperceptible to the naked eye. It is actually impossible to achieve perfect measurement, yet it is possible to get it to within specified bounds. These bound are often referred to as error bounds. The importance of the situation will limit the accuracy required. There is no point in spending millions of dollars to get a kitchen table an exact length to within on thousandth of a millimeter, yet it is essential that such money be spent when the context is building planes or space shuttles. If there were substantial errors, then the consequences could be disastrous.

Clearly such accuracy cannot be achieved using the simple ruler. It is virtually impossible to measure to within one millimeter without using magnification techniques. It was Sir Isaac Newton who, in 1672, stumbled upon a method called interferometry, which is related to the use of light to accurately measure at microscopic level.

EVALUATING ERRORS IN MEASUREMENT AND QUALITY CONTROL

Being aware of the error involved in real-life production can make calculations to work out the worst-case scenario possible. For instance, if a rod of 10 cm width is required to fit into a hole, a tight fit is not required. It's possible that the rod itself has errors. As an example, this could mean that it could be as wide as 10.05 cm or as low as 9.95 cm. Logic dictates that the hole needs to be larger than 10.05 cm for a guaranteed fit. This leads to the next possible problem: the hole was produced using a drill size that also has an error. If a drill with a size of 10.05 cm is used, there is always the chance that it will create a hole that is too small.

It is for all these reasons that mathematicians are employed to reduce the chance of waste and potential problems. Indeed, the problem often boils down to one involving statistics and probability; it is the job of quality control to reduce the chance of errors occurring, while maximizing profit and ensuring the equipment actually works as required.

ENGINEERING

Engineers are required to produce and measure important objects for complex design projects. It may take only one defective piece for the whole project to fail. These engineering works will often be integral parts of society. Most of the things taken for granted, such as water, food production, and good health, are direct consequences of engineering projects.

Measuring the Height of Everest

It was during the 1830s that the Great Trigonometrical Survey of The Indian sub-continent was undertaken by William Lambdon. This expedition was one of remarkable human resilience and mathematical application. The aim was to accurately map the huge area, including the Himalayans. Ultimately, they wanted not only the exact location of the many features, but to also evaluate the height above sea level of some of the world's tallest mountains, many of which could not be climbed at that time. How could such a mammoth task be achieved?

Today, it is relatively easy to use trigonometry to estimate how high an object stands. Then, if the position above sea level is known, it takes simple addition to work out the object's actual height compared to Earth's surface. Yet, the main problem for the surveyors in the 1830s was that, although they got within close proximity of the mountains and hence estimated the relative heights, they did not know how high they were above sea level. Indeed they were many hundreds of miles from the nearest ocean.

The solution was relatively simple, though almost unthinkable. Starting at the coast the surveyors would progressively work their way across the vast continent, continually working out heights above sea level of key points on the landscape. This can be referred to in mathematics as an inductive solution. From a simple starting point, repetitions are made until the final solution is found. This method is referred to as triangulation because the key points evaluated formed a massive grid of triangles. In this specific case, this network is often referred to as the great arc.

Eventually, the surveyors arrived deep in the Himalayas and readings from known places were taken; the heights of the mountains were evaluated without even having to climb them! It was during this expedition that a mountain, measured by a man named Waugh, was recorded as reaching the tremendous height of 29,002 feet (recently revised; 8,840 m). That mountain was dubbed Everest, after a man named George Everest who had succeeded Lambdon halfway through the expedition. George Everest never actually saw the mountain.

ARCHAEOLOGY

Archaeology is the study of past cultures, which is important in understanding how society may progress in the future. It can be extremely difficult to explore ancient sites and extract information due to the continual shifting and changing of the surface of the earth. Very few patches of ground are ever left untouched over the years.

While exploring ancient sites, it is important to be able to make accurate representations of the ground. Most items are removed to museums, and so it is important to retain a picture of the ground as originally discovered. A mathematical technique is employed to do so accurately. The distance and depth of items found are measured and recorded, and a map is constructed of the relative positions. Accurate measurements are essential for correct deductions to be made about the history of the site.

ARCHITECTURE

The fact that the buildings we live in will not suddenly fall to the ground is no coincidence. All foundations and structures from reliable architects are built on strict principles of mathematics. They rely upon accurate construction and measurement. With the pressures of deadlines, it is equally important that materials with insufficient accuracy within their measurements are not used.

COMPUTERS

The progression of computers has been quite dramatic. Two of the largest selling points within the computer industry are memory and speed. The speed of a computer is found by measuring the number of calculations that it can perform per second.

BLOOD PRESSURE

When checking the health of a patient, one of the primary factors considered is the strength of the heart, and how it pumps blood throughout the body. Blood pressure measurements reveal how strongly the blood is pumped and other health factors. An accurate measure of blood pressure could ultimately make the difference between life and death.

DOCTORS AND MEDICINE

Doctors are required to perform accurate measurements on a day-to-day basis. This is most evident during surgery where precision may be essential. The administration of drugs is also subject to precise controls. Accurate amounts of certain ingredients to be prescribed could determine the difference between life and death for the patient.

Doctors also take measurements of patients' temperature. Careful monitoring of this will be used to assess the recovery or deterioration of the patient.

CHEMISTRY

Many of the chemicals used in both daily life and in industry are produced through careful mixture of required substances. Many substances can have lethal consequences if mixed in incorrect doses. This will often require careful measurement of volumes and masses to ensure correct output.

Much of science also depends on a precise measurement of temperature. Many reactions or processes require an optimal temperature. Careful monitoring of temperatures will often be done to keep reactions stable.

NUCLEAR POWER PLANTS

For safety reasons, constant monitoring of the output of power plants is required. If too much heat or dangerous levels of radiation are detected, then action must be taken immediately.

MEASURING TIME

Time drives and motivates much of the activity across the globe. Yet it is only recently that we have been able to measure this phenomenon and to do so consistently. The nature of the modern world and global trade requires the ability to communicate and pass on information at specified times without error along the way.

The ancients used to use the Sun and other celestial objects to measure time. The sundial gave an approximate idea for the time of the day by using the rotation of the Sun to produce a shadow. This shadow then pointed towards a mark/time increment. Unfortunately, the progression of the year changes the apparent motion of the Sun. (Remember, though, that it is due to the change in Earth's orbit around the Sun, not the Sun moving around Earth.) This does not allow for accurate increments such as seconds.

It was Huygens who developed the first pendulum clock. This uses the mathematical principal that the length of a pendulum dictates the frequency with which the pendulum oscillates. Indeed a pendulum of approximately 39 inches will oscillate at a rate of one second. The period of a pendulum is defined to be the time taken for it to do a complete swing to the left, to the right, and back again. These however were not overly accurate, losing many minutes across one day. Yet over time, the accuracy increased.

It was the invention of the quartz clock that allowed much more accurate timekeeping. Quartz crystals vibrate (in a sense, mimicking a pendulum) and this can be utilized in a wristwatch. No two crystals are alike, so there is some natural variance from watch to watch.

THE DEFINITION OF A SECOND

Scientists have long noted that atoms resonate, or vibrate. This can be utilized in the same way as pendulums. Indeed, the second is defined from an atom called cesium. It oscillates at exactly 9,192,631,770 cycles per second.

MEASURING SPEED, SPACE TRAVEL, AND RACING

In a world devoted to transport, it is only natural that speed should be an important measurement. Indeed, the quest for faster and faster transport drives many of the nations on Earth. This is particularly relevant in long-distance travel. The idea of traveling at such speeds that space travel is possible has motivated generations of filmmakers and science fiction authors. Speed is defined to be how far an item goes in a specified time. Units vary greatly, yet the standard unit is meters traveled per second. Once distance and time are measured, then speed can be evaluated by dividing distance by time.

All racing, whether it involves horses or racing cars, will at some stage involve the measuring of speed. Indeed, the most successful sportsperson will be the one who, overall, can go the fastest. This concept of overall speed is often referred to as average speed. For different events, average speed has different meanings.

A sprinter would be faster than a long-distance runner over 100 meters. Yet, over a 10,000-meter race, the converse would almost certainly be true. Average speed gives the true merit of an athlete over the relevant distance. The formula for average speed would be average speed = total distance/total time.

NAVIGATION

The ability to measure angles and distances is an essential ingredient in navigation. It is only through an accurate measurement of such variables that the optimal route can be taken. Most hikers rely upon an advanced knowledge of bearings and distances so that they do not become lost. The same is of course true for any company involved in transportation, most especially those who travel by airplane or ship. There are no roads laid out for

To make a fair race, the tracks must be perfectly spaced. RANDY FARIS/CORBIS.

them to follow, so ability to measure distance and direction of travel are essential.

SPEED OF LIGHT

It is accepted that light travels at a fixed speed through a vacuum. A vacuum is defined as a volume of space containing no matter. Space, once an object has left the atmosphere, is very close to being such. This speed is defined as the speed of light and has a value close to 300,000 kilometers per second.

HOW ASTRONOMERS AND NASA MEASURE DISTANCES IN SPACE

When it comes to the consideration of space travel, problems arise. The distances encountered are so large that if we stick to conventional terrestrial units, the numbers become unmanageable. Distances are therefore expressed as light years. In other words, the distance between two celestial objects is defined to be the time light would take to travel between the two objects.

SPACE TRAVEL AND TIMEKEEPING

The passing of regular time is relied upon and trusted. We do not expect a day to suddenly turn into a year, though psychologically time does not always appear to pass regularly. It has been observed and proven using a branch of mathematics called relativity that, as an object accelerates, so the passing of time slows down for that particular object.

An atomic clock placed on a spaceship will be slightly behind a counterpart left on Earth. If a person could actually travel at speeds approaching the speed of light, they would only age by a small amount, while people on Earth would age normally.

Indeed, it has also been proven mathematically that a rod, if moved at what are classed as relativistic velocities (comparable to the speed of light), will shorten. This is known as the Lorentz contraction. Philosophically, this leads to the question, how accurate are measurements? The simple answer is that, as long as the person and the object are moving at the same speed, then the problem does not arise.

Distance in Three Dimensions

In mathematics it is important to be able to evaluate distance in all dimensions. It is often the case that only the coordinates of two points are known and the distance between them is required. For example, a length of rope needs to be laid across a river so that it is fully taut. There are two trees that have suitable branches to hold the rope on either side. The width of the river is 5 meters. The trees are 3 meters apart widthwise. One of the branches is 1 meter higher than the other. How much rope is required?

The rule is to use an extension of Pythagoras in three dimensions: $a^2 + b^2 = h^2$. An extension to this in three dimensions is: $a^2 + b^2 + c^2 = h^2$. This gives us width, depth, and height. Therefore, $5^2 + 3^2 + 1^2 = h^2 = 35$. Therefore h is just under 6. So at least 6 m of rope is needed to allow for the extra required for tying the knots.

WHY DON'T WE FALL OFF EARTH?

As Isaac Newton sat under a tree, an apple fell off and hit him upon the head. This led to his work on gravity. Gravity is basically the force, or interaction, between Earth and any object. This force varies with each object's mass and also varies as an object moves further away from the surface of Earth.

This variability is not a constant. The reason astronauts on the moon seem to leap effortlessly along is due to the lower force of gravity there. It was essential that NASA was able to measure the gravity on the moon before landing so that they could plan for the circumstances upon arrival.

How is gravity measured on the moon, or indeed anywhere without actually going there first? Luckily, there is an equation that can be used to work it out. This formula relies on knowing the masses of the objects involved and their distance apart.

MEASURING THE SPEED OF GRAVITY

Gravity has the property of speed. Earth rotates about the Sun due to the gravitational pull of the Sun. If the Sun were to suddenly vanish, Earth would continue its orbit until gravity actually catches up with the new situation. The speed of gravity, perhaps unsurprisingly, is the speed of light.

Stars are far away, and we can see them in the sky because their light travels the many light years to meet our retina. It is natural that, after a certain time, most stars end their life often undergo tremendous changes. Were a star to explode and vanish, it could take years for this new reality to be evident from Earth. In fact, some of the stars viewable today may actually have already vanished.

MEASURING MASS

A common theme of modern society is that of weight. A lot of television airplay and books, earning authors millions, are based on losing weight and becoming healthy. Underlying the whole concept of weighing oneself is that of gravity. It is actually due to gravity that an object can actually be weighed.

The weight of an object is defined to be the force that that object exerts due to gravity. Yet these figures are only relevant within Earth's gravity. Interestingly, if a person were to go to the top of a mountain, their measurable weight will actually be less than if they were at sea level. This is simply because gravity decreases the further away an object is from Earth's surface, and so scales measure a lower force from a person's body.

Potential applications

People will continue to take measurements and use them across a vast spectrum of careers, all derived from applications within mathematics. As we move into the future, the tools will become available to increase such measurements to remarkable accuracies on both microscopic and macroscopic levels.

Advancements in medicine and the ability to cure diseases may come from careful measurements within cells and how they interact. The ability to measure, and do so accurately, will drive forward the progress of human society.

Where to Learn More

Periodicals

Muir, Hazel. "First Speed of Gravity Measurement Revealed." *New Scientist.com.*

Web sites

Keay, John. "The Highest Mountain in the World." The Royal Geographical Society. 2003. <http://imagingeverest.rgs.org/Concepts/Virtual_Everest/-288.html> (February 26, 2005).

Medical Mathematics

Overview

Mathematics finds wide applications in medicine and public health. Epidemiology, the scientific discipline that investigates the causes and distribution of disease and that underlies public health practice, relies heavily on mathematical data and analysis. Mathematics is also a critical tool in clinical trials, the cornerstone of medical research supporting modern medical practice, which are used to establish the efficacy and safety of medical treatments. As medical technology and new treatments rely more and more on sophisticated biological modeling and technology, medical professionals will draw increasingly on their knowledge of mathematics and the physical sciences.

There are three major ways in which researchers and practitioners apply mathematics to medicine. The first and perhaps most important is that they must use the mathematics of probability and statistics to make predictions in complex medical situations. The most important example of this is when people try to predict the outcome of illnesses, such as AIDS, cancer, or influenza, in either individual patients or in population groups, given the means that they have to prevent or treat them.

The second important way in which mathematics can be applied to medicine is in modeling biological processes that underlie disease, as in the rate of speed with which a colony of bacteria will grow, the probability of getting disease when the genetics of Mendelian inheritance is known, or the rapidity with which an epidemic will spread given the infectivity and virulence of a pathogen such as a virus. Some of the most commercially important applications of bio-mathematical modeling have been developed for life and health insurance, in the construction of life tables, and in predictive models of health premium increase trend rates.

The third major application of mathematics to medicine lies in using formulas from chemistry and physics in developing and using medical technology. These applications range from using the physics of light refraction in making eyeglasses to predicting the tissue penetration of gamma or alpha radiation in radiation therapy to destroy cancer cells deep inside the body while minimizing damage to other tissues.

While many aspects of medicine, from medical diagnostics to biochemistry, involve complex and subtle applications of mathematics, medical researchers consider epidemiology and its experimental branch, clinical trials, to be the medical discipline for which mathematics is indispensable. Medical research, as furthered by these two disciplines, aims to establish the causes of disease and prove treatment efficacy and safety based on quantitative (numerical) and logical relationships among observed and recorded data. As such, they comprise the "tip of the iceberg" in the struggle against disease.

The mathematical concepts in epidemiology and clinical research are basic to the mathematics of biology, which is after all a science of complex systems that respond to many influences. Simple or nonstatistical mathematical relationships can certainly be found, as in Mendelian inheritance and bacterial culturing, but these are either the most simple situations or they exist only under ideal laboratory conditions or in medical technology that is, after all, based largely on the physical sciences. This is not to downplay their usefulness or interest, but simply to say that the budding mathematician or scientist interested in medicine has to come to grips with statistical concepts and see how the simple things rapidly get complicated in real life.

Noted British epidemiologist Sir Richard Doll (1912–) has referred to the pervasiveness of epidemiology in modern society. He observed that many people interested in preventing disease have unwittingly practiced epidemiology. He writes, "Epidemiology is the simplest and most direct method of studying the causes of disease in humans, and many major contributions have been made by studies that have demanded nothing more than an ability to count, to think logically and to have an imaginative idea."

Because epidemiology and clinical trials are based on counting and constitute a branch of statistical mathematics in their own right, they require a rather detailed and developed treatment. The presentation of the other major medical mathematics applications will feature explanations of the mathematics that underlie familiar biological phenomena and medical technologies.

Fundamental Mathematical Concepts and Terms

The most basic mathematical concepts in health care are the measures used to discover whether a statistical association exists between various factors and disease. These include rates, proportions, and ratios. Mortality (death) and morbidity (disease) rates are the "raw material" that researchers use in establishing disease causation. Morbidity rates are most usefully expressed in terms of disease incidence (the rate with which population or research sample members contract a disease) and prevalence (the proportion of the group that has a disease over a given period of time).

Beyond these basic mathematical concepts are concepts that measure disease risk. The population at risk is the group of people that could potentially contract a disease, which can range from the entire world population (e.g., at risk for the flu), to a small group of people with a certain gene (e.g., at risk for sickle-cell anemia), to a set of patients that are randomly selected to participate in groups to be compared in a clinical trial featuring alternative treatment modes. Finally, the most basic measure of a population group's risk for a disease is relative risk (the ratio of the prevalence of a disease in one group to the prevalence in another group).

The simplest measure of relative risk is the odds ratio, which is the ratio of the odds that a person in one group has a disease to the odds that a person in a second group has the disease. Odds are a little different from the probability that a person has a disease. One's odds for a disease are the ratio between the number of people that have a disease and the number of people that do not have the disease in a population group. The probability of disease, on the other hand, is the proportion of people that have a disease in a population. When the prevalence of disease is low, disease odds are close to disease probability. For example, if there is a 2%, or 0.02, probability that people in a certain Connecticut county will contract Lyme disease, the odds of contracting the disease will be $2/98 = 0.0204$.

Suppose that the proportion of Americans in a particular ethnic or age group (group 1) with type II diabetes in a given year is estimated from a study sample to be 6.2%, while the proportion in a second ethnic or age group (group 2) is 4.5%. The odds ratio (OR) between the two groups is then: $OR = (6.2/93.8)/(4.5/95.5) = 0.066/0.047 = 1.403$.

This means that the relative risk of people in group 1 developing diabetes compared to people in group 2 is 1.403, or over 40% higher than that of people in group 2.

The mortality rate is the ratio of the number of deaths in a population, either in total or disease-specific, to the total number of members of that population, and is usually given in terms of a large population denominator, so that the numerator can be expressed as a whole number. Thus in 1982 the number of people in the United States was 231,534,000, the number of deaths from all causes was 1,973,000, and therefore the death rate from all causes of 852.1 per 100,000 per year. That same year there were 1,807 deaths from tuberculosis, yielding a disease-specific mortality rate of 7.8 per million per year.

Assessing disease frequency is more complex because of the factors of time and disease duration. For example, disease prevalence can be assessed at a point in time (point prevalence) or over a period of time (period prevalence), usually a year (annual prevalence). This is the prevalence that is usually measured in illness surveys that are reported to the public. Researchers can also measure prevalence over an indefinite time period, as in the case of lifetime prevalence. Researchers calculate this time period by asking every person in the study sample whether or not they have ever had the disease, or by checking lifetime health records for everybody in the study sample for the occurrence of the disease, counting the occurrences, and then dividing by the number of people in the population.

The other critical aspect of disease frequency is incidence, which is the number of cases of a disease that occur in a given period of time. Incidence is an extremely critical statistic in describing the course of a fast-moving epidemic, in which medical decision-makers must know how quickly a disease is spreading. The incidence rate is the key to public health planning because it enables officials to understand what the prevalence of a disease is likely to be in the future. Prevalence is mathematically related to the cumulative incidence of a disease over a period of time as well as the expected duration of a disease, which can be a week in the case of the flu or a lifetime in the case of juvenile onset diabetes. Therefore, incidence not only indicates the rate of new disease cases, but is the basis of the rate of change of disease prevalence.

For example, the net period prevalence of cases of disease that have persisted throughout a period of time is the proportion of existing cases at the beginning of that period plus the cumulative incidence during that period of time minus the cases that are cured, self-limited, or that die, all divided by the number of lives in the population at risk. Thus, if there are 300 existing cases, 150 new cases, 40 cures, and 30 deaths in a population of 10,000 in a particular year, the net period (annual) prevalence for that year is $(300 + 150 - 40 - 30) / 10{,}000 = 380/10{,}000 = 0.038$. The net period prevalence for the year in question is therefore nearly 4%.

A crucial statistical concept in medical research is that of the research sample. Except for those studies that have access to disease mortality, incidence, and prevalence rates for the entire population, such as the unique SEER (surveillance, epidemiology and end results) project that tracks all cancers in the United States, most studies use samples of people drawn from the population at risk either randomly or according to certain criteria (e.g., whether or not they have been exposed to a pathogen, whether or not they have had the disease, age, gender, etc.). The size of the research sample is generally determined by the cost of research. The more elaborate and detailed the data collection from the sample participants, the more expensive to run the study.

Medical researchers try to ensure that studying the sample will resemble studying the entire population by making the sample representative of all of the relevant groups in the population, and that everyone in the relevant population groups should have an equal chance of getting selected into the sample. Otherwise the sample will be biased, and studying it will prove misleading about the population in general.

The most powerful mathematical tool in medicine is the use of statistics to discover associations between death and disease in populations and various factors, including environmental (e.g., pollution), demographic (age and gender), biological (e.g., body mass index, or BMI), social (e.g., educational level), and behavioral (e.g., tobacco smoking, diet, or type of medical treatment), that could be implicated in causing disease.

Familiarity with basic concepts of probability and statistics is essential in understanding health care and clinical research and is one of the most useful types of knowledge that one can acquire, not just in medicine, but also in business, politics, and such mundane problems as interpreting weather forecasts.

A statistical association takes into account the role of chance. Researchers compare disease rates for two or more population groups that vary in their environmental, genetic, pathogen exposure, or behavioral characteristics, and observe whether a particular group characteristic is associated with a difference in rates that is unlikely to have occurred by chance alone.

How can scientists tell whether a pattern of disease is unlikely to have occurred by chance? Intuition plays a role, as when the frequency of disease in a particular population group, geographic area, or ecosystem is dramatically out of line with frequencies in other groups or settings. To confirm the investigator's hunches that some kind of statistical pattern in disease distribution is emerging, researchers use probability distributions.

Probability distributions are natural arrays of the probability of events that occur everywhere in nature. For example, the probability distribution observed when one flips a coin is called the binomial distribution, so-called because there are only two outcomes: heads or tails, yes or no, on or off, 1 or 0 (in binary computer language). In the binomial distribution, the expected frequency of heads and tails is 50/50, and after a sufficiently long series of coin flips or trials, this is indeed very close to the proportions of heads and tails that will be observed. In medical research, outcomes are also often binary, i.e., disease is

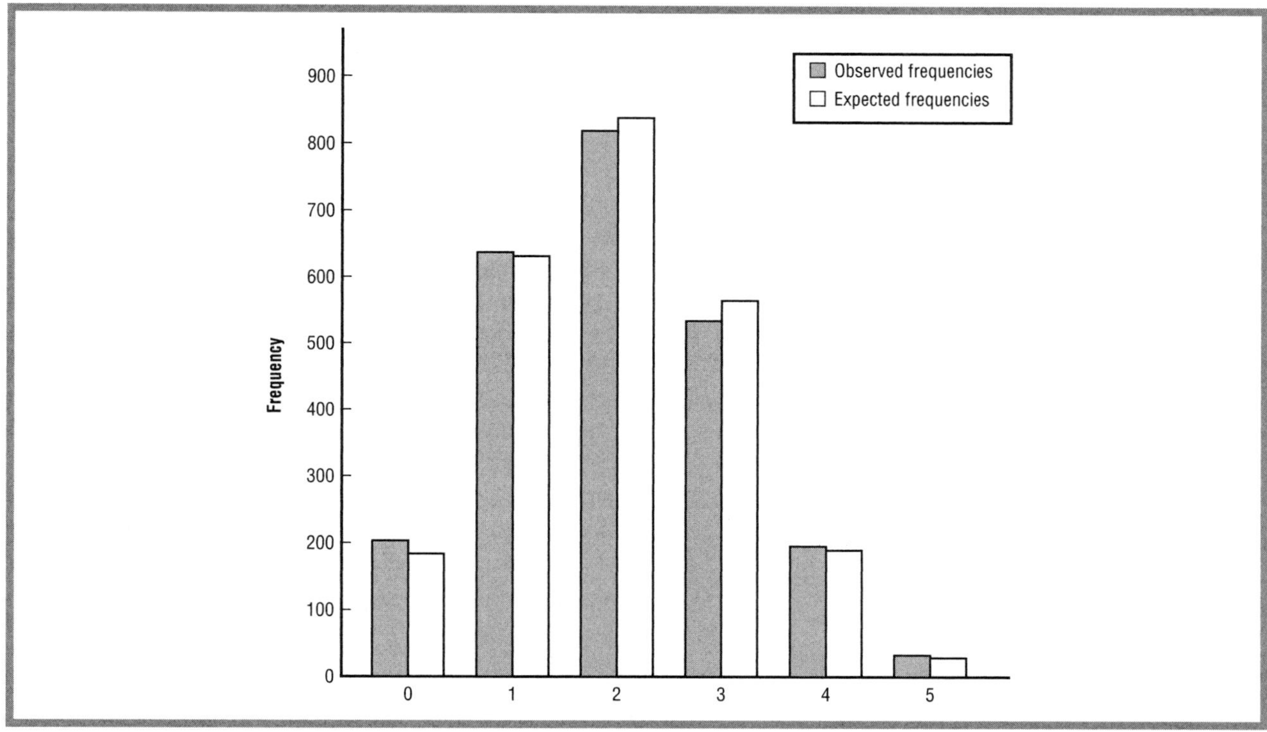

Figure 1: Binomial distribution.

present or absent, exposure to a virus is present or absent, the patient is cured or not, the patient survives or not.

However, people almost never see exactly 50/50, and the shorter the series of coin flips, the bigger the departure from 50/50 will likely be observed. The binomial probability distribution does all of this coin-flipping work for people. It shows that 50/50 is the expected odds when nothing but chance is involved, but it also shows that people can expect departures from 50/50 and how often these departures will happen over the long run. For example, a 60/40 odds of heads and tails is very unlikely if there are 30 coin tosses (18 heads, 12 tails), but much more likely if one does only five coin tosses (e.g., three heads, two tails). Therefore, statistics books show binomial distribution tables by the number of trials, starting with n = 5, and going up to n = 25. The binomial distribution for ten trials is a "stepwise," or discrete distribution, because the probabilities of various proportions jump from one value to another in the distribution. As the number of trials gets larger, these jumps get smaller and the binomial distribution begins to look smoother. Figure 1 provides an illustration of how actual and expected outcomes might differ under the binomial distribution.

Beyond n = 30, the binomial distribution becomes very cumbersome to use. Researchers employ the normal distribution to describe the probability of random events in larger numbers of trials. The binomial distribution is said to approach the normal distribution as the number of trials or measurements of a phenomenon get higher. The normal distribution is represented by a smooth bell curve. Both the binomial and the normal distributions share in common that the expected odds (based on the mean or average probability of 0.5) of "on-off" or binary trial outcomes is 50/50 and the probabilities of departures from 50/50 decrease symmetrically (i.e., the probability of 60/40 is the same as that of 40/60). Figure 2 provides an illustration of the normal distribution, along with its cumulative S-curve form that can be used to show how random occurrences might mount up over time.

In Figure 2, the expected (most frequent) or mean value of the normal distribution, which could be the average height, weight, or body mass index of a population group, is denoted by the Greek letter μ, while the standard deviation from the mean is denoted by the Greek letter σ. Almost 70% of the population will have a measurement that is within one standard deviation

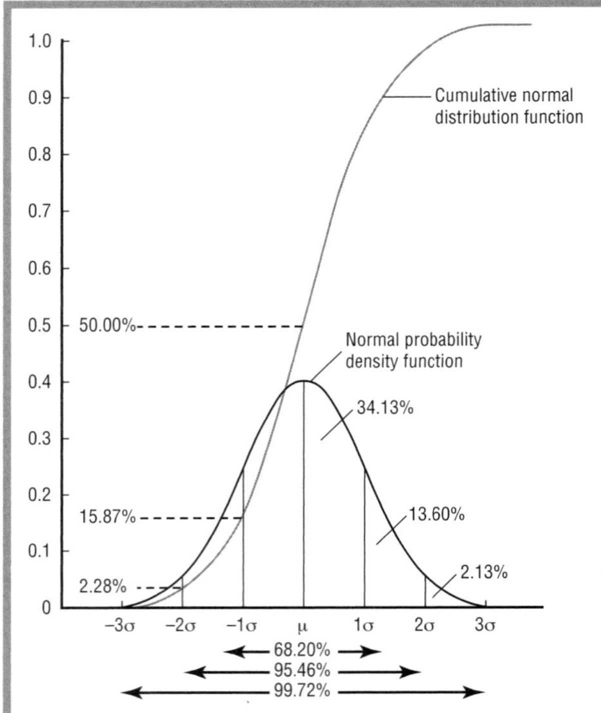

Figure 2: Population height and weight.

from the mean; on the other hand, only about 5% will have a measurement that is more than two standard deviations from the mean. The low probability of such measurements has led medical researchers and statisticians to posit approximately two standard deviations as the cutoff point beyond which they consider an occurrence to be significantly different from average because there is only a one in 20 chance of its having occurred simply by chance.

The steepness with which the probability of the odds decreases as one continues with trials determines the width or variance of the probability distribution. Variance can be measured in standardized units, called standard deviations. The further out toward the low probability tails of the distribution the results of a series of trials are, the more standard deviations from the mean, and the more remarkable they are from the investigator's standpoint. If the outcome of a series of trials is more than two standard deviations from the mean outcome, it will have a probability of 0.05 or one chance in 20. This is the cutoff, called the alpha (α) level beyond which researchers usually judge that the outcome of a series of trials could not have occurred by chance alone. At that point they begin to consider that one or more factors are causing the observed pattern. For example, if the frequency pattern of disease is similar to the frequencies of age, income, ethnic groups, or other features of population groups, it is usually a good bet that these characteristics of people are somehow implicated in causing the disease, either directly or indirectly.

The normal distribution helps disease investigators decide whether a set of odds (e.g., 10/90) or a probability of 10% of contracting a disease in a subgroup of people that behave differently from the norm (e.g., alcoholics) is such a large deviation (usually, more than two standard deviations) from the expected frequency that the departure exceeds the alpha level of a probability of 0.05. This deviation would be considered to be statistically significant. In this case, a researcher would want to further investigate the effect of the behavioral difference. Whether or not a particular proportion or disease prevalence in a subgroup is statistically significant depends on both the difference from the population prevalence as well as the number of people studied in the research sample.

Real-life Applications

VALUE OF DIAGNOSTIC TESTS

Screening a community using relatively simple diagnostic tests is one of the most powerful tools that health care professionals and public health authorities have in preventing disease. Familiar examples of screening include HIV testing to help prevent AIDS, cholesterol testing to help prevent heart disease, mammography to help prevent breast cancer, and blood pressure testing to help prevent stroke. In undertaking a screening program, authorities must always judge whether the benefits of preventing the illness in question outweigh the costs and the number of cases that have been mistakenly identified, called false positives.

Every diagnostic or screening test has four basic mathematical characteristics: sensitivity (the proportion of identified cases that are true cases), specificity (the proportion of identified non-cases that are true non-cases), positive predictive value (PV^+, the probability of a positive diagnosis if the case is positive), and negative predictive value (PV^-, the probability of a negative diagnosis if the case is negative). These values are calculated as follows. Let a = the number of identified cases that are real cases of the disease (true positives), b = the number of identified cases that are not real cases (false positives), c = the number of true cases that were not identified by the test (false negatives), and d = the number of individuals identified as non-cases that were true non-cases (true negatives). Thus, the number of true cases is a + c,

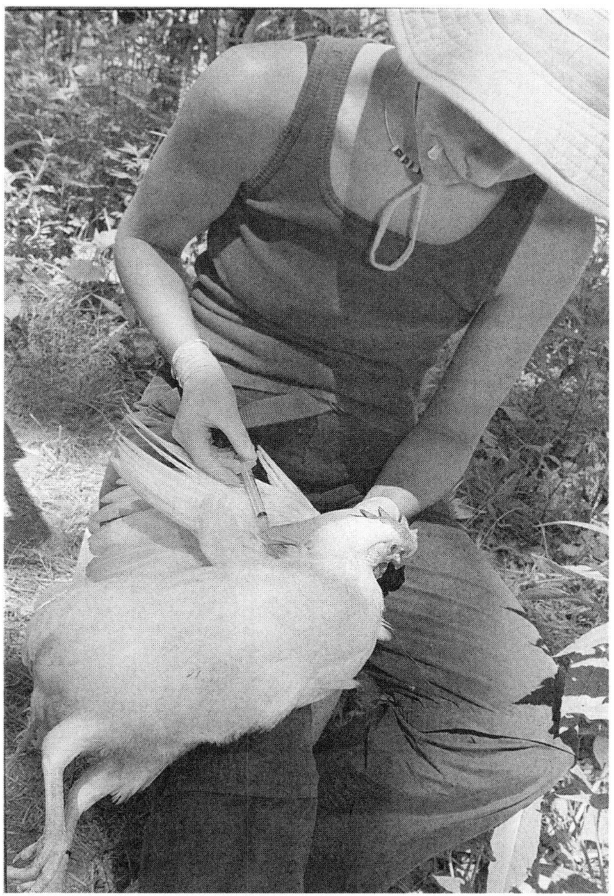

A researcher collects blood from a "sentinel" chicken from an area being monitored for the West Nile virus. FADEK TIMOTHY/CORBIS SYGMA.

the number of true non-cases is b + d, and the total number of cases is a + b + c + d. The four test characteristics or parameters are thus Sensitivity = a/a + b; Specificity = d/b + d; PV^+ = a/a + b; PV^- = d/c + d. These concepts are illustrated in Table 1 for a mammography screening study of nearly 65,000 women for breast cancer.

Calculating the four parameters of the screening test yields: Sensitivity = 132 / 177 = 74.6%; Specificity = 63,650 / 64, 633 = 98.5%; PV^+ = 132 / 1,115 = 11.8%; PV^- = 63,650 / 63,695 = 99.9%.

These parameters, especially the ability of the test to identify true negatives, make mammography a valuable prevention tool. However, the usefulness of the test is proportional to the disease prevalence. In this case, the disease prevalence is very low: (a + c)/(b + d) = 177/64,683 ≈ 0.003, and the positive predictive value is less than 12%. In other words, the actual cancer cases identified are a small minority of all of the positive cases. As the prevalence of breast cancer rises, as in older women, the proportion of actual cases rises. This makes the test much more cost effective when used on women over the age of 50 because the proportion of women that undergo expensive biopsies that do not confirm the mammography results is much lower than if mammography was administered to younger women or all women.

CALCULATION OF BODY MASS INDEX (BMI)

The body mass index (BMI) is often used as a measure of obesity, and is a biological characteristic of individuals that is strongly implicated in the development or etiology of a number of serious diseases, including diabetes and heart disease. The BMI is a person's weight, divided by his or her height squared: BMI = weight/$height^2$. For example, if a man is 1.8 m tall and weighs 85 kg, his body mass index is: 85 kg^2/1.8 m = 26.2. For BMIs over 26, the risk of diabetes and coronary artery disease is elevated, according to epidemiological studies. However, a more recent study has shown that stomach girth is more strongly related to diabetes risk than BMI itself, and BMI may not be a reliable estimator of disease risk for athletic people with more lean muscle mass than average.

STANDARD DEVIATION AND VARIANCE FOR USE IN HEIGHT AND WEIGHT CHARTS

Concepts of variance and the standard deviation are often depicted in population height and weight charts.

Suppose that the average height of males in a population is 1.9 meters. Investigators usually want to know more than just the average height. They might also like to know the frequency of other heights (1.8 m, 2.0 m, etc.). By studying a large sample, say 2,000 men from the population, they can directly measure the men's heights and calculate a convenient number called the sample's standard deviation, by which they could describe how close or how far away from the average height men in this population tend to be. To get this convenient number, the researchers simply take the average difference from the mean height. To do this, they would first sum up all of these differences or deviations from average, and then divide by the number of men measured. To use a simple example, suppose five men from the population are measured and their heights are 1.8 m, 1.75 m, 2.01 m, 2.0 m, and 1.95 m. The average or mean height of this small sample in meters = (1.8 + 1.75 + 2.01 + 2.0 + 1.95)/5 = 1.902. The difference of each man's height from the average height of the sample, or the deviation from average. The sample standard deviation is simply the average

Counting calories is a practice of real-life mathematics that can have a dramatic impact on health. A collection of menu items from opposite ends of the calorie spectrum including a vanilla shake from McDonald's (1,100 calories); a Cuban Panini sandwich from Ruby Tuesday's (1,164 calories), and a six-inch ham sub, left, from Subway (290 calories). All the information for these items is readily available at the restaurants that serve them. AP/WIDE WORLD PHOTOS. REPRODUCED BY PERMISSION.

of the deviations from the mean. The deviations are $1.8 - 1.902 = -0.102$, $1.75 - 1.902 = -0.152$, $2.01 - 1.902 = 0.108$, $2.0 - 1.902 = 0.008$, and $1.95 - 1.902 = 0.048$. Therefore, the average deviation for the sample is $(-1.02 - 0.152 + 0.108 + 0.008 + 0.048)/5 = -0.2016$ m.

However, this is a negative number that is not appropriate to use because the standard deviation is supposed to be a directionless unit, as is an inch, and because the average of all of the average deviations will not add up to the population average deviation. To get the sample standard deviation to always be positive, no matter which sample of individuals that is selected to be measured, and to ensure that it is a good estimator of the population average deviation, researchers go through additional steps. They sum up the squared deviations, calculate the average squared deviation (mean squared deviation), and take the square root of the sum of the squared deviations (the root mean squared deviation or RMS deviation). They then add a correction factor of -1 in the denominator.

So the sample standard deviation in the example is

$$s = \sqrt{\frac{(-.102)^2 + (-.152)^2 + (.108)^2 + (.008)^2 + (.048)^2}{4}} \cong 0.109$$

Note that the sample average of 1.902 m happens in this sample to be close to the known population average, denoted as μ, of 1.9 m. The sample standard deviations might or might not be close to the population standard deviation, denoted as σ. Regardless, the sample average and standard deviation are both called estimators of the population average and standard deviation. In order for any given sample average or standard deviation to be considered to be an accurate estimator for the population average and standard deviation, a small correction factor is applied to these estimators to take into account that a sample has already been drawn, which puts a small constraint (eliminates a degree of freedom) on the estimation of μ and σ for the population. This is done so that after many samples are examined, the mean of all the sample means and the average of all of the sample standard deviations approaches the true population mean and standard deviation.

GENETIC RISK FACTORS: THE INHERITANCE OF DISEASE

Nearly all diseases have both genetic (heritable) and environmental causes. For example, people of Northern European ancestry have a higher incidence of skin cancer from sun exposure in childhood than do people of Southern European or African ancestry. In this case, Northern Europeans' lack of skin pigment (melanin) is the heritable part, and their exposure to the sun to the point of burning, especially during childhood, is the environmental part. The proportion of risk due to inheritance and the proportion due to the environment are very difficult to figure out. One way is to look at twins who have the same genetic background, and see how often various environmental differences that they have experienced have resulted in different disease outcomes.

However, there is a large class of strictly genetic diseases for which predictions are fairly simple. These are diseases that involve dominant and recessive genes. Many genes have alternative genotypes or variants, most of which are harmful or deleterious. Each person receives

Real-life sensitivity and specificity in cancer screening

Screening test (mammography)	Cancer confirmed	Cancer not confirmed	Total
Positive	a = 132	b = 983	**a + b = 1,115**
Negative	c = 45	d = 63,650	**c + d = 63,695**
Total	**a + c = 177**	**b + d = 64,683**	**a + b + c + d = 64,810**

Table 1.

one of these gene variants from each parent, so he or she has two variants for each gene that vie for expression as one grows up. People express dominant genes when the variant contributed by one parent overrides expression of the other parent's variant (or when both parents have the same dominant variant). Some of these variants make the fetus a "non-starter," and result in miscarriage or spontaneous abortion. Other variants do not prevent birth and may not express disease until middle age. In writing about simple Mendelian inheritance, geneticists can use the notation AA to denote homozygous dominant (usually homozygous normal), Aa to denote heterozygous recessive, and aa to denote homozygous recessive.

One tragic example is that of Huntington's disease due to a dominant gene variant, in which the nervous system deteriorates catastrophically at some point after the age of 35. In this case, the offspring can have one dominant gene (Huntington's) and one normal gene (heterozygous dominant), or else can be homozygous dominant (both parents had Huntington's disease, but had offspring before they started to develop symptoms). Because Huntington's disease is caused by a dominant gene, the probability of the offspring developing the disease is 100%.

When a disease is due to a recessive gene allele or variant, one in which the normal gene is expressed in the parents, the probability of inheriting the disease is slightly more complicated. Suppose that two parents are heterozygous recessive (both are Aa). The pool of variants contributed by both parents that can be distributed to the offspring, two at a time, are thus A, A, a, and a. Each of the four gene variant combinations (AA, Aa, aA, aa) has a 25% chance of being passed on to an offspring. Three of these combinations produce a normal offspring and one produces a diseased offspring, so the probability of contracting the recessive disease is 25% under the circumstances.

In probability theory, the probability of two events occurring together is the product of the probability of each of the two events occurring separately. So, for example, the probability of the offspring getting AA is $\frac{1}{2} \times \frac{1}{2} = \frac{1}{4}$ (because half of the variants are A), the probability of getting Aa is $2 \times \frac{1}{4} = \frac{1}{2}$ (because there are two ways of becoming heterozygous), and the probability of getting aa is $\frac{1}{4}$ (because half of the variants are a). Only one of these combinations produces the recessive phenotype that expresses disease.

Therefore, if each parent is heterozygous recessive (Aa), the offspring has a 50% chance of receiving aa and getting the disease. If only one parent is heterozygous normal (Aa) and the other is homozygous recessive (aa), and the disease has not been devastatingly expressed before childbearing age, then the offspring will have a 75% chance of inheriting the disease. Finally, if both parents are homozygous recessive, then the offspring will have a 100% chance of developing the disease.

Some diseases show a gradation between homozygous normal, heterozygous recessive, and homozygous recessive. An example is sickle-cell anemia, a blood disease characterized by sickle-shaped red blood cells that do not efficiently convey oxygen from the lungs to the body, found most frequently in African populations living in areas infested with malaria carried by the tsetse fly. Let AA stand for homozygous for the normal, dominant genotype, Aa for the heterozygous recessive genotype, and aa for the homozygous recessive sickle-cell genotype. It turns out that people living in these areas with the normal genotype are vulnerable to malaria, while people carrying the homozygous recessive genotype develop sickle-cell anemia and die prematurely. However, the heterozygous individuals are resistant to malaria and rarely develop sickle-cell anemia; therefore, they actually have an advantage in surviving or staying healthy long enough to bear children in these regions. Even though the sickle-cell variant leads to devastating disease that prevents an individual from living long enough to reproduce, the population in the tsetse fly regions gets a great benefit from having this variant in the gene pool. Anthropologists cite the distribution of sickle-cell anemia as evidence of how environmental conditions influence the gene pool in a population and result in the evolution of human traits.

The inheritance of disease becomes more and more complicated as the number of genes involved increase. At

How Simple Counting has Come to be the Basis of Clinical Research

The first thinker known to consider the fundamental concepts of disease causation was none other than the ancient Greek physician Hippocrates (460–377 B.C.), when he wrote that medical thinkers should consider the climate and seasons, the air, the water that people use, the soil and people's eating, drinking, and exercise habits in a region. Subsequently, until recent times, these causes of diseases were often considered but not quantitatively measured. In 1662 John Graunt, a London haberdasher, published an analysis of the weekly reports of births and deaths in London, the first statistical description of population disease patterns. Among his findings he noted that men had a higher death rate than women, a high infant mortality rate, and seasonal variations in mortality. Graunt's study, with its meticulous counting and disease pattern description, set the foundation for modern public health practice.

Graunt's data collection and analytical methodology was furthered by the physician William Farr, who assumed responsibility for medical statistics for England and Wales in 1839 and set up a system for the routine collection of the numbers and causes of deaths. In analyzing statistical relationships between disease and such circumstances as marital status, occupations such as mining and working with earthenware, elevation above sea level, and imprisonment, he addressed many of the basic methodological issues that contemporary epidemiologists deal with. These include defining populations at risk for disease and the relative disease risk between population groups, and considering whether associations between disease and the factors mentioned above might be caused by other factors, such as age, length of exposure to a condition, or overall health.

A generation later, public health research came into its own as a practical tool when another British physician, John Snow, tested the hypothesis that a cholera epidemic in London was being transmitted by contaminated water. By examining death rates from cholera, he realized that they were significantly higher in areas supplied with water by the Lambeth and the Southwark and Vauxhall companies, which drew their water from a part of the Thames River that was grossly polluted with sewage. When the Lambeth Company changed the location of its water source to another part of the river that was relatively less polluted, rates of cholera in the areas served by that company declined, while no change occurred among the areas served by the Southwark and Vauxhall. Areas of London served by both companies experienced a cholera death rate that was intermediate between the death rates in the areas supplied by just one of the companies. In recognizing the grand but simple natural experiment posed by the change in the Lambeth Company water source, Snow was able to make a uniquely valuable contribution to epidemiology and public health practice.

After Snow's seminal work, epidemiologists have come to include many chronic diseases with complex and often still unknown causal agents; the methods of epidemiology have become similarly complex. Today researchers use genetics, molecular biology, and microbiology as investigative tools, and the statistical methods used to establish relative disease risk draw on the most advanced statistical techniques available.

Yet reliance on meticulous counting and categorizing of cases and the imperative to think logically and avoid the pitfalls in mathematical relationships in medical data remain at the heart of all of the research used to prove that medical treatments are safe and effective. No matter how high technology, such as genetic engineering or molecular biology, changes the investigations of basic medical research, the diagnostic tools and treatments that biochemists or geneticists propose must still be adjudicated through a simple series of activities that comprise clinical trials: random assignments of treatments to groups of patients being compared to one another, counting the diagnostic or treatment outcomes, and performing a simple statistical test to see whether or not any differences in the outcomes for the groups could have occurred just by chance, or whether the newfangled treatment really works. Many hundreds of millions of dollars have been invested by governments and pharmaceutical companies into ultra-high technology treatments only to have a simple clinical trial show that they are no better than placebo. This makes it advisable to keep from getting carried away by the glamour of exotic science and technologies when it comes to medicine until the chickens, so to speak, have all been counted.

a certain point, it is difficult to determine just how many genes might be involved in a disease—perhaps hundreds of genes contribute to risk. At that point, it is more useful to think of disease inheritance as being statistical or quantitative, although new research into the human genome holds promise in revealing how information about large numbers of genes can contribute to disease prognosis and treatment.

CLINICAL TRIALS

Clinical trials constitute the pinnacle of Western medicine's achievement in applying science to improve human life. Many professionals find trial work very exciting, even though it is difficult, exacting, and requires great patience as they anxiously await the outcomes of trials, often over periods of years. It is important that the sense of drama and grandeur of the achievements of the trials should be passed along to young people interested in medicine. There are four important clinical trials currently in the works, the results of which affect the lives and survival of hundreds of thousands, even millions, of people, young and old.

The first trial was a rigorous test of the effectiveness of condoms in HIV/AIDS prevention. This was a unique experiment reported in 1994 in the *New England Journal of Medicine* that appears to have been under-reported in the popular press. Considering the prestige of the Journal and its rigorous peer-review process, it is possible that many lives could be saved by the broader dissemination of this kind of scientific result. The remaining three trials are a sequence of clinical research that have had a profound impact on the standard of breast cancer treatment, and which have resulted in greatly increased survival. In all of these trials, the key mathematical concept is that of the survival function, often represented by the Kaplan-Meier survival curve, shown in Figure 4 below.

Clinical trial 1 was a longitudinal study of human immunodeficiency virus (HIV) transmission by heterosexual partners Although in the United States and Western Europe the transmission of AIDS has been largely within certain high-risk groups, including drug users and homosexual males, worldwide the predominant mode of HIV transmission is heterosexual intercourse. The effectiveness of condoms to prevent it is generally acknowledged, but even after more than 25 years of the growth of the epidemic, many people remain ignorant of the scientific support for the condom's preventive value.

A group of European scientists conducted a prospective study of HIV negative subjects that had no risk factor for AIDS other than having a stable heterosexual relationship with an HIV infected partner. A sample of 304 HIV negative subjects (196 women and 108 men) was followed for an average of 20 months. During the trial, 130 couples (42.8%) ended sexual relations, usually due to the illness or death of the HIV-infected partner. Of the remaining 256 couples that continued having exclusive sexual relationships, 124 couples (48.4%) consistently used condoms. None of the seronegative partners among these couples became infected with HIV. On the other hand, among the 121 couples that inconsistently used condoms, the seroconversion rate was 4.8 per 100 person-years (95% confidence interval, 2.5–8.4). This means that inconsistent condom-using couples would experience infection of the originally uninfected partner between 2.5 and 8.4 times for every 100 person-years (obtained by multiplying the number of couples by the number of years they were together during the trial), and the researchers were confident that in 95 times out of a 100 trials of this type, the seroconversion rate would lie in this interval. The remaining 11 couples refused to answer questions about condom use. HIV transmission risk increased among the inconsistent users only when infected partners were in the advanced stages of disease ($p < 0.02$) and when the HIV negative partners had genital infections ($p < 0.04$).

Because none of the seronegative partners among the consistent condom-using couples became infected, this trial presents extremely powerful evidence of the effectiveness of condom use in preventing AIDS. On the other hand, there appear to be several main reasons why some of the couples did not use condoms consistently. Therefore, the main issue in the journal article shifts from the question of whether or not condoms prevent HIV infection—they clearly do—to the issue of why so many couples do not use condoms in view of the obvious risk. Couples with infected partners that got their infection through drug use were much less likely to use condoms than when the seropositive partner got infected through sexual relations. Couples with more seriously ill partners at the beginning of the study were significantly more likely to use condoms consistently. Finally, the couples who had been together longer before the start of the trial were positively associated with condom use.

Clinical trial 2 investigated the survival value of dense-dose ACT with immune support versus ACT given in three-week cycles Breast cancer is particularly devastating because a large proportion of cases are among young and middle-aged women in the prime of life. The majority of cases are under the age of 65 and the most aggressive cases occur in women under 50. The very most aggressive cases occur in women in their 20s, 30s, and 40s. The development of the National Cancer Care Network (NCCN) guidelines for treating breast cancer is the result

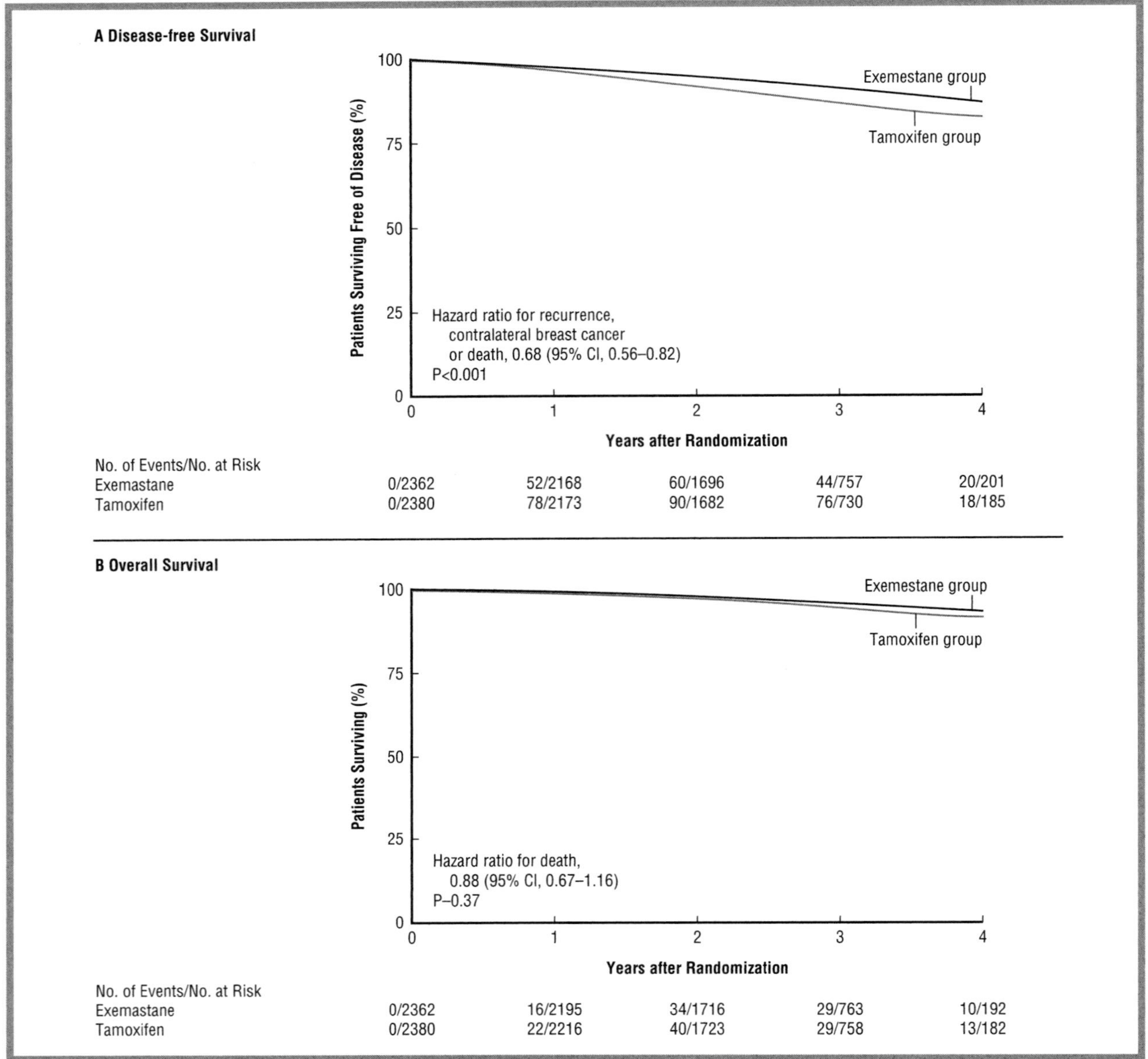

Figure 4: Cancer survival data.

of an accumulation of clinical trial evidence over many years. At each stage of the NCCN treatment algorithm, the clinician must make a treatment decision based on the results of cancer staging and the evidence for long-term (generally five-year) survival rates from clinical trials.

A treatment program currently recommended in the guidelines for breast cancer that is first diagnosed is that the tumor is excised in a lumpectomy, along with any lymph nodes found to contain tumor cells. Some additional nodes are usually removed in determining how far the tumor has spread into the lymphatic system. The tumor is tested to see whether it is stimulated by estrogen or progesterone. If so, the patient is then given chemotherapy with a combination of doxorubicin (Adriamycin) plus cyclophosphamide (AC) followed by paclitaxel (Taxol, or T) (the ACT regimen). In the original protocol, doctors administered eight chemotherapy infusion cycles (four AC and four T) every three weeks to give the patient's immune system time to recover. The patient then receives radiation therapy for six weeks. After radiation, the patient receives either Tamoxifen or an aromatase inhibitor for years as secondary preventive treatment.

Oncologists wondered whether compressing the three-week cycles to two weeks (dense dosing) while supporting the immune system with filgrastim, a white cell growth factor, would further improve survival. They speculated that dense dosing would reduce the opportunity for cancer cells to recover from the previous cycle and continue to multiply. Filgrastim was used between cycles because a patient's white cell count usually takes about three weeks to recover spontaneously from a chemotherapy infusion, and this immune stimulant has been shown to shorten recovery time.

The researchers randomized 2,005 patients into four treatment arms: 1) A-C-T for 36 weeks, 2) A-C-T for 24 weeks, 3) AC-T for 24 weeks, and 4) AC-T for 16 weeks. The patients in the dense dose arms (2 and 4) received filgrastim. These patients were found to be less prone to infection than the patients in the other arms (1 and 3).

After 36 months of follow-up, the primary endpoint of disease-free survival favored the dense dose arms with a 26% reduction in the risk of recurrence. The probability of this result by chance alone was only 0.01 ($p = 0.01$), a result that the investigators called exciting and encouraging. Four-year disease-free survival was 82% in the dense-dose arms versus 75% for the other arms. Results were also impressive for the secondary endpoint of overall survival. Patients treated with dense-dose therapy had a mortality rate 31% lower than those treated with conventional therapy ($p = 0.013$). They had an overall four-year survival rate of 92% compared with 90% for conventional therapy. No significant difference in the primary or secondary endpoints was observed between the A-C-T patients versus the AC-T patients: only dense dosing made a difference. The benefit of the AC-T regimen was that patients were able to finish their therapy eight weeks earlier, a significant gain in quality of life when one is a cancer patient.

One of the salient mathematical features of this trial is that it had enough patients (2,005) to be powered to detect such a small difference (2%) in overall survival rate. Many trials with fewer than 400 patients in total are not powered to detect differences with such precision. Had this difference been observed in a smaller trial, the survival difference might not have been statistically significant.

Clinical trial 3 studied the treatment of patients over 50 with radiation and tamoxifen versus tamoxifen alone. Some oncologists have speculated that women over 50 may not get additional benefit by receiving radiation therapy after surgery and chemotherapy. A group of Canadian researchers set up a clinical trial to test this hypothesis that ran between 1992 and 2000 involving women 50 years or older with early stage node-negative breast cancer with tumors 5 cm in diameter or less. A sample of 769 women was randomized into two treatment arms: 1) 386 women received breast irradiation plus tamoxifen, and 2) 383 women received tamoxifen alone. They were followed up for a median of 5.6 years.

The local recurrence rate (reappearance of the tumor in the same breast) was 7.7% in the tamoxifen group and 0.6% in the tamoxifen plus radiation group. Analysis of the results produced a hazard ratio of 8.3 with a 95% confidence interval of [3.3, 21.2]. This means that women in the tamoxifen group were more than eight times as likely to have local tumor recurrences than the group that received irradiation, and the researchers were confident that in 95 times out of a 100 trials of this type, the hazard ratio would at least be over three times as great and as much as 21.2 times as great, given the role of random chance fluctuations. The probability of this result was that it could occur by chance alone only once in a 1,000 trials ($p < 0.001$).

As mentioned above, clinical trials are the interventional or experimental application of epidemiology and constitute a unique branch of statistical mathematics. Statisticians that are specialists in such studies are called trialists. Clinical trial shows how the rigorous pursuit of clinical trial theory can result in some interesting and perplexing conundrums in the practice of medicine.

In this trial, they studied the secondary prevention effectiveness of tamoxifen versus Exemestane. For the past 20 years, the drug tamoxifen (Nolvadex) has been the standard treatment to prevent recurrence of breast cancer after a patient has received surgery, chemotherapy, and radiation. It acts by blocking the stimulatory action of estrogen (the female hormone estrogen can stimulate tumor growth) by binding to the estrogen receptors on breast tumor cells (the drug is an estrogen imitator or agonist). The impact of tamoxifen on breast cancer recurrence (a 47% decrease) and long-term survival (a 26% increase) could hardly be more striking, and the life-saving benefit to hundreds of thousands of women has been one of the greatest success stories in the history of cancer treatment. One of the limitations of tamoxifen, however, is that after five years patients generally receive no benefit from further treatment, although the drug is considered to have a "carryover effect" that continues for an indefinite time after treatment ceases.

Nevertheless, over the past several years a new class of endocrine therapy drugs called aromatase inhibitors (AIs) that have a different mechanism or mode of action from that of tamoxifen have emerged. AIs have an even more complete anti-estrogen effect than tamoxifen, and

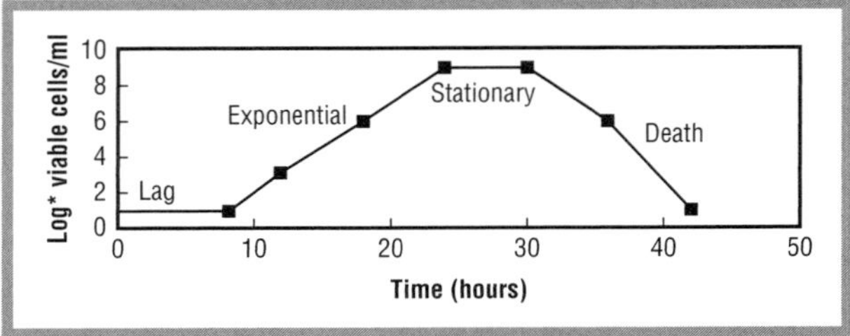

Figure 5: Bacterial growth curve for viable (living) cells.

showed promise as a treatment that some patients could use after their tumors had developed resistance to tamoxifen. As recently as 2002 the medical information company WebMD published an Internet article reporting that some oncologists still preferred the tried-and-true tamoxifen to the newcomer AIs despite mounting evidence of their effectiveness.

However, the development of new "third generation" aromatase inhibitors has spurred new clinical trials that now make it likely that doctors will prescribe an AI for new breast cancer cases that have the most common patient profile (stages I–IIIa, estrogen sensitive) or for patients that have received tamoxifen for 2–5 years. A very large clinical trial reported in 2004 addressed switching from tamoxifen to an AI. A large group of 4,742 postmenopausal patients over age 55 with primary (nonmetastatic) breast cancer that had been using tamoxifen for 2–3 years was recruited into the trial between February 1998 and February 2003. About half (2,362) were randomly assigned (randomized) into the exemestane group and the remainder (2,380) were randomized into the tamoxifen group (the group continuing their tamoxifen therapy). Disease-free survival, defined as the time from the start of the trial to the recurrence of the primary tumor or occurrence of a contralateral (opposite breast) or a metastatic tumor, was the primary trial endpoint.

In all, 449 first events (new tumors) were recorded, 266 in the tamoxifen group and 183 in the exemestane group, by June 30, 2003. This large excess of events in the tamoxifen group was highly statistically significant ($p < 0.0004$, known as the O'Brien-Fleming stopping boundary), and the trial's data and safety-monitoring committee, a necessary component of all clinical trials, recommended an early halt to the trial. Trial oversight committees always recommend an early trial ending when preliminary results are so statistically significant that continuing the trial would be unethical. This is because continuation would put the lives of patients in one of the trial arms at risk because they were not receiving medication that had already shown clear superiority.

The unadjusted hazard ratio for the exemestane group compared to the tamoxifen group was 0.62 (95% confidence interval 0.56–0.82, $p < 0.00005$, corresponding to an absolute benefit of 4.7%). Disease-free survival in the exemestane group was 91.5% (95% confidence interval 90.0–92.7%) versus 86.8% for the tamoxifen group (95% confidence interval 85.1–88.3%). The 95% confidence interval around the average disease-free survival rate for each group is a band of two standard errors (related to the standard deviation) on each side. If these bands do not overlap, as these did not, the difference in disease-free survival for the two groups is statistically significant.

The advantage of exemestane was even greater when deaths due to causes other than breast cancer were censored (not considered in the statistical analysis) in the results. One important ancillary result, however, was that at the point the trial was discontinued; there was no statistically significant difference in overall survival between the two groups. This prompted an editorial in the *New England Journal of Medicine* that raised concern that many important clinical questions that might have been answered had the trial continued, such as whether tamoxifen has other benefits, for instance osteoporosis and cardiovascular disease prevention effects, in breast cancer patients, now could not be and perhaps might never be addressed.

RATE OF BACTERIAL GROWTH

Under the right laboratory conditions, a growing bacterial population doubles at regular intervals and the growth rate increases geometrically or exponentially (2^0, 2^1, 2^2, $2^3 \ldots 2^n$) where n is the number of generations. It should be noted that this explosive growth is not really representative of the growth pattern of bacteria in nature, but it illustrates the potential difficulty presented when a patient has a runaway infection, and is a useful tool in diagnosing bacterial disease.

When a medium for culturing bacteria captured from a patient in order to determine what sort of infection might be causing symptoms is inoculated with a certain number of bacteria, the culture will exemplify a growth curve similar to that illustrated below in Figure 5. Note that the growth curve is set to a logarithmic scale in order to straighten the steeply rising exponential growth curve. This works well because $\log 2^2 = 2x$ is a formula for a straight line in analytic geometry.

The bacterial growth curve displays four typical growth phases. At first there is a temporary lag as the bacteria take time to adapt to the medium environment. An exponential growth phase as described above follows as the bacteria divide at regular intervals by binary fission. The bacterial colony eventually runs out of enough nutrients or space to fuel further growth and the medium

Key Terms

Exponential growth: A growth process in which a number grows proportional to its size. Examples include viruses, animal populations, and compound interest paid on bank deposits.

Probability distribution: The expected pattern of random occurrences in nature.

becomes contaminated with metabolic waste from the bacteria. Finally, the bacteria begin to die off at a rate that is also geometric, similar to the exponential growth rate. This phenomenon is extremely useful in biomedical research because it enables investigators to culture sufficient quantities of bacteria and to investigate their genetic characteristics at particular points on the curve, particularly the stationary phase.

Potential Applications

One of the most interesting future developments in this field will likely be connected to advances in knowledge concerning the human genome that could revolutionize understanding of the pathogenesis of disease. As of 2005, knowledge of the genome has already contributed to the development of high-technology genetic screening techniques that could be just the beginning of using information about how the expression of thousands of different genes impacts the development, treatment, and prognosis of breast and other types of cancer, as well as the development of cardiovascular disease, diabetes, and other chronic diseases.

For example, researchers have identified a gene-expression profile consisting of 70 different genes that accurately predicted the prognosis for a group of breast cancer patients into poor prognosis and good prognosis groups. This profile was highly correlated with other clinical characteristics, such as age, tumor histologic grade, and estrogen receptor status. When they evaluated the predictive power of their prognostic categories in a ten-year survival analysis, they found that the probability of remaining free of distant metastases was 85.2% in the good prognosis group, but only 50.6% in the poor prognosis group. Similarly, the survival rate at ten years was 94.6% in the good prognosis group, but only 54.6% in the poor prognosis group. This result was particularly valuable because some patients that had positive lymph nodes that would have been classified as having a poor prognosis using conventional criteria were found to have good prognoses using the genetic profile.

Physicians and scientists involved in medical research and clinical trials have made enormous contributions to the understanding of the causes and the most effective treatment of disease. The most telling indicator of the impact of their work has been the steadily declining death rate throughout the world. Old challenges to human survival continue, and new ones will certainly emerge (e.g., AIDS and the diseases of obesity). The mathematical tools of medical research will continue to be humankind's arsenal in the struggle for better health.

Where to Learn More

Books

Hennekens, C.H., and J.E. Buring. *Epidemiology in Medicine.* Boston: Little, Brown & Co., 1987.

Periodicals

Coombes, R., et al. "A Randomized Trial of Exemestane after Two to Three Years of Tamoxifen Therapy in Postmenopausal Women with Primary Breast Cancer." *New England Journal of Medicine.* (March 11, 2004) 350(11): 1081–1092.

De Vincenzi,Isabelle. "A Longitudinal Study of Human Immunodeficiency Virus Transmission by Heterosexual Partners." *New England Journal of Medicine.* (August 11, 1994) 331:6: 341–346.

Fyles, A.W., et al. "Tamoxifen with or without Breast Irradiation in Women 50 Years of Age or Older with Early Breast Cancer." *New England Journal of Medicine* (2004) 351(10): 963–970.

Shapiro, S., et al. "Lead Time in Breast Cancer Detection and Implications for Periodicity of Screening." *American Journal of Epidemiology* (1974) 100: 357–366.

Van't Veer, L., et al. "Gene Expression Profiling Predicts Clinical Outcome of Breast Cancer." *Nature.* (January 2002) 415: 530–536.

Web sites

"Significant improvements in disease free survival reported in women with breast cancer." First report from The Cancer and Leukemia Group B (CALGB) 9741 (Intergroup C9741) study. December 12, 2002 (May 13, 2005). <http://www.prnewswire.co.uk/cgi/news/release?id=95527>.

"Old Breast Cancer Drug Still No. 1." WebMD, May 20, 2002. (May 13, 2005.) <http://my.webmd.com/content/article/16/2726_623.htm>.

Modeling

Overview

A model is a representation that mimics the important features of a subject. A mathematical model uses mathematical structures such as numbers, equations, and graphs to represent the relevant characteristics of the original. Mathematical models rely on a variety of mathematical techniques. They vary in size from graphs to simple equations, to complex computer programs. A variety of computer coding languages and software programs have been developed to aid in computer modeling. Mathematical models are used for an almost unlimited range of subjects including agriculture, architecture, biology, business, design, education, engineering, economics, genetics, marketing, medicine, military, planning, population genetics, psychology, and social science.

Fundamental Mathematical Concepts and Terms

There are three fundamental components of a mathematical model. The first includes the things that the model is designed to reflect or study. These are often referred to as the output, the dependent variables, or the endogenous variables. The second part is referred to as input, parameters, independent variables, or exogenous variables. It represents the features that the model is not designed to reflect or study, but which are included in or assumed by the model. The last part is the things that are omitted from the model.

Consider a marine ecologist who wants to build a model to predict the size of the population of kelp bass (a species of fish) in a certain cove during a certain year. This number is the output or the dependent variable. The ecologist would consider of all the factors that might influence the fish population. These might include the temperature of the water, the concentration of food for the kelp bass, population of kelp bass from the previous year, the number of fishermen who use the cove, and whatever else he considers important. These items are the input or the dependent variables. The things that might be excluded from the model are those things that do not influence the size of the kelp bass population. These might include the air temperature, the number of sunny days per year, the number of cars that are licensed within a 5-mile (8 km) radius of the cove, and anything else that does not have a clear, direct impact on the fish population.

Once the model is built, it can often serve a variety of purposes and the variables in the model can change depending on the model's use. Imagine that the same model of kelp bass populations is used by an officer at the Department of Fish and Wildlife to set fishing regulations. The officer cares a lot about how many fishermen use the cove and he can set regulations controlling the number of licenses granted. For the regulator, the number of fisherman changes to the independent variable and the population of fish is a dependent variable.

Building mathematical models is somewhat similar to creating a piece of artwork. Model building requires imagination, creativity, and a deep understanding of the process or situation being modeled. Although there is no set method that will guarantee a useful, informative model, most model building requires, at the very least, the following four steps.

First, the problem must be formulated. Every model answers some question or solves a problem. Determining the nature of the problem or the fundamentals involved in the question are basic to building the model. This step can be the most difficult part of model building.

Second, the model must be outlined. This includes choosing the variables that will be included and omitted. If parameters that have no impact on the output are included in the model, it will not work well. On the other hand, if too many variables are included in the model, it will become exceedingly complex and ineffective. In addition, the dependent and independent variables must be determined and the mathematical structures that describe the relationships between the variables must be developed. Part of this step involves making assumptions. These assumptions are the definitions of the variables and the relationships between them. The choice of assumptions plays a large role in the reliability of a model's predictions.

The third step of building a model is assessing its usefulness. This step involves determining if the data from model are what it was designed to produce and if the data can be used to make the predictions the model was intended to make. If not, then the model must be reformulated. This may involve going back to the outline of the model and checking that the variables are appropriate and their relationships are structured properly. It may even require revisiting the formulation of the problem itself.

The final step of developing a model is testing it. At this point results, from the model are compared against measurements or common sense. If the predictions of the model do not agree with the results, the first step is to check for mathematical errors. If there are none, then fixing the model may require reformulations to the mathematical structures or the problem itself. If the predictions of the model are reasonable, then the range of variables for which the model is accurate should be explored. Understanding the limits of the model is part of the testing process. In some cases it may be difficult to find data to compare with predictions from the model. Data may be difficult, or even impossible, to collect. For example, measurements of the geology of Mars are quite expensive to gather, but geophysical models of Mars are still produced. Experience and knowledge of the situation can be used to help test the model.

After a model is built, it can be used to generate predictions. This should always be done carefully. Models usually only function properly within certain ranges. The assumptions of a model are also important to keep in mind when applying it.

Models must strike a balance between generality and specificity. When a model can explain a broad range of circumstances, it is general. For example, the normal distribution, or the bell curve, predicts the distribution of test scores for an average class of students. However, the distribution of test scores for a specific professor might vary from the normal distribution. The professor may write extremely hard tests or the students may have had more background in the material than in prior years. A U-shaped or linear model may better represent the distribution of test scores for a particular class. When a model more specific to a class is used, then the model loses its generality, but it better reflects reality. The trade-offs between these values must be considered when building and interpreting a model.

There are a variety of different types of mathematical models. Analytical models or deterministic models use groups of interrelated equations and the result is an exact solution. Often advanced mathematical techniques, such as differential equations and numerical methods, are required to solve analytical models. Numerical methods usually calculate how things change with time based on the value of a variable at a previous point in time. Statistical or stochastic models calculate the probability that an event will occur. Depending on the situation, statistical models may have an analytical solution, but there are situations in which other techniques such as Bayesian methods, Markov random models, cluster analysis, and Monte Carlo methods are necessary. Graphical models are extremely useful for studying the relationships between variables, especially when there are only a few variables or when several variables are held constant. Optimization is an entire field of mathematical modeling that focuses on maximizing (or minimizing) something, given a group of constraining conditions. Optimization often relies on graphical techniques. Game theory and catastrophe theory can also be used in modeling. A relatively new branch of mathematics called chaos theory has been used to model many phenomena in nature such as the growth of trees and ferns and weather patterns. String theory has been used to model viruses.

Computers are obviously excellent tools for building and solving models. General computer coding languages have the basic functions for building mathematical models. For example, JAVA, Visual Basic and C^{++} are commonly used to build mathematical models. However, there are a number of computer programs that have been developed with the particular purpose of building mathematical models. Stella II is an object oriented modeling program. This means that variables are represented by boxes and the relationships between the variables are represented by different types of arrows. The way in which the variables are connected automatically generates the mathematical equations that build the model. MathCad, MatLab and Mathematica are based on built-in codes that automatically perform mathematical functions and can solve complex equations. These programs also include a variety of graphing capabilities. Spreadsheet programs like Microsoft Excel are useful for building models, especially ones that depend on numerical techniques. They include built-in mathematical functions that are commonly used in financial, biological, and statistical models.

Real-life Applications

Mathematical models are used for an almost unlimited range of purposes. Because they are so useful for understanding a situation or a problem, nearly any field of study or object that requires engineering has had a mathematical model built around it. Models are often a less expensive way to test different engineering ideas than using larger construction projects. They are also a safer and less expensive way to experiment with various scenarios, such as the effects of wave action on a ship or wind action on a structure. Some of these fields that commonly rely on mathematical modeling are agriculture, architecture, biology, business, design, education, engineering, economics, genetics, marketing, medicine, military, planning, population genetics, psychology, and social science. Two classic examples of mathematical modeling from the vast array of mathematical models are presented below.

ECOLOGICAL MODELING

Ecologists have relied on mathematical modeling for roughly a century, ever since ecology became an active field of research. Ecologists often deal with intricate systems in which many of the parts depend on the behavior of other parts. Often, performing experiments in nature is not feasible and may also have serious environmental consequences. Instead, ecologists build mathematical models and use them as experimental systems. Ecologists can also use measurements from nature and then build mathematical models to interpret these results.

A fundamental question in ecology concerns the size of populations, the number of individuals of a given species that live in a certain place. Ecologists observe many types of fluctuations in population size. They want to understand what makes a population small one year and large the next, or what makes a population grow quickly at times and grow slowly at other times. Population models are commonly studied mathematical models in the field of ecology.

When a population has everything that it needs to grow (food, space, lack of predators, etc.), it will grow at its fastest rate. The equation that describes this pattern of growth is $\Delta N/\Delta t = rN$. The number of organisms in the population is N, time is t, and the rate of change in the number of organisms is r. The Δ is the Greek letter delta and it indicates a change in something. The equation says that the change in the number of organisms (ΔN) during a period of time (Δt) is equal to the product of the rate of change (r) and the number of organisms that are present (N).

If the period of time that is considered is allowed to become very small and the equation is integrated, it becomes $N = N_0e^{rt}$, where N_0 is the number of organisms at an initial point in time. This is an exponential equation, which indicates that the number of organisms will increase extremely fast. Because the graph of this exponential equation shoots upward very quickly, it has a shape that is similar to the shape of the letter "J". This exponential growth is sometimes called "J-shaped" growth.

J-shaped growth provides a good model of the growth of populations that reproduce rapidly and that have few limiting resources. Think about how quickly mosquitoes seem to increase when the weather warms up in the spring. Other animals with J-shaped growth are many insects, rats, and even the human population on a global scale. The value of r varies greatly for these different species. For example, the value of r for the rice weevil (an insect) is about 40 per year, for a brown rat about 5 per year and for the human population about 0.2 per year. In addition, environmental conditions, such as temperature, will influence the exponential rate of increase of a population.

In reality, many populations grow very quickly for some time and then the resources they need to grow become limited. When populations become large, there may be less food available to eat, less space available for each individual or predators may be attracted to the large food supply and may start to prey on the population. When this happens the population growth stops increasing so quickly. In fact, at some point, it may stop increasing at all. When this occurs, the exponential growth model, which produces a J-shaped curve, does not represent the population growth very well.

Another factor must be added to the exponential equation to better model what happens when limited resources impact a population. The mathematical model, which expresses what happens to a population limited by its resources, is $\Delta N/\Delta t = rN(1 - N/K)$. The variable K is sometimes called the carrying capacity of a population. It is the maximum size of a population in a specific environment. Notice that when the number of individuals in the population is near 0 ($N = 0$), the term $1-N/K$ is approximately equal to 1. When this is the case, the model will behave like an exponential model; the population will have rapid growth. When the number of individuals in the population is equal to the carrying capacity ($N = K$), then the term $1 - N/K$ becomes $1 - K/K$, or 0. In this case the model predicts that the changes in the size of the population will be 0. In fact, when the size of a population approaches its carrying capacity, it stops growing.

The graph of a population that has limited resources starts off looking like the letter J for small population sizes and then curves over and becomes flat for larger population sizes. It is sometimes called a sigmoid growth curve or "S-shaped" growth. The mathematical model $\Delta N/\Delta t = rN(1-N/K)$ is referred to as the logistic growth curve.

The logistic growth curve is a good approximation for the population growth of animals with simple life histories, like microorganisms grown in culture. A classic example of logistic growth is the sheep population in Tasmania. Sheep were introduced to the island in 1800 and careful records of their population were kept. The population grew very quickly at first and then reached a carrying capacity of about 1,700,000 in 1860.

Sometimes a simple sigmoidal shape is not enough to clearly represent population changes. Often populations will overshoot their carrying capacity and then oscillate around it. Sometimes, predators and prey will exhibit cyclic oscillations in population size. For example the population sizes of Arctic lynx and hare increase and decrease in a cycle that lasts roughly 9–10 years.

Ecologists have often wondered whether modeling populations using just a few parameters (such as the rate of growth of the population, the carrying capacity) accurately portrays the complexity of population dynamics. In 1994, a group of researchers at Warwick University used a relatively new type of mathematics called chaos theory to investigate this question.

A mathematical simulation model of the population dynamics between foxes, rabbits and grass was developed. The computer screen was divided into a grid and each square was assigned a color corresponding to a fox, a rabbit, grass, and bare rock. Rules were developed and applied to the grid. For example, if a rabbit was next to grass, it moved to the position of the grass and ate it. If a fox was next to a rabbit, it moved to the position of the rabbit and ate it. Grass spread to an adjacent square of bare rock with a certain probability. A fox died if it did not eat in six moves, and so on.

The computer simulation was played out for several thousand moves and the researchers observed what happened to the artificial populations of fox, rabbits, and grass. They found that nearly all the variability in the system could be accounted for using just four variables, even though the computer simulation model contained much greater complexity. This implies that the simple exponential and logistic modeling that ecologists have been working with for decades may, in fact, be a very adequate representation of reality.

MILITARY MODELING

The military uses many forms of mathematical modeling to improve its ability to wage war. Many of these models involve understanding new technologies as they are applied to warfare. For example, the army is interested in the behavior of new materials when they are subjected to extreme loads. This includes modeling the conditions under which armor would fail and the mechanics of penetration of ammunition into armor. Building models of next generation vehicles, aircraft and parachutes and understanding their properties is also of extreme importance to the army.

The military places considerable emphasis on developing optimization models to better control everything from how much energy a battalion in the field requires to how to get medical help to a wounded soldier more effectively. Special probabilistic models are being developed to try to detect mine fields in the debris of war. These models incorporate novel mathematical techniques such as Bayesian methods, Markov random models, cluster analysis, and Monte Carlo simulations. Simulation models are used to develop new methods for fighting wars. These types of models make predictions about the outcome of war since it has changed from one of battlefield combat to one that incorporates new technologies like smart weapon systems.

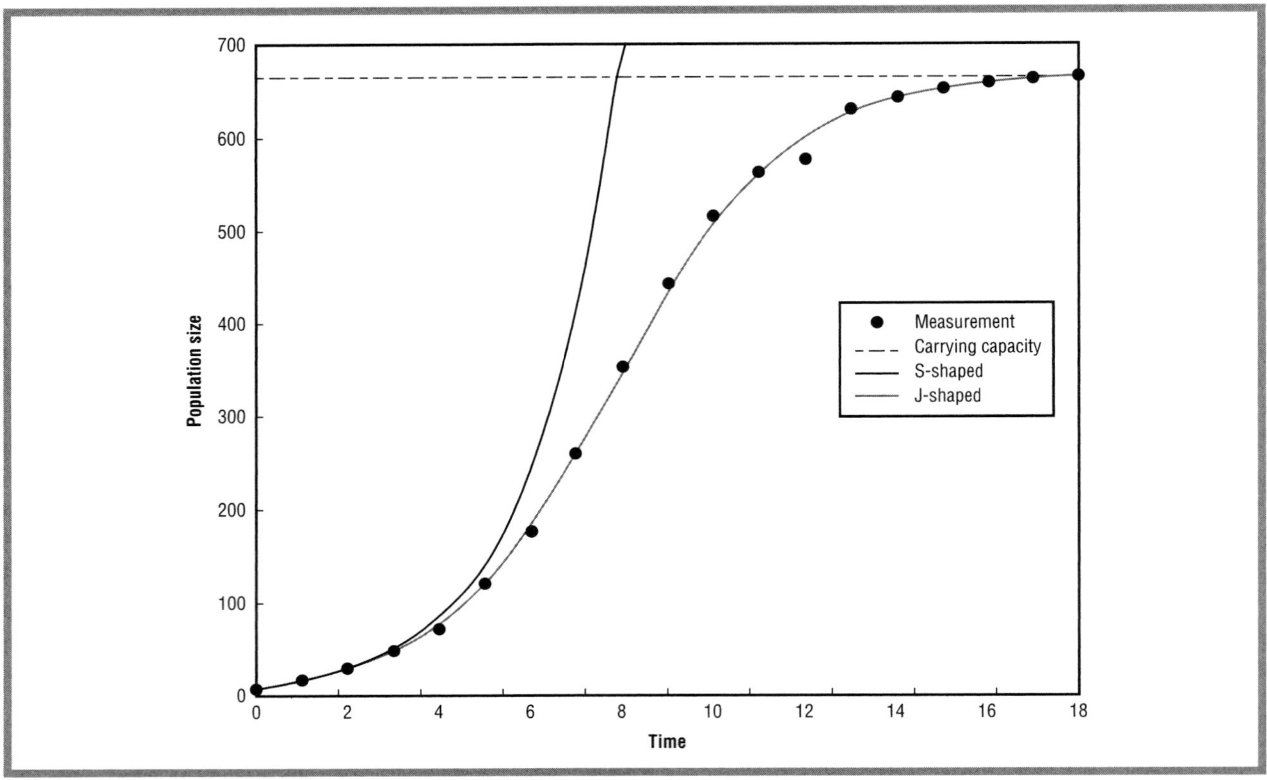

Figure 1: Examples of population growth models. The dots are measurements of the size of a population of yeast grown in a culture. The dark line is an exponential growth curve showing J-shaped growth. The lighter line is a sigmoidal or logistic growth curve showing S-shaped growth. The dashed line shows the carrying capacity of the population.

Game theory was developed in the first half of the twentieth century and applied to many economic situations. This type of modeling attempts to use mathematics to quantify the types of decisions a person will make when confronted with a dilemma. Game theory is of great importance to the military as a means for understanding the strategy of warfare. A classic example of game theory is illustrated by the military interaction between General Bradley of the United States Army and General von Kluge of the German Army in August 1944, soon after the invasion of Normandy.

The U.S. First Army had advanced into France and was confronting the German Ninth Army, which outnumbered the U.S. Army. The British protected the U.S. First Army to the North. The U.S. Third Army was in reserve just south of the First Army.

General von Kluge had two options; he could either attack or retreat. General Bradley had three options concerning his orders to the reserves. He could order them to the west to reinforce the First Army; he could order them to the east to try to encircle the German Army; or he could order them to stay in reserve for one day and then order them to reinforce the First Army or strike eastward against the Germans.

In terms of game theory, six outcomes result from the decisions of the two generals and a payoff matrix is constructed which ranks each of the outcomes. The best outcome for Bradley would be for the First Army's position to hold and to encircle the German troops. This ranks 6, or the highest in the matrix and it would occur if von Kluge attacks and the First Army and Bradley holds the Third Army in reserve one day to see if the First Army needed reinforcement and if not he could then order them to the east to encircle the German troops. The worst outcome for Bradley is a 1 and it would occur if von Kluge orders an attack and at the same time Bradley ordered the reserve troops eastward. In this case, the Germans could possibly break through the First Army's position and there would be no troops available for reinforcement.

Game theory suggests that the best decision for both generals is one that makes the most of their worst possible

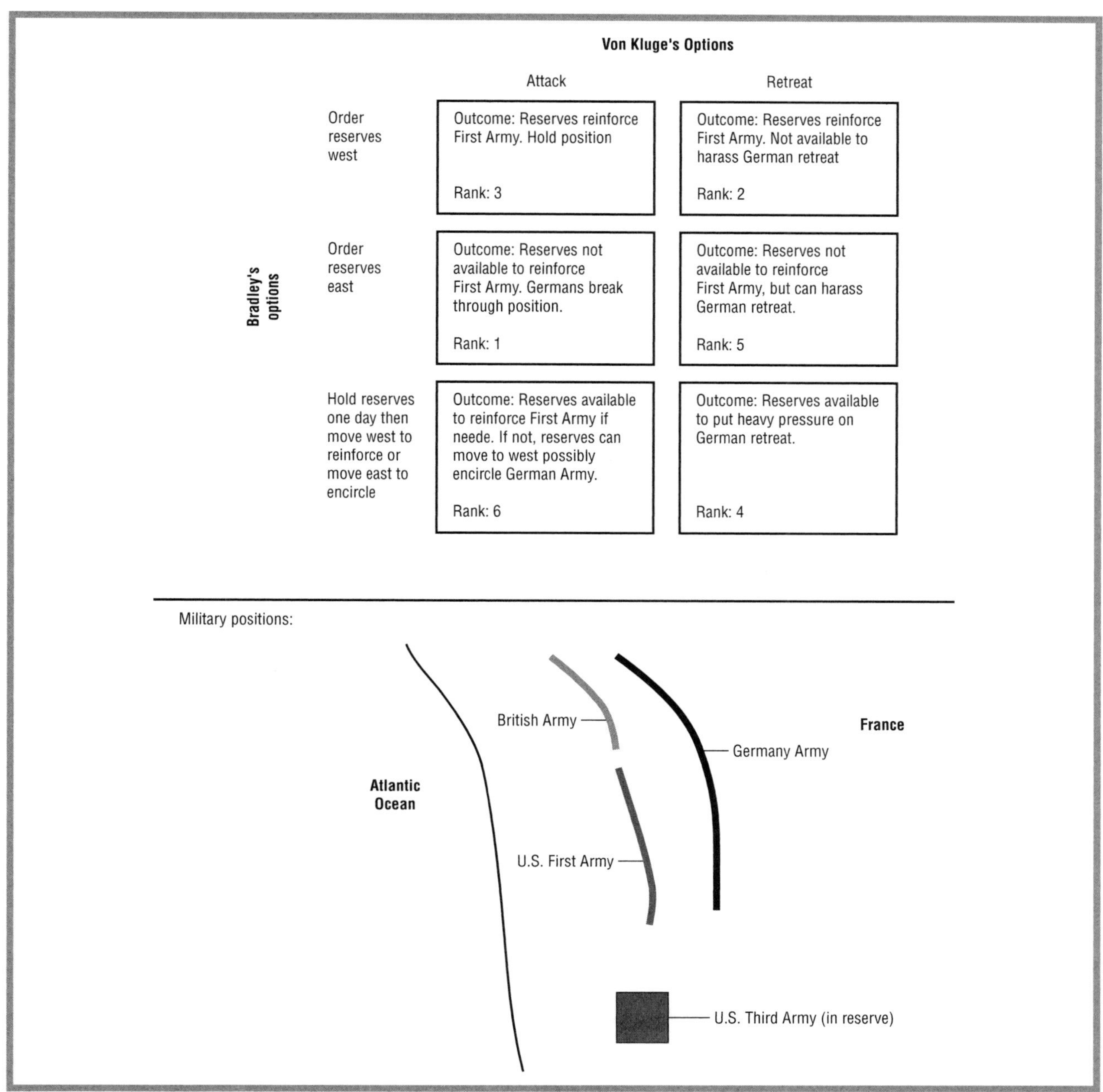

Figure 2: Payoff matrix for the various scenarios in the battle between the U.S. Army and the German Army in 1944. If possible add graphic of military positions as well. Caption should read: Military positions of the U.S. and German Armies during the battle. The U.S. and British forces held positions to the west of the German Army. The U.S. Third Army was in reserve to the south of the U.S. First Army.

outcome. Given the six scenarios, this results in von Kluge deciding to withdraw and Bradley deciding to hold the Third Army in reserve for one day, a 4 in the matrix. The expected outcome of this scenario is that the Third Army would be one day late in moving to the east and could only put moderate pressure on the retreating German Army. On the other hand, they would not be committed to the wrong action. From the German point of view, the Army does not risk being encircled and cut off by the Allies, and it avoids excessive harassment during its retreat.

Interestingly, the two generals decided to follow the action suggested by game theory. However, after van Kluge decided to withdraw, Hitler ordered him to attack. The U.S. First Army held their position on the first day of

Key Terms

Dependent variable: What is being modeled; the output.

Exponential growth: A growth process in which a number grows proportional to its size. Examples include viruses, animal populations, and compound interest paid on bank deposits.

Independent variable: Data used to develop a model, the input.

Input: What is used to develop a model, the independent variables.

Model: A system of theoretical ideas, information, and inferences presented as a mathematical description of an entity or characteristic.

Output: What is being modeled; the dependent variables.

the battle and Bradley ordered the Third Army to the east to encircle the Germans. Hitler unwittingly generated the best possible outcome for Bradley, the 6th or highest rank in the matrix.

Where to Learn More

Books

Beltrami, Edward. *Mathematical Models in the Social and Biological Sciences.* Boston: Jones and Bartlett Publishers, 1993.

Bender, Edward A. *An Introduction to Mathematical Modeling.* Mineola NY: Dover Publications, 2000.

Burghes, D.N., and A.D. Wood. *Mathematical Models in the Social, Management and Life Sciences.* Chichester: Ellis Horwood Limited, 1980.

Harte, John. *Consider a Spherical Cow.* Sausalito CA: University Science Books, 1988.

Odum, Eugene P. *Fundamentals of Ecology.* Philadelphia: Saunders College Publishing, 1971.

Skiena, Steven.*Calculated Bets: Computers, Gambling, and Mathematical Modeling to Win.* Cambridge: Cambridge University Press, 2001.

Stewart, Ian. *Nature's Numbers.* New York: BasicBooks, 1995.

Web sites

Carlton College. "Mathematical Models." Starting Point: Teaching Entry Level Geoscience. January 15, 2004. <http://serc.carleton.edu/introgeo/models/mathematical/> (April 18, 2005).

Department of Mathematical SciencesUnited States Military Academy. "Military Mathematical Modeling (M3)" May 1, 1998. <http://www.dean.usma.edu/departments/math/pubs/mmm99/default.htm> (April 18, 2005).

Multiplication

Overview

Multiplication is a method of easily adding various quantities of identical numbers without performing each addition equation individually.

Fundamental Mathematical Concepts and Terms

In a multiplication equation, the two values being multiplied are called coefficients or factors, while the result of a multiplication equation is labeled the product. Several forms of notation can be used to designate a multiplication operation. The most common symbol for multiplication in arithmetic is $\times$. In algebra and other forms of mathematics where letters substitute for unknown quantities, the $\times$ is often omitted, so that the expression $3x + 7y$ is understood to mean $3 \times x + 7 \times y$. In other cases, parentheses can be used to express multiplication, as in 5(2), which is mathematically identical to 5×2, or 10.

For both subtraction and division, the order of the values being operated on has a significant impact on the final answer; in multiplication, the order has no effect on the result. The commutative property of multiplication states that $x \times y$ gives the same result as $y \times x$ for any values of x and y, making the order of the factors irrelevant to the product. Another property of multiplication is that any value multiplied times 0 produces a product of 0, while any number multiplied times 1 gives the starting number. The signs of the factors also affect the product; multiplying two numbers with the same sign (either two positives or two negatives) will produce a positive result, while multiplying numbers with differing signs will produce a negative value.

A Brief History of Discovery and Development

As an extension of the basic process of addition, multiplication's origins are lost in ancient history, and early merchants probably learned to perform basic multiplication operations long before the system was formalized. The first formal multiplication tables were developed and used by the Babylonians around 1800 B.C. One of these earliest tables was created to process simple calculations of the area of a square farm field, using the length and width as data and allowing a user to look up the area in the table body. These early tables function identically to

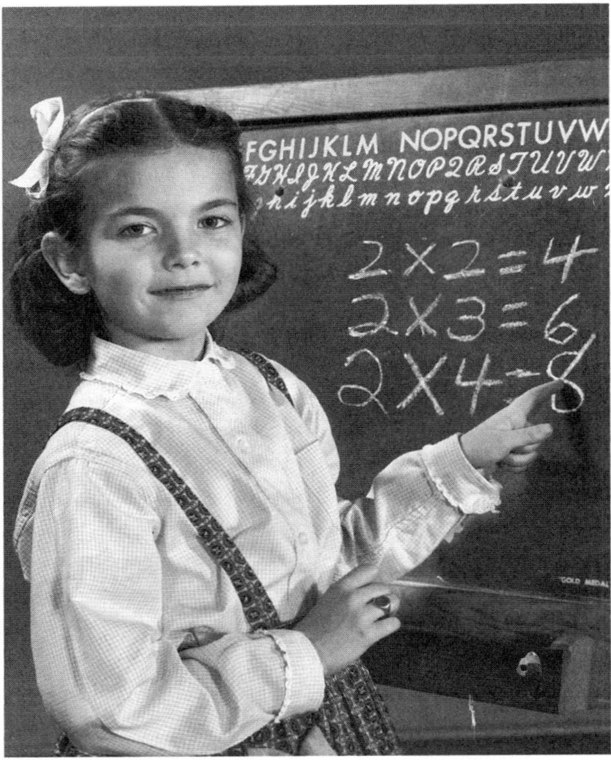

Girl executing simple multiplication problems. Lambert/Getty Images.

today's multiplication tables, meaning that the tables which modern elementary school students labor to memorize have actually been in use for close to forty centuries.

Moving past the basic single digit equations of the elementary school multiplication table, long multiplication can become a time-consuming, complex process, and many different techniques for performing long multiplication have been developed and used. In the thirteenth century, educated Europeans used a multiplication technique known as lattice multiplication. This somewhat complicated method involved drawing a figure resembling a garden lattice, then writing the two factors above and to the right of the figure. Following a step by step process of multiplying, adding, and summing values, this method allowed one to reliably multiply large numbers.

An earlier, more primitive method of long multiplication was devised by the early Egyptians, and is described in a document dating to 1700 B.C. The Egyptian system seems rather unusual, due largely to the Egyptian perspective on numbers. Whereas modern mathematics views numbers as independent, discrete entities with an inherent value, ancient Egyptians thought of numbers only in terms of collections of concrete objects. In other words, to an ancient Egyptian, the number nine would have no inherent meaning, but would always refer to a specific collection of objects, such as nine swords or nine cats.

For this reason, Egyptian math generally did not attempt to deal with extremely large quantities, as these calculations offered little practical value. Instead, the Egyptians devised a method of multiplication which could be accomplished by a complex series of manipulations using nothing more than simple addition. Due to its complexity and limited utility, this method does not appear to have gained favor outside Egypt. As an interesting side note, elements of the Egyptian method actually involve binary mathematics, the system which forms the basis of modern computer logic systems.

A similar, binary-based system was developed and used in Russia. This so-called peasant method of multiplication involved repeatedly doubling and halving the two values to be multiplied until an answer was produced. While tedious to apply, this method involved little more than removing the right-most value at each step until the result was produced. Like the previously discussed methods, this technique seems remarkably slow in modern terms; however, in a context in which one might only need to perform a single multiplication problem each week or each month, such techniques would have been useful.

Given the complexity of performing long multiplication manually, numerous inventors attempted to create mechanical multiplying machines. Far more difficult than creating a simple adding machine, this task was first successfully completed by Gottfried Wilhelm Von Leibniz (1646–1716), a German philosopher and mathematician who also invented differential calculus. This device, which Von Leibniz called the Stepped Reckoner, used a series of mechanical cranks, drums, and gears to evaluate multiplication equations, as well as division, addition, and subtraction problems. Only two of these machines were ever built; both survive and are housed in German museums. Von Leibniz apparently shared the somewhat common dislike of calculating by hand; he is quoted as saying that the process of performing hand calculations squanders time and reduces men to the level of slaves. Unfortunately, his bulky, complex mechanical calculator never came into widespread use.

Additional attempts were made to construct multiplying machines, and various mechanical and electromechanical versions were created during the ensuing centuries. However the first practical hand-held tools for performing multiplication did not appear until the 1970s, with the introduction of microprocessors and handheld calculators by firms such as Hewlett Packard and Texas Instruments. Today, using these inexpensive tools or spreadsheet software, long multiplication is no more difficult or time-consuming to perform than simple addition.

Real-life Applications

EXPONENTS AND GROWTH RATES

Growth rates describe the application of simple multiplication many times to a starting value. In cases where the growth rate is constant over time, a special notation is used to define the projected value; this notation is called an exponent, and its value conveys how many times the starting value is to be multiplied by itself. For example, the expression 3×3 can also be written 3^2, which is read "Three to the second power," or simply "Three squared." As the sequence progresses, the values become more cumbersome to work with, and exponents greatly simplify the process. For instance, the expression $3 \times 3 \times 3 \times 3 \times 3 \times 3 \times 3 \times 3 \times 3 \times 3$ can be easily written as 3^{10}, and when evaluated produces a value of 59,049.

INVESTMENT CALCULATIONS

One common application of exponents deals with growth rates. For example, assume that an investment of $100 will earn 7% over the course of a year, yielding a total of $107 at year-end. This process can be continued indefinitely; at the end of two years, this $107 will have earned another $7.49, making the total after two years $114.49.

Using exponents, we can easily determine how much the original $100 will have earned after any specific number of years; in this example, we will find the total value after nine years. First, we note that the growth rate is 7%, meaning that the starting value must be multiplied by 1.07 in order to find the new value after one year. In order to find the multiplier, or value we would apply to our starting number to find the final total, we simply multiply 1.07 times itself until we account for all nine years of the calculation. Expressed in long terms, this equation would be $1.07 \times 1.07 \times 1.07 \times 1.07 \times 1.07 \times 1.07 \times 1.07 \times 1.07 \times 1.07 = 1.84$. Expressing this value exponentially we write the expression as 1.07^9. We can now multiply our original investment value by our calculated multiplier to find the final value of the investment: $\$100 \times 1.84 = \$184$. Further, if we wish to recalculate the size of the investment over a shorter or longer period of time, we simply change the exponent to reflect the new time period.

Two unusual situations occur when using exponents. First, by convention, the value of any number raised to the power 0 is 1; so $4^0 = 1{,}26^0 = 1$, and $995^0 = 1$. While mathematicians offer lengthy explanations of why this is so, a more intuitive explanation is simply that moving from one exponent to the next lower one requires a division by the base value; for example, to move from 3^4 to 3^3, we divide by 3, or in expanded terms, we divide 81 by 3 to get to 27. If we follow this sequence to its natural progression, we will eventually reach 3^1, and if we divide this value (3) by 3, we find a result of 1. Since this sequence will end with 1 for any base value, then any value raised to the power 0 will equal 1.

A second curiosity of exponents occurs in the case of negative values, either in the exponent or in the base value. In some situations, base values are raised to a negative power, as in the expression 5^{-3}. By convention, this

How Much Wood Could a Woodchuck Chuck, if a Woodchuck Could Chuck Wood?

This nursery rhyme tongue-twister has puzzled children for years, and has in fact inspired numerous online discussions regarding the specific details of the riddle and how to solve it. Using a simple formula, we can take the amount the rodent chucks per hour, multiply it times the number of working hours each day, then multiply again by 365 to get a total per year. This, multiplied by the animal's lifespan in years would give us a total amount chucked, which one online estimate places at somewhere around 26 tons.

Like all such estimations, in which a single event is multiplied repeatedly to predict performance over a long period of time, this estimate is fraught with assumptions, any of which can cause the final estimate to be either too high or too low. For example, even a small error in estimating how much can be chucked per hour could throw the final total off by a ton or more. Another major source of error is found in the variability of the woodchuck's work; unlike mechanical wood chuckers, woodchucks work faster some days than others. Also unlike machines, rodents frequently spend the winter hibernating, significantly reducing the actual volume of wood chucked. To sum up, the question of how much wood can be chucked remains difficult to answer, given the number of assumptions required; the most generally correct answer may simply be "Quite a lot."

expression is evaluated as the inverse of this expression with the exponent sign made positive, or $1/5^3 = 1/125$. A related complication arises when the base value is itself negative, as in the case of $(-5)^3$. Multiplying negative and positive values is accomplished according to a simple set of rules: if the signs are the same, the final value is positive, otherwise the final value is negative. So 4×4 and -4×-4 produce the same result, a value of 16. However 4×-4 produces a value of -16. In the case of a negative base being raised to a specific power, a related set of rules apply: if the exponent is even, the final value is positive, otherwise it is negative. Following this rule, $(-5)^3 = -125$, while $(-5)^2 = 25$.

CALCULATING EXPONENTIAL GROWTH RATES

One ancient myth is based on the concept of an exponential growth rate. The legend of Hercules describes a series of twelve great works which the hero was required to perform; one of these assignments was to slay the Hydra, a horrible beast with nine heads. While Hercules was unimaginably strong, he quickly found that traditional tactics would not work against the Hydra; each time Hercules cut off one of the Hydra's heads, two new heads grew in its place, meaning that as soon as he turned from dispatching one head, he quickly found himself being attacked by even more heads than before. Hercules eventually triumphed by discovering how to cauterize the stumps of the severed heads, preventing them from regenerating. While this story is ancient, it illustrates a simple principle which frequently holds true: in stopping an exponentially growing system, the best solution is typically to interrupt the growth cycle, rather than trying to keep up with it in this case was to prevent or interrupt the growth in the first place, rather than trying to keep up with it as it occurs.

While some animals are able to regenerate severed body parts, no real-life animal is able to do so as quickly as the mythical Hydra. However, some animal populations do multiply at an alarming rate, and in the right circumstances can rapidly reach plague proportions. Mice, for example, can produce offspring about every three weeks, and each litter can include up to eighteen young. To simplify this equation, we can assume one litter per month, and 16 young per litter. We also assume, for simplicity, that the mice only live to be 1 month old, so only their offspring live on into the next month. Beginning with a single pair of healthy mice on New Year's Day, by the end of January, we will have eight pair. Thus, over the course of the first month, the mouse population will have grown by a factor of eight.

While this first month's performance is impressive, the process becomes even more startling as the months pass. At the end of February, the eight pair from month one will have each given birth to another sixteen young (eight pair), making the new population $8 \times 8 = 64$ pair. This number will continue to increase by a factor of eight each month, meaning that by the end of May, more than 3,000 pair of mice will exist. By the end of December, the total mouse population will be almost 70 billion, or about 10 times the human population of Earth.

Obviously, mice have lived on Earth for eons without ever taking over, so this conclusion raises some question about the validity of the math involved, as well as pointing out some potential problems with the methodology used. First, the calculation assumes that mice can begin breeding immediately after birth, which is incorrect. Also, it assumes that all the mice in each generation survive to reproduce, when in fact many mice do not. Additionally, it assumes that adequate food exists for all the mice to continue eating, which would also be a near-impossibility. Finally, it assumes that the mouse's natural predators, including humans, would sit idly by and watch this takeover occur. Since these limitations all impact the growth rate of mouse populations in real life, a mouse population explosion of the size described here is unlikely to occur. Nevertheless, the high multiplication rate of mice and other rodents helps explain why they are so difficult to eradicate, and can so quickly infest large buildings.

While the final result of the mouse calculation is somewhat unrealistic, similar population explosions have actually occurred. A small number of domestic rabbits were released in Australia during the 1800s; with adequate food and few natural predators, they quickly multiplied and began destroying the natural vegetation. During the 1950s, government officials began releasing the Myxoma virus, which killed 99% of animals exposed to it. However, resistant animals quickly replenished the population, and by the mid-1990s, parts of the Australian rangeland were inhabited by more than 3,000 rabbits per square kilometer. Rabbit control remains an issue in Australia today; the country boasts the world's longest rabbit fence, which extends more than 1,000 kilometers. As of 1991, the estimated rabbit population of Australia was approximately 300 million, or about fifteen times the human population of the continent.

SPORTS MULTIPLICATION CALCULATING A BASEBALL ERA

Comparing the performance of baseball pitchers can be difficult. In a typical major league game three, four, or more pitchers all work for the same goal, but only one is awarded a win or loss. To help compare pitching performance on more even basis, baseball analysts frequently discuss a pitcher's earned run average, or ERA. The ERA is used to evaluate what might happen if pitchers could pitch entire games, providing a basis for comparison among multiple players.

Calculating a pitcher's ERA is fairly simple, and involves just a few values. The process begins with the number of earned runs scored on the pitcher during his time in the game. This value is then multiplied by nine (the assumed number of innings in a full game), and that total is divided by the number of innings actually pitched. For example, if a pitcher plays three innings and allows two runs, his ERA would be calculated as $2 \times 9/3 = 6$. Like most projections, this one is subject to numerous other factors, but suggests that if this pitcher could maintain his performance at this level, he would allow six runs in a typical full game.

The ERA calculation becomes more complex when a pitcher is removed from a game during an inning. In such cases, the number of innings pitched will be measured in thirds, with each out equaling one third of an inning. If the pitcher who allows two runs is removed after one out has been made in the fourth inning, he would have pitched 3 1/3 innings. Historically, major league ERAs have risen and fallen as the rules of the game have changed. Today, a typical professional pitcher will have an ERA around 4.50, while league leaders often post single-season ERAs of 2.00 or less. One of a coach's more difficult challenges is recognizing when a pitcher has reached the end of his effectiveness and should be removed from a game. Fatigue typically leads to poorer performance and a rapidly rising ERA.

RATE OF PAY

An old joke says that preachers hold the most lucrative jobs, since they are paid for a week's labor but only work one day of each week. Using this arguably flawed logic, professional rodeo cowboys might be considered some of the highest paid athletes today, since they spend so little time actually "working." A bull rider's working weekend typically consists of a two day competition. Each competitor rides one bull the first night, and a second the following night. If he is able to stay on each bull for the full eight seconds, and scores enough style points for his riding ability, he then qualifies for a third ride in the final round of competition.

Because each ride lasts only eight seconds, a bull rider's complete work time for each event is only 24 seconds, not counting time spent settling into the saddle and the inevitable sprint to escape after the ride ends. Multiplying this 24 seconds of work times the 31 events in an entire professional season produces a total working time each year of about 13 minutes. Because a top professional rider earns over $250,000 per season, this rider's income works out to an amazing $19,230 per minute, or $1,153,846 per hour. Unfortunately, this average does not include the enormous amounts of time spent practicing, traveling, and healing from injuries, and in many cases, professional bull riders win only a few thousand dollars per season. But even for the wages paid to top riders, few people are willing to strap themselves atop an angry animal that weighs more than some small cars.

MEASUREMENT SYSTEMS

Some sports have their own unique measurement systems. Horse racing is a sport in which races are frequently measured in furlongs; since a furlong is approximately 66 feet, a 50 furlong race would be 3,300 feet long, or around .6 miles. Furlongs can be converted to feet by multiplying by 66, or converted to miles by dividing by 80. Horses themselves are frequently measured in an arcane measurement unit, the hand. A hand equals approximately four inches, and hands can be converted to feet by multiplying the number of hands by 3, or to inches by multiplying the number of hands by .25. Like many other traditional units of measurement, the hand is a standardized version of an ancient method of measurement, in which the width of four fingers serves as a standard measurement tool.

ELECTRONIC TIMING

Electronic timing has made many sports more exciting to watch, with Olympic medallists often separated from also-rans by mere thousandths of a second. In some events, split times are calculated, such as a time at the halfway mark of a downhill ski race. Along with providing an assessment of how well a skier is performing on the top half of the course, these measurements can also be used to predict the final time by simply doubling the mid-point time to predict the final. While this method is not foolproof, it is close enough to give fans an idea of whether a skier will be chasing a world record or simply trying to reach the bottom of the hill without falling down.

MULTIPLICATION IN INTERNATIONAL TRAVEL

Despite enormous growth in international trade, the United States still uses the imperial measurement system, rather than the more common and simpler metric system.

Because of this disparity, conversions between the two systems are sometimes necessary. While the 2-liter soft drink is one of the few common uses of the metric system in America today, a short trip to Canada would reveal countless situations in which converting between the two systems would be necessary.

While packing for the trip, an important consideration would be the weather forecast for Canada, which would normally be given in degrees Celsius. The conversion from Celsius to the Fahrenheit system used in the U.S. requires multiplication and division, using this formula: $F = 9/5 \times C + 32$. To get a ballpark figure (a rough estimate), simply double the Celsius reading and add 30. Obviously, this difference in measurement systems means that a frigid sounding temperature of 40 degrees Celsius is in fact quite hot, equal to 104 degrees Fahrenheit. Converting Fahrenheit to Celsius is equally simple: just reverse the process, subtracting 32 and multiplying by 5/9. No conversion is necessary at –40, because this is the point at which both scales read the same value.

Driving in Canada would also require mathematical conversions; while Canadians drive on the right-hand side of the highway, they measure speed in kilometers per hour (km/h), rather than the U.S. traditional miles per hour (mph) system. Because one mile equals 1.6 kilometers, the kilometer values for a given speed are larger than the mile values; the typical highway speed of 55 mph in the U.S. is approximately equal to 88 km/h in Canada, and mph can be converted to km/h using a multiplication factor of 1.6.

Gasoline in Canada is often more expensive than in the United States; however prices there are not posted in gallons, but in liters, meaning the posted price may appear exceptionally low. One gallon equals 3.8 liters, and gallons are converted to liters by multiplying by this value. Soft drinks are often sold in 2-liter bottles in the U.S., making this one of the few metric quantities familiar to Americans. Also, smaller volumes of liquid are measured not in ounces, quarts, or pints, but in deciliters and milliliters.

One of the greatest advantages of the metric system is its simplicity, with unit conversions requiring only a shift of the decimal point. For example, under the U.S. system, converting miles to yards requires one to multiply by 1,760, and converting to feet requires multiplication by 5,280. Liquids are even more confusing, with gallons to quarts using a factor of 4, and quarts to ounces using 32. Weights are similarly inconsistent, with pounds equaling 16 ounces. Using the metric system, each conversion is based on a factor of ten: multiplying by ten, one hundred, or one thousand allows conversions among kilometers, meters, and millimeters for distance, liters, deciliters, and milliliters for volume, and kilograms, decigrams, and milligrams for weight.

OTHER USES OF MULTIPLICATION

Multiplication is frequently used to find the area of a space; as previously discussed, one of the oldest known multiplication tables was apparently created to calculate the total area of pieces of farm property based on only the side dimensions. The area of a square or rectangle is found by multiplying the length times the width; for a field 40 feet long and 20 feet wide, the total area would be $40 \times 20 = 800$ square feet. Other shapes have their own formulae; a triangle's area is calculated by multiplying the length of the base by the height, then multiplying this total by 0.5; a triangle with a 40 foot base and a 20 foot height would be half the size of the previously described rectangle, and its area would be $40 \times 20 \times 0.5 = 400$ square feet.

Formulas also exist for determining the area of more complex shapes. While simple multiplication will suffice for squares, rectangles, and triangles, additional information is needed to find the area of a circle. One of the best-known and most widely used mathematical constants is the value pi, which is approximately 3.14. Pi was first calculated by the ancient Babylonians, who placed its value at 3.125; in 2002, researchers calculated the value of pi to the 1.2 trillionth decimal place.

Pi's value lies in its use in calculating both the circumference and the area of a circle. The circumference, or distance around the perimeter, of a circle, is found by multiplying pi times the diameter; for a circle with diameter of 10 inches, the circumference would be 3.14×10, or 31.4 inches. The area of this same circle can be found by multiplying pi times the radius squared; for a circle with diameter of 10 and radius of 5, the formula would be $3.14 \times 5 \times 5$, giving an area of 78.5 square inches.

Other techniques can be used to calculate the area of irregular shapes. One approach involves breaking an irregular shape into a series of smaller shapes such as rectangles and triangles, finding the area of each smaller shape, and adding these values together to produce a total; this method is frequently used when calculating the number of shingles needed to cover an irregularly shaped roof.

A branch of mathematics called calculus can be used to calculate the area under a curve using only the formula which describes the curve itself. This technique is fundamentally similar to the previously described method, in that it mathematically slices the space under the curve into extremely thin sections, then finds the area of each and sums the results. Calculus has numerous applications in fields such as engineering and physics.

CALCULATING MILES PER GALLON

As the price of gasoline rises and occasionally falls, one common question deals with how to reduce the cost of fuel. The initial part of this question involves determining how much gas a car uses in the first place. Some cars now have mileage computers which calculate this automatically, but for most drivers, dividing the number of miles driven (a figure taken from the trip odometer) by the number of gallons added (a figure on the fuel pump) will provide a simple measure of miles per gallon. Using this figure along with the capacity of the fuel tank allows a calculation of a vehicle's range, or how far it can travel before refueling.

In general, larger vehicles will travel fewer miles per gallon of gas, making them more expensive to operate. However, these vehicles also typically have larger fuel tanks, making their range on a single tank equal to that of a smaller car. For example, a 2003 Hummer H2 has a 30-gallon fuel tank and gets around 12 miles per gallon, giving it a theoretical range of 360 miles on a full tank. In comparison, the fuel-sipping 2004 Toyota Prius hybrid sedan has only a 12 gallon tank. However, when combined with the car's mileage rating of more than 50 miles per gallon, this vehicle can travel around 600 miles per tank, and could conceivably travel more than 1,500 miles on the Hummer's oversized 30-gallon fuel load. In general, most cars are built to allow a 300–500-mile driving range between fill-ups, however the price of the fill-up varies widely depending on the car's efficiency and tank size.

SAVINGS

Small amounts of money can often add up quickly. Consider a convenience store, and a student who stops there each morning to purchase a soft drink. These drinks sell for $1.00, but by reusing his cup from previous days, the student could save 32 cents per day, since the refill price is only 68 cents. While this amount of money seems trivial when viewed alone, consider the implications over time.

Over the course of just one week, this small savings rapidly adds up; multiplying the savings times five days gives a total savings of $1.60, or enough to buy two more refills. Multiplying this weekly savings times four gives us a monthly savings of around $6.40, and multiplying the weekly savings by 52 yields a total annual savings of $83.20, enough to pay for a tank or two of gas or perhaps a nice evening out. Perhaps more amazing is the result when a consumer decides to save small amounts wherever possible; saving this same tiny amount on ten items each day would yield annual savings of $832.00, a significant amount of savings for doing little more than paying attention to how the money is being spent.

Potential Applications

One increasingly popular marketing technique illustrates the use of exponential growth for practical use. Traditional marketing practices work largely by addition: as more advertisements are run, the number of potential customers grows and a percentage of those potential customers eventually buy the product or service. As advertising markets have become more fragmented and audiences have grown harder to reach, one emerging technique is called viral marketing.

SPAM AND EMAIL COMMUNICATIONS

Viral marketing refers to a marketing technique in which information is passed from the advertiser to one generation of customers who then pass it to succeeding generations in rapidly expanding waves. In the same way that the rabbit population in Australia expanded by several times as each generation was born, viral marketing depends on people's tendency to pass messages they find amusing or thought-provoking to a long list of friends.

The growth of e-mail in particular has helped spur the rise of viral marketing, since forwarding a funny email is as simple as clicking an icon. In the same way that viruses rapidly multiply, viral e-mail messages can expand so rapidly that they clog company e-mail servers. Some companies have begun taking advantage of this phenomenon by intentionally producing and releasing viral marketing messages, such as humorous parodies of television commercials. Viral marketing can be an exceptionally inexpensive technique, as the material is distributed at no cost to the originating firm.

Where to Learn More

Web sites

Brain Bank. "A History of Measurement and Metrics." <http://www.cftech.com/BrainBank/OTHERREFERENCE/WEIGHTSandMEASURES/MetricHistory.html> (April 9, 2005).

A Brief History of Mechanical Calculators. "Leibniz Stepped Drum." <http://www.xnumber.com/xnumber/mechanical1.htm> (April 5, 2005).

Centre for Experimental and Constructive Mathematics. "Table of Computation of Pi from 2000 B.C. to Now." <http://

Key Terms

Exponential growth: A growth process in which a number grows proportional to its size. Examples include viruses, animal populations, and compound interest paid on bank deposits.

Integral calculus: A branch of mathematics used for purposes such as calculating such values as volumes displaced, distances traveled, or areas under a curve.

www.cecm.sfu.ca/projects/ISC/Pihistory.html> (April 5, 2005).

Ferel feast!. "History of Rabbits in Australia." <http://library.thinkquest.org/03oct/00128/en/rabbits/history.htm> (April 7, 2005).

History of Electronics. "Calculators." <http://www.etedeschi.ndirect.co.uk/museum/concise.history.htm> (April 6, 2005).

Homerun Web. "How to Calculate Earned Run Average (ERA)." <http://www.homerunweb.com/era.html> (April 8, 2005).

Information Technology at St. Lawrence University. "Early Dynastic Mathematics." <http://it.stlawu.edu/~dmelvill/mesomath/3Mill/ED.html> (April 6, 2005).

The Math Forum at Drexel. "Egyptian Method of Multiplication." <http://mathforum.org/library/drmath/view/57542.html> (April 5, 2005).

The Math Forum at Drexel. "Lattice Multiplication." <http://mathforum.org/library/drmath/view/52468.html> (April 4, 2005).

Music and Mathematics

Overview

Mathematics and music are basic elements of cultures and civilizations. They are fundamental. Along with moving, speaking, and reading, basic mathematics is one of the key early developmental skills parents try to instill in their children. Even before children are born, their parents may play music for them. Music is said to help babies' brains develop, and music made specifically for unborn or newborn babies can be found ranging from classical to reggae. Mathematics and music are often combined in books, toys, or songs, as musical counting can be easier for children to remember and is more fun.

Fundamental Mathematical Concepts and Terms

Mathematics is the study of mathematics. Music is experienced and created as music. These statements may seem obvious and trivial at first, but consider the study of physics without mathematics, or the study of economics without statistics, or the creation of poetry without language. Most subjects draw on tools from other disciplines, but mathematics and music can be studied as pure forms. In music, the form and the medium are the same. In mathematics, the methods and the subject are the same.

What is a number? What is a minor key? Both mathematics and music have invented special symbols to represent their seemingly abstract concepts. Everyone thinks they know what a number is, but defining the concept of a number is very difficult. One may know what a minor key is, but defining the concept rigorously is again difficult.

However, these are casual similarities between mathematics and music. There are more fundamental and formal ways in which the two disciplines interact, and the realization of this dates back to at least the Ancient Greeks.

A Brief History of Discovery and Development

PYTHAGORAS AND STRINGS

Pythagoras (fifth to sixth century B.C.) is chiefly remembered for discovering the method of calculating the length of one side of a right-angled triangle when the lengths of the other sides are known. He was a Greek philosopher who came to believe that mathematics was the most important discipline to study, and that nature was, at its deepest level, mathematical. Very little is known about Pythagoras, and what writings there are have come from his later followers and contain many inventions and exaggerations, often treating Pythagoras as a god-like being with divine powers. However, even if nothing his supporters tell is true, the legend and legacy of Pythagoras had a profound effect on both mathematics and music in the Western tradition.

Supposedly, one day Pythagoras was walking by a blacksmith's shop and was distracted by the sounds of the hammers falling on the anvil. Several hammers were being used, producing distinct notes, the notes separated in regular musical intervals that Pythagoras found pleasing to hear. The hammers were of different weights, and Pythagoras wondered if the ratio of weights might be related to the notes they played. Then Pythagoras is said to have done something unusual for a Greek philosopher; he experimented to see if his observations had a physical basis, by playing with a string.

The string was a monochord, which is a taut string stretched between two supports on the top of a hollow sounding box, much like a single string guitar. Plucking the string makes it vibrate, which produces a note dependent on the length and tension of the string. A stopper, or bridge, can move up and down the string, changing the length of the string that vibrates, and thereby changing the note played. The shorter the string, the higher the note.

Only certain combinations of notes seem to go together to the ear. Pythagoras had found the music of the hammers pleasing. The notes they played when striking the anvil seemed to complement each other. The monochord, like many stringed instruments, can play notes that seem 'sweet' and those that seem 'sour' to the ear, as well as those that seem to go together in harmony and those that seem discordant. The Greeks, like many other cultures, had developed rules for the playing of the good notes that produced harmonious listening. In modern Western terminology, the notes were collected into scales and octaves. What Pythagoras is said to have discovered is that there is a mathematical relationship between such notes. On the monochord, the relationship could be seen in the length of the string to be played. If the string is played without a stopper, so that its full length vibrates, a certain note is played. If the stopper is then placed halfway, the note that results from only half the string vibrating is an octave higher than the original. If the string were to be doubled in length, it would play a note an octave lower.

In the Western tradition, the separation of these notes is called an octave because there were originally eight notes placed in this interval. However, that is just one possible way of dividing the interval, and different cultures have produced different divisions. For example, in the Chinese tradition there are five notes, while in the Arabic there are 17. Yet while the number of notes placed into an "octave" is variable, all cultures use the same interval. The notes an octave apart are the same note, and they sound the same, just pitched higher or lower. Musicians in all cultures have recognized this interval as a natural phenomenon. What the Pythagoreans showed was that this interval had a mathematical basis, and could be expressed in relation to the length of a monochord string by the ratio 2:1.

Playing two notes an octave apart either together or immediately after one another sounds harmonious. It is said two musical notes are harmonious when they sound pleasing together, as opposed to discordant notes that sound bad together and can make listeners wince or block their ears. Musicians in all cultures had, through trial and error, discovered those notes that seemed to go together well in harmony, simply by listening. The Pythagoreans revealed a mathematical relationship between these harmonious notes.

The modern notes C and G sound pleasing when played close together. If a monochord is set up so the full length of the string plays a C, then to play a G the stopper is moved two-thirds along (a third from the end). The ratio of the lengths of C and G is 3:2. The modern musical term is that these notes are separated by a perfect fifth. Other simple ratios of the string also give rise to harmonious notes. Some notes, however, produce a discordant sound together. Playing a C with an F sharp, an interval known as an augmented fourth with the ratio of 45:32, does not sound good.

The reason for these different sounds, it was discovered centuries later, is the frequency, or number of vibrations per second, at which the strings vibrate. A string of half the length vibrates at twice the frequency of a full string, whereas two-thirds of a string, the perfect fifth ratio, vibrates at one and a half times the speed of the full string. The Pythagoreans concluded that musical harmony was a mathematical property, and occurred when the ratios between the notes were simple.

Pythagoras and his followers believed that the ratios they had discovered using the monochord could not only be applied to other musical instruments, but to the whole of nature. They developed a theory that linked music, mathematics, and the motion of the planets. The Greek view of the universe placed the Earth at the center, with the sun, the moon, the five known planets, and the stars all rotating in fixed crystalline spheres around it. The Pythagoreans applied the same principle of ratios to the supposed orbits of the planets, and concluded that they had the same properties as a musical scale. The crystal spheres, they said, must make a sound as they moved, and these sounds must be harmonious. Other Greek philosophers, such as Plato, became enamored with this rather beautiful notion of a singing universe, and added to it. The idea of the music of the spheres would remain in fashion for hundreds of years, heavily influencing the study of astronomy.

MEDIEVAL MONKS

The Greeks linked music and mathematics so tightly that the study of music was considered a branch of mathematics, along with arithmetic, geometry, and astronomy. The Greek ideas were, however, almost lost to the Western world after the fall of the Roman Empire. Monks cloistered in monasteries were the only ones with the education and time to translate and copy the surviving writings of the Greeks. Anicis Manlius Severinus Boethius (c. 480–524) translated and copied the ideas of the Pythagoreans. Boethius did more than just copy ancient texts, he also drew together many sources into coherent books. Also, like many other monastic copyists, he edited sections so that they conformed to his own beliefs. Boethius was an excellent translator and copyist, but his grasp of mathematics was poor, and his writings were often hard to follow or misrepresented their sources. However, his copies, mistakes and all, were copied by others and followed blindly, even when they contradicted real-life experiences.

Whatever the limitations of Boethius' copies, they had a profound and lasting effect on Western teaching. Copies of his compilation spread across Europe, and influenced the way music and mathematics were thought of and studied. Music remained a kind of sub-discipline of mathematics, and was taught in European universities as part of the quadrivium ("the four ways"), the same set of subjects that Pythagoras had grouped together: arithmetic, geometry, astronomy, and music.

Under the influence of Boethius, the science of harmonics was the main focus of musical study. The medieval scholars categorized music in three ways, the actual making of music (*musica instrumentalis*), the harmony of the human soul (*musica humana*), and the music of the spheres (*musica mundane*). In this scheme, music was seen as part of the basic nature of the universe, while the playing of music was merely the lowliest part of musical study.

Fixed styles of singing, based heavily on the theories of Pythagoras, became entrenched in the musical practice of the times. Most formal music was made for religious purposes, and just what forms of music were appropriate were written into canon law as the "Ecclesiastical modes." This provided a unifying structure to European music, but also limited innovation. Monophonic chants with one melody line without an accompaniment, where all the singers sing the same words at the same pace, were the standard form. Harmonies were slowly introduced into Gregorian chants by singing the same words at different frequencies at the same time, and these were usually limited to an octave, a fifth, or a fourth apart.

Then a new kind of song type called polyphony, where two or more melodies would be sung or played at the same time, began to take over. The composer had to be careful to make the separate melodies blend together in a harmonious manner, which was rather hard to do. Even harder was trying to write such a song down on paper. Monophonic songs only required the composer to write the pitch or note that was to be sung. A polyphonic song required a method of recording the time each note was to take for all the melodies.

QUANTIFICATION OF MUSIC

In monophonic plainchant, the singers all start together, sing the same words, and stop together, so the level of notation that is required is simple. However, in polyphonic songs, different melodies must be sung, and without a method of knowing the time each note is to take, the whole process slides into anarchy and discordance.

The rise of polyphony, therefore, coincided with advances in musical notation, which also coincided with the development of mathematical notation. Early musical notation developed in the West as a way of fixing songs into a specific form. In the sixth century A.D., Saint Isidore had complained that "[u]nless sounds are remembered by man, they perish," and wondered how reliable were the memories of some singers. However, the early notation could not express the concept of time in music, such as how long to sing a note.

Polyphony required such tools, so musical notation evolved into a way of quantifying time. The melodies a composer had ticking in his head could be transmitted to performers in different places, and even later times, by writing them down using the special code of music.

One of the most interesting developments in musical notation was the ability to note a specific period of silence with rests, which were introduced in the thirteenth century. Around the same time the Hindu-Arabic numerals were being popularized in Europe, and with them the strange symbol for zero. While the ideas of silence and zero may be taken for granted, they were both revolutionary in their respective fields. The musical rests enabled even more complex melodies to be linked harmonically, and the zero opened up new ways of thinking about numbers in mathematics.

DISCORDANCE OF THE SPHERES

Innovation in music, mathematics, and other fields led to a realization among the scholars of the age that they were doing new things, going beyond the ancients ideas of the Greeks and Romans. In music, the new styles came to be called the *Ars Nova* (The New Art). However, the ideas of the ancients were difficult for many to abandon. The idea of the harmony of the spheres remained a central part of scholarly thinking, partly because it meshed nicely with Christian beliefs and implied an order to creation.

Just as there had been great changes in music and mathematics, the field of astronomy also went through a revolution. In 1543, Nicolas Copernicus (1473–1543) challenged the accepted wisdom of the ancients by publishing a book that suggested the Sun, not the Earth, was at the center of the universe. One scholar who embraced this new idea was Johannes Kepler (1571–1630), who attempted to merge the new Copernican universe with the ancient idea of the harmony of the spheres.

Because the study of astronomy was almost always accompanied with musical study, many important figures in the development of astronomy were also keenly interested in music, and Kepler was no exception. Kepler tried to piece together a model of the universe that used the musical and geometrical ideas of Pythagoras and the new theory of Copernicus, to finally reveal just what notes each planet sang as they moved through the heavens in their crystalline spheres. However, he could just not get the theories to fit together.

After a number of unsuccessful early attempts, Kepler was fortunate enough to inherit a huge collection of accurate planetary observations upon the death of his master, Tycho Brahe (1546–1601). Kepler tried once more, using the observations to guide him in recreating the motion of the planets. He found a regular order in the motion of the planets, but to his great surprise it was an entirely different type of motion than expected, and contradicted all the theories he was attempting to unify. Kepler realized the observation figures he had been given, and those he had made himself, showed that the planets moved in elliptical orbits. While this was not the philosophical order he had been searching for, it was a mathematical order, and he produced equations that predicted the planetary motion with unprecedented accuracy. His insights slowly spread and gained acceptance, shattering the crystalline spheres, and bringing to an end the scientific search for the harmony of the spheres.

WELL-TEMPERED TONES

Music composers grew more and more innovative and daring as the years progressed, experimenting with new styles and ideas such as modulations of scale, just as innovators in other fields experimented with new theories and devices. However, some of the new compositions began to push the limits of the music instruments and tunings of the times. The musical notes that had been evolved from the Pythagorean theories used "just tones," that is, intervals between the notes that were derived from integer ratios. However, this only allowed for perfect tuning in one key at a time. If the composer wanted a piece of music to change keys, then either the musicians had to re-tune on the fly, use different instruments, or sound out of tune.

The problem was that the notes in the just scale were not equally spaced in the octave. When key changes were made, there was a need for new notes to fill in the gaps in the scale. For example, in the diatonic scale the pure (or just) ratios between the notes are either a tone or a semitone apart. The semitones all have the interval 15:16, and the tones are either separated by 8:9 or 9:10, just to make it more complicated. Because the intervals between the notes are different, playing in a different key (which can be thought of as the note started with) means every key has different patterns of intervals between the notes. Also, if the key chosen is too far from the instrument's tuning, so-called "wolf-tones" are heard, where discordant sounds shrill and howl for notes that theoretically should sound harmonious.

Many possible solutions were suggested and tried, and compromise tunings, or temperings, were attempted with various degrees of success. The mean-tone temperament used only major thirds and minor sixth intervals between the notes, effectively averaging out the scale. This meant that all the fifths and fourths that could be played were a little out of tune, but barely. This system worked well for six major and three minor keys, but outside of those, was very discordant.

Eventually the well-tempered tuning was introduced. In this tuning, the interval between each note is made equal, so it is often referred to as the equal temperament. The octave was simply divided into 12 equal parts, giving 12 semitones. In mathematical terms, since the interval ratio of an octave is 2:1, the interval between semitones had to be that number that equals two when multiplied by itself 12 times, which is 1.0595. This meant that changing keys was no longer a problem, as every key has the same interval between notes. However, the beautiful and precise Pythagorean ratios between the notes were now lost, so that all the notes in the octave interval were slightly out of just tuning. The equal temperament sacrificed accuracy for flexibility, precision for practicality, but in doing so allowed for much more innovation and experimentation in music.

The well, or equal, temperament was suggested as early as 1550, but was slow to be accepted. It became championed by Johann Sebastian Bach (1685–1750), and in 1722 he published a set of 24 preludes and fugues in the 12 major and 12 minor keys of the well tempering, naming them *The Well Tempered Clavier.* Anyone wishing to play these pieces (and Bach's music was quite popular) had to tune their instruments to the equal temperament. Other composers also adopted the new tuning, and slowly it became the standard tuning for Western music, confusing music students ever since by the fact the modern octave now consists of 12 notes.

Real-life Applications

MATHEMATICAL ANALYSIS OF SOUND

A vibrating string produces a fundamental note, the note that is heard when it is played. However, in reality it also produces other sounds called overtones that add complexity to the sound. This is why strings made from different materials will make different sounds even when everything else is equal, or why different kinds of musical instruments can play the same note yet give out a unique tone. However, overtones also add complexity to the analysis of sound. A pure tone, it was discovered, can be represented as a simple wave, but when overtones are introduced, the analysis becomes much more difficult.

The mathematical work of Jean-Baptiste-Joseph Fourier (1765–1830) led to a method for analyzing, and eventually recreating, sound. Fourier discovered that any periodic oscillation, of which sound waves were later shown to be one type, can be broken up into a set of simple sine curves. The sine function is one of the basic functions of trigonometry. A sine curve or wave is defined by the function $y = \sin(x)$, and can be considered as the modeling of a pure tone without overtones. What Fourier's work showed was that complex waves can be thought of as the addition of a number of sine waves of different frequencies and amplitudes. Fourier analysis can take a complex sound wave and break it apart into a collection of simple sine curves. The way the sine curves interact, canceling out in some places and combining in others, means that sound waves, even strange artificial curves such as square or triangular waves, can be broken down and analyzed with some simple mathematics.

The oscilloscope allows regular vibrations, including sound waves, to be displayed in real-time. An oscilloscope, in essence, draws a graph on the screen, displaying the signal as a waveform, which can be broken down into a sum of sine curves. Oscilloscopes are used by scientists, television and automotive repair technicians, in medical research, and to measure and analyze diverse phenomena from stress in buildings and brain waves, and, of course, sound waves.

If sound can be deconstructed into simple sine curves, then simple sine waves can be generated, then added together to reconstruct, or synthesize, any sound. This is the principle of modern electronic music synthesizers. Theoretically, the quality of the sound these instruments can produce depends only on the accuracy of the original analysis of the sound to be reconstructed and the quality of the sine waves that can be produced.

ELECTRONIC INSTRUMENTS

In the last 100 years, the electrification of music has changed the way music is produced and listened to. Electric instruments were first introduced as a means of making a louder sound. For example, the Hawaiian-style, or slide, guitar was a popular instrument in the 1930s, but because it was played horizontally it projected most of its sound upwards, rather than toward the audience, limiting the size of the audience it could reach. In the 1930s, the Gibson company successfully placed an electronic amplifier inside an otherwise acoustic Hawaiian guitar, and the electric guitar was born.

Distortions introduced in the amplification process limited the volume of early instruments and their acceptance. However, by the 1950s electric guitars produced cleaner, louder sound, and could sustain notes for a longer time than any acoustic model. The rising popularity of the electric guitar ushered in a whole new range of sound, such as the electronic manipulations of the sound, using fuzz-boxes, wah-wah peddles, and many other devices. The loudness of the electric guitar had to be matched with louder supporting instruments, and so all the instruments had to be wired up and their sound amplified.

Early electronic instruments still relied on a natural vibrator, such as string or drum skin, even though the final sound might bear little resemblance to the natural sound of the vibrating element. However, since the mathematical analysis of sound had shown that sound waves could be thought of as the sum of simple oscillations, then it must be possible to build sounds from a simple electronic source, and so the synthesizer was born. The Hammond electric organ, which debuted in 1935, can be considered the first of the electronic synthesizers, although the technology it used relied on many moving parts. Later electronic keyboards use a myriad of mathematical algorithms to reproduce the sounds and rhythms of other musical instruments. The synthesizer contains an oscillator to produce the basic frequency, which is then amplified, mixed with other frequencies to add richness to the sound, filtered to remove unwanted noises, and can then be submitted to a host of other modulations to give the final desired sound. Early synthesizers sounded artificial, with sounds that mimicked actual instruments, but did not sound as natural. Better mathematical analysis of sound, combined with better programming and electronics, has produced much richer sounding electronic instruments, with the ability to not just sound like an genre of instruments but an actual specific instrument, such as a particular Steinway grand piano or an individual Stradivarius violin.

Personal computers (PCs) have added another dimension to electronic music, with many add-on components and programs that turn the PC into a recording studio. A single instrument can, through the manipulation of computer algorithms, produce the tracks for a complete band or orchestra. While early programs were little more than novelties, producing bleeps and beats, it has become possible to record professional quality music on a PC, and a number of professional musicians have done so. The flexibility and sophistication of computer-generated music has even allowed musicians to experiment with scales other than the well tempered scale, and many alternate tunings that use the just or Pythagorean scales have again become popular, as their limitations can be overcome electronically.

ACOUSTIC DESIGN

Making sound louder by electronic amplification is one method of making sounds easier to hear. However, for some styles of music, amplification is not an option, such as classical or acoustic concerts. Either the number of people who can hear the music will be very limited, or some other method of boosting the strength of sound to help it arrive to a listener's ear must be found. Acoustic design is a branch of architecture that attempts to build concert halls, sound stages, and auditoriums in such a manner as to maximize the amount of sound that reaches the audience.

Sound quickly fades in strength as one moves from the source, in proportion to the inverse square law. So if one moves twice as far away from a sound, one hears a quarter of the original strength. However, sound can be reflected with the right surface. Ancient Greek amphitheaters used hard rocks in large amounts to help large audiences hear the voices of actors, using clever angles to get the most reflection and the best focus on the listeners as they could. The designs for modern concert halls and stadiums are computer modeled to maximize the sound reflection, while removing unwanted reverberations. Geometry plays a key role in such design, as well as the reflective and absorbent qualities of the materials used. By integrating different materials with careful design, the sound of a room can be crafted to give a warm, rich sound or a clean, intimate sound, or many other desired results.

For many buildings, it is sound reduction or diffusion that is important. By using materials that absorb sound and angles that diffuse rather than concentrate, architects can design rooms where sound does not travel clearly, for example, to avoid secrets being overheard. Sound insulation and soundproofing can effectively isolate the sound of a room from rooms or areas close by.

DIGITAL MUSIC

A digital revolution in the late twentieth century changed the way music is recorded, stored, and listened to. Early music recording and storage devices were analog devices, such as LPs (long-playing records) and audiotape. Analog means a continuous property that varies, such as the bumps on a wax cylinder, the groove of a record, or the magnetic alignment of grains on an audiotape. Analog recordings are very susceptible to errors, such as a scratch on a record, or being dropped or jolted, and are inefficient, needing a large surface to record small amounts of information. Digital music was introduced as a way to eliminate errors, and as a more compact medium. The compact disc is much smaller than an old LP, yet can store almost twice as much music, while a computer hard drive can have an entire music collection stored in digital form.

Digital simply means the information is recorded as binary numbers, ones and zeros. The sound to be encoded is sampled a large number of times each second, and a value is given for the wave height at that moment. Compact discs (CDs) were designed for a sample rate of 44,100 times per second (44.1 kHz). Going from analog sound to digital data is an inexact process, and even with a high sample rate some information will be lost or simplified, because the number of samples is limited, and the acceptable values for the wave height are also restricted.

Like building a staircase, sampling sound has to consider the distance between the steps (the sample rate) and the height difference between the steps (the allowed values of the sampling, referred to as the bit depth). The wider the steps, the harder it is to walk up them, and by analogy, the sound is poorer in quality, because the sampling is too far between steps. The taller the steps, the harder they are to walk up, and by analogy, the less accurate the sample estimations are. The solution is to make short, low steps, to sample frequently, and to have many small increments in the values the estimates can take. However, there is a practical limit to how small these steps can be made and still be usable. In the case of digital music, the limit is how much storage space the information requires.

COMPRESSING MUSIC

CDs use comparatively large amounts of space to store their digitally encoded information. Binary information on the CD is stored in groups of 16 ones and zeros (16-bit samples). There are 8 bits per byte, so 44,100 16-bit samples per second equals 88,200 bytes per second. That's for one channel; twice as much is needed for a stereo recording. Consequently, 176,400 bytes per second is about 10 megabtyes (10,584,000 bytes) per minute for stereo recording.

One method to reduce storage size is to take less-frequent samples (lower sample rate) or less accurate values (less bit-depth), but this leads to a poorer sound quality. So compressing the data is the preferred option, used in formats like the mp3. A range of mathematical topics contributes to the field of data compression, including algebra, statistics, graph theory, calculus, Fourier analysis, and fractals. Basic compression takes redundant information that can be simplified to the point where it takes less binary numbers to encode.

Another compression technique relies on the fact that the human ear is an imperfect listening device. The ear cannot detect all the complexities of sound, and will often be unable to tell the difference between a complex noise and a simplified substitute. Mathematical models of the human ear have helped define methods for dramatically reducing sound files in this manner, essentially tricking the ears of those who listen.

ERROR CORRECTION

Whenever a compact disc skips or an mp3 file gives out an unexpected sharp squawk it is because of an error. These errors can be from physical substances, such as dust, dirt, or scratches, from jolting a music player, or from recording or manufacturing mistakes. Because errors are so common, music players must use error-correction algorithms to fill in missing information where possible.

The mathematics for correcting these errors were discovered many years before any use for them existed. The work of Claude Shannon (1916–2001), Warren Weaver (1894–1978), and others on information theory (IT) showed that information transmission errors could be corrected by mathematical algorithms, and later work by other mathematicians and engineers provided working applications. The basic ideas of error correction are to add extra, or redundant, information to the signal so that there is a better chance of one of the bits of information being read correctly, and the information encoded in such a way that it can self-check as it goes along, so that it literally knows what to expect next. The mathematics of error correction is also used in linguistics, psychology, cryptography, and the elimination of noise in the transmission of data.

USING RANDOMNESS

Mathematical ideas have often been incorporated into composition techniques. Chance music, also called algorithmic, aletoric, random, or stochastic music, allows for unexpected structures to be introduced as defined by a set of rules, for example, by tossing a coin. Some chance pieces have strict rules that provide a high degree of structure, while others are so flexible in the choices that can be made as to be almost unpredictable.

Wolfgang Amadeus Mozart (1756–1791) created a musical game, *Musikalisches Wurfelspiel* (Dice Music), in 1787, where a minuet is formed by rolling dice, which determines the order of pre-written measures of music, sort of like cutting and pasting. The background music for the 1938 film, *Alexander Nevsky*, composed by Sergei Prokofiev (1891–1953), used the landscape in the film as a pattern for the notes. John Cage's (1912–1992) *Atlas Eclipticalis* (1961) was composed by placing the score over star-charts, with the position and brightness of the stars visible through the paper determining the notes, while his Reunion (1968) is performed by playing chess on a chessboard equipped with photo-receptors, each move determining the series of sounds to be played. Many of Cage's pieces allow for great freedom of interpretation, including how many instruments, or what instruments, are used, and the pitch, duration, intonation, and loudness of the notes to be played. Another example is the composition of Brian Eno's (1948–) *Music for Airports* (1978), where pieces of prerecorded audio tape were cut up and several played simultaneously in loops. These laborious cut-and-paste techniques of composition were later revolutionized by personal computers.

COMPUTER-GENERATED MUSIC

Human composers may incorporate random or chance elements in their work, but wisely tend to let such processes guide them rather than dictate the final results. Getting a computer to compose music means letting go of that human input, and gives rise to many problems. A number of attempts to have a computer compose music

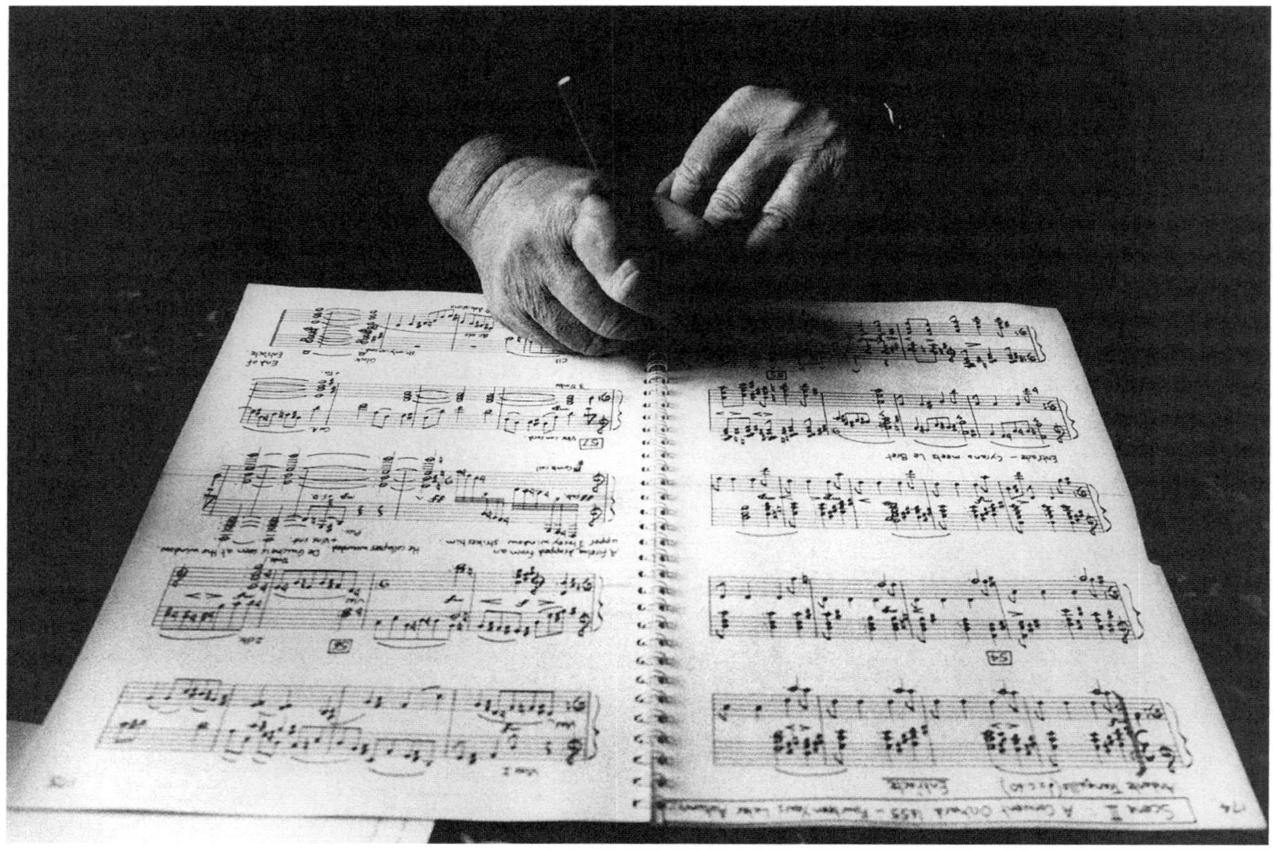

Musical scores convey information with almost mathematical elegance. JOHN GARRETT/CORBIS.

were made as early as the 1950s, with limited success. Sometimes the resulting pieces had snatches of tuneful music, but were overall disappointing, while other attempts produced listenable, but extremely simple tunes. In 1959 the ILLIAC computer, an early supercomputer, was programmed with the rules of composition that had been written during the period of Baroque music. The *Illiac Suite for String Quartet* was performed successfully, but the composition was not considered to be of high quality.

The problems with computer composition were that when there are few rules for generating the notes, the resulting tunes are too random in structure, resembling white noise. The music generated has no direction or coherence. If the rules are stricter, then some coherence can be found in some tunes, because rather than white noise the use of strict algorithms often produces brown noise. Brown noise, named after Robert Brown (1773–1858), refers to a type of process that can be thought of as a random walk. The starting point can be set, but after that the walker may end up moving forward or back, to the sides, or some combination. Over time, the walker will have progressed somewhere and how they will get there cannot be determined. Not surprisingly, brown-noise computer compositions sometimes have a direction, of sorts, but can be rambling and dull.

Fractal mathematics, popularized by Benoit Mandelbrot (1924–) in the 1970s, offer a different type of structure for computer-generated music. Fractals are curves, surfaces, and objects that have non-integer dimensions. A point has a dimension of zero, a line a dimension of one, a square two, and a cube three. However, fractals have dimensions that lie between these; for example, a geometric construction called Koch's cube has a dimension of about 1.26. Surprisingly, a number of natural phenomena displays fractal properties, such as clouds, coastlines, landscapes, plants, and many more. Fractional dimensions produce some interesting patterns that usually have a degree of self-similarity in them; that is, the large-scale pattern resembles smaller structures within the pattern, which in turn resemble smaller structures within themselves. For example, a tree has a growth pattern that resembles that of a single branch, and within a branch there can be smaller branches with the same pattern.

Compositions that use fractal formulas sound more coherent that other types of computer-generated chance music. The property of self-similarity means that fractal compositions repeat themes in complex ways, which closely mimics a common property of human composition. Without knowing it, composers like Prokofiev and Cage had already used fractals in their music by using landscapes and star patterns to determine notes. However, most computer-generated music still sounds aimless and flat in comparison with human compositions. While many musicians have embraced the flexibility and inspiration that computer generation can give to a composition, computers are in no danger of taking over the writing of music just yet. There is still something human choices can give to music that cannot be fully simulated.

Randomness was originally seen as a negative when it entered music as unexpected or unwanted noise, yet when harnessed in the right manner it has produced many innovative and important pieces of music. Randomness has also entered into the way music is listened to, from CD-shuffling stereos to mp3 players with a random song selection, the old structures of albums and playlists are often sidestepped.

FREQUENCY OF CONCERT A

Sometimes mathematics and music have come into conflict, such as the long debates over the correct frequency of the notes in the Western scale. Orchestras and many other musicians often tune their instruments to the note known as concert A, and from that all the other notes in the scale are then set by their musical intervals from that frequency. However, the choice of this frequency is arbitrary. At first, Western music had no standard frequency for concert A, as there was very little communication across medieval Europe. Different regions sang and performed with their own pitches, because they had their own frequencies for the same notes. However, as contact between musicians increased across Europe, a rough standard was introduced and in the eighteenth century, concert A, as estimated by music historians, was about 420–425 Hz (Hertz, or cycles per second).

Once sound frequencies became better understood, and methods of measuring frequency were available, there were attempts to introduce a more specific and universal standard for concert A, although national pride and politics got in the way. The French and English set different frequencies, of 435 Hz (cycles per second) for the French and 439 Hz for the English.

Then in 1939, an international standard of 440 Hz was introduced, but against the will of a mathematical lobby that wished concert A to be set at 426.7 Hz, so that middle C would be at 256 Hz. This was called the philosophical pitch, as 256 is 2^8 (two to the power of eight, or two multiplied by itself eight times), and so seemed to the mathematicians a more formal, even Pythagorean, derivation of the note. The musicians, however, did not want such a dramatic change to the pitch of the music they played, as such a low number for concert A would have altered the sound of all existing music.

MATH-ROCK

Most musicians do not consider the mathematics that lie behind their music. Music can be composed and performed extremely well without any mathematical input from those involved. However, with the introduction of electronic instruments, it has become easier to introduce mathematical concepts into music. There is even a genre of rock music that calls itself math-rock, and is categorized by the creative use of time signatures.

A time or meter signature can be thought of as the number of beats in a measure of music, or in basic rhythmic terms, the number of drum beats in a set period of time. Normally, all the instruments in a piece of music will play in the same time, as this makes it easy to keep together, and usually sounds better. If one instrument plays in a different meter than the others, the result is usually unpleasant to the listeners. In math-rock, however, the musicians play in different meters on purpose. For example, in the Frank Zappa (1940–1993) instrumental *Toads of the Short Forest* from the 1970 album "Weasels Ripped My Flesh" there are two drummers, one playing in 7/8 time, while the other plays in 3/4. At the same time the organist plays in 5/8, creating an effect known as polyrhythm, or polymeter. The roots of polyrhythmic music go back to Indian and African music, as well as Latin music.

Music performance and composition are art forms, and many have called mathematics at its highest levels more art than science. Yet, when the basics of mathematics or music are learned, they both must start with simple rules and learned by rote, memorizing the building blocks of the subject until they become second nature. As learning progresses, the rules become more complex, and the effort needed to master them increases. For some people the effort is too great, or the rules too complex. Only a few people master a branch of mathematics or a genre of music, and at the highest levels the rules do not seem to be so important. They are still there, underpinning everything, but can be used in new ways, or stretched, or combined with unexpected results. For those people on the outside looking in, these highest workings of music and mathematics may be fascinating,

Key Terms

Analogue: A continuously variable medium, for use as a method of storing, processing, or transmitting information.

Frequency: Number of times that a repeated event occurs in a given time period, usually one second.

Scale: The ratio of the size of an object to the size of its representation.

spellbinding, even beautiful, but are strange and unexplainable; they are to be enjoyed, but never fully understood.

Potential Applications

Mathematics and music have become more entwined than ever before. The ways music is made, produced, transmitted, and listened to all rely heavily on mathematics, and many practical applications in music have come from abstract mathematical concepts. The shift to digital formats for music has been accompanied by continuing work in compression, error correction, and improving quality. Work in the field of music has produced mathematical tools and applications for other areas, and will continue to do so. In turn, mathematical ideas in other fields have successfully been transposed into musical applications. The experimental ethos that is at the heart of musical expression also exists in the field of musical instrumentation and engineering, and new devices are constantly being created.

Where to Learn More

Books

Garland, Trudi Hammel, and Charity Vaughan Kahn. *Math and Music: Harmonious Connections.* Palo Alto, CA: Dale Seymour Publications, 1995.

Helmholtz, Hermann. *The Sensations of Tone.* Mineola, NY: Dover Publications, 1954.

Johnston, Ian. *Measured Tones: The Interplay of Physics and Music*, 2nd Ed. Bristol, England: Institute of Physics Publishing, 2002.

Roederer, Juan G. *The Physics and Psychophysics of Music: An Introduction*, Third Edition. New York: Springer Publishing, 2001.

Web sites

Mozart's Musikalisches Wurfelspiel (Musical Dice Game) November 1995. <http://sunsite.univie.ac.at/Mozart/dice> (May 25, 2005).

Music and Mathematics: Dave Benson. 2003. <http://www.math.uga.edu/~djb/html/math-music.html> (May 25, 2005).

Nature and Numbers

Overview

Scientists depend on numbers and mathematics to explain many of the patterns that they observe in nature. For example, scientists find the same numbers repeated in many different places in nature. Fibonacci numbers and the Golden section are found in the geometry of mollusk shells and in the petal and leaf patterns of plants. Scientists also use special mathematical formulas called models to predict patterns that occur in nature. Mathematical models are used to explain how horses move and where phytoplankton grow most quickly in the ocean. New fields of mathematics have also resulted in insights into nature. Fractals are mathematical rules that can be used to represent complex shapes in nature and knot theory has the potential for providing an improved understanding of viral infections. Both numbers themselves and a variety of mathematical techniques greatly improve our understanding of the natural world.

Real-life Applications

FIBONACCI NUMBERS AND THE GOLDEN RATIO

Fibonacci (1175–1250) was an Italian mathematician credited with introducing the decimal system to Europe. In 1225 he took part in a tournament to solve a mathematical problem: a pair of rabbits produces another pair of rabbits every month, but it takes two months before any pair can produce their first offspring. Assuming no rabbits die, at the end of a year, how many rabbits will there be? Fibonacci answered the question by writing out a series of numbers. The first month there is one pair. Since the pair cannot reproduce until after the second month, the second month there is still just one pair. The third month, the pair reproduces so there are two pairs. The fourth month, the original pair reproduces, but the younger pair is still too young to reproduce; there are three pairs of rabbits. Continuing with this logic, Fibonacci showed that at the end of 12 months, there would be 144 rabbits. The series of numbers that answered the question is 1, 1, 2, 3, 5, 8, 13, 21, 34, 55, 89, 144 . . . Fibonacci noticed that this series could be extended infinitely by adding the previous two numbers to produce the next number in the series.

The Fibonacci numbers are repeatedly found in nature, particularly in plants. The number of petals on most flowers and leaves on many plants are Fibonacci numbers. For example, calla lilies have one petal, irises and lilies have three petals, buttercups, roses and columbine have five petals, delphiniums have eight petals, some daisies and marigolds have 13 petals. The numbers of leaves on only a few plants are not Fibonacci numbers: consider how rare

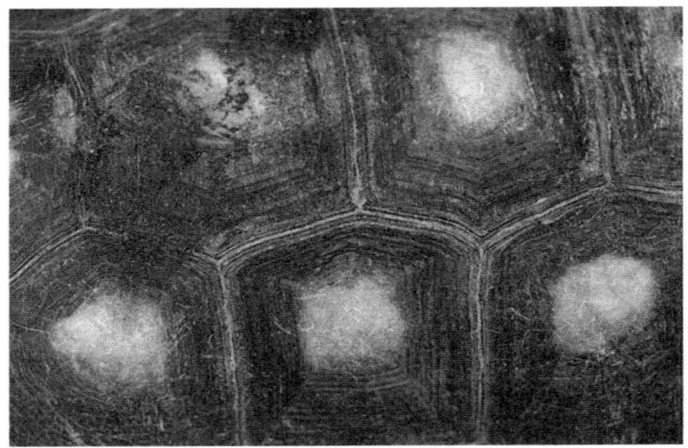

Note the geometric detail of a tortoise shell. ROYALTY-FREE/CORBIS.

four leaf clovers are. The center of many flowers, like sunflowers, is called the seedhead and it consists of seeds packed together in spirals. These spirals are organized in both clockwise and counterclockwise directions and they are interlocking. On a flower's seedhead the numbers of spirals in both directions are also Fibonacci numbers. For example, the seedheads of coneflowers have 55 clockwise spirals and 34 counterclockwise spirals; sunflowers have 89 clockwise and 55 counterclockwise. It turns out that this arrangement of seeds allows for optimal packing: seeds are not crowded and the most seeds possible can fit on the seedhead. The same pattern of interlocking spirals also appears on cauliflower heads, broccoli heads, pinecones and pineapples. The numbers of spirals on these structures are also always Fibonacci numbers. Additionally, the branching patterns of many plants can be expressed using Fibonacci numbers. The patterns of leaf growth around plant stems, called phyllotaxis, uses Fibonacci numbers.

Dividing every number of the Fibonacci series by its previous number results in another series: 1, 2, 1.5, 1.66, 1.6, 1.625, 1.61538 . . . , which eventually converges to 1.618034. This number is called the Golden ratio, the Golden section or the Golden number. It is often represented by the Greek letter Phi, ϕ. The part of the number to the right of the decimal, 0.618034 is called phi (with a lowercase p) and is equal to $1/\phi$. Both Phi and phi are observed in many places in nature. For example, many spiral mollusk shells incorporate Phi in their growth pattern. The nautilus shell is created as the mollusk adds larger and larger chambers to the outside of its shell. Drawing a line from the center of the spiral to the outside, the ratio of the distance between one turn of the spiral and the next turn of the spiral along that line is Phi. The DNA molecule, which contains the genetic information for all organisms, has the shape of a double helix. Each turn of the helix measures 34 angstroms in length. The width of the DNA molecule is 21 angstroms. These numbers are successive Fibonacci numbers and their ratio is close to the Golden ratio. The Golden ratio is observed in ratio of the width to the height of a dolphin's fin. The ratio of the length of the bones at the tips of the human fingers to the length of the bones between the first and second knuckle is the Golden ratio. Similarly, the ratio of the length of bones between the first and second knuckle to the length of the bone between the second knuckle and the hand is also the Golden ratio. The Golden ratio also turns up in the ratio of the distance between the top of the human head and the navel and the distance between the navel and the feet. In fact, some consider the Golden ratio fundamental to art, architecture and music.

MATHEMATICAL MODELING OF NATURE

A group of equations that are combined together in order to make a prediction is called a mathematical model. Scientists and mathematicians often use mathematical models to better understand their observations of nature. Mathematical models explain a variety of processes in the world, from how horses move to where phytoplankton are found in the ocean.

Biomechanics is a field of research that investigates how the materials in living organisms behave and mathematical models are fundamental to this research. For example, biomechanists have developed a mathematical model that describes the stresses and strains on the bones and muscles in horses' legs. These models lead to predictions of how horses move. At rest, the stress on a horse's bones is about 25% of the stress it would take to break the bones. When a horse starts walking the stress increases and it continues to increase as the horse walks more quickly. At the point where the stress reaches 30% of the breaking point, the horse switches to a trot, relieving the stress on the bones. Again the stress increases as the horse trots more quickly. The model predicts that at the point when the stress is about 30% of the breaking point, the horse should switch its behavior. Indeed at the speed where the stress on the bones is 30% of the breaking point, the horse switches its motion and begins to canter and the stress decreases. Finally, when the canter speed increases to the point that stress reaches 30% the model suggests another change in behavior. This speed is, in fact, the speed at which the horse begins to gallop decreasing the stress once again.

Phytoplankton are microscopic plants that live in the ocean. They perform photosynthesis, which is responsible for producing most of the oxygen on Earth and for removing a great deal of the greenhouse gas, carbon dioxide, from the atmosphere. In addition, phytoplankton are the base of the oceanic food chain, so where phytoplankton are, there are usually fish to be caught. Phytoplankton

Key Terms

Fibonacci numbers: The numbers in the series, 1, 1, 2, 3, 5, 8, 13, 21, 34, 55, 89, 144 . . . , which are formed by adding the two previous numbers together.

Fractal: A self-similar shape that is repeated over and over to form a complex shape.

Golden ratio: The number 1.61538 that is found in many places in nature.

Knot theory: A branch of mathematics that studies the way that knots are formed.

grow by dividing into two individuals, so when they grow faster, they are found in higher concentration. Phytoplankton growth requires both light and nutrients and it usually occurs more quickly when water temperature is warmer. Given information about light levels and nutrient concentrations, scientists construct mathematical model that predict where phytoplankton are found and in what concentration they will be found throughout the ocean. In fact, scientist can use their models with satellite imagery of the ocean to predict concentrations of phytoplankton throughout the oceans. These predictions help estimate the effects of carbon dioxide on the planet and help regulators develop fisheries policy.

Mathematical modeling is not limited to these two examples. Practically any observation or a pattern in nature, such as describing how leopards get their spots, why monkeys could never grow as large as King Kong, where hurricanes are likely to make landfall and even the height of the tides, can be better understood by developing mathematical models.

USING FRACTALS TO REPRESENT NATURE

In the 1970s a new branch of mathematics emerged that demonstrates how complex shapes in nature result from the repetition of the same pattern over and over. These repeating patterns, or self-similar shapes, are called fractals. Notice that a branch of a tree has roughly the same shape as the tree itself. Likewise, a piece of a cloud looks similar to an entire cloud. These complicated shapes are built by writing computer codes with very simple rules, or fractals, and then letting the computer repeat the rules over and over. The result is computer simulations of shapes that look like forms from nature: the complicated branches of trees, the clouds in the sky or a sprig of lilacs on a branch. Fractals have also been used to help forecast weather patterns. Fractals have helped mathematicians and biologists understand that complex-looking forms in nature may not really be that complicated. Instead these shapes are intricate repetitions of simple patterns.

Potential Applications

SPECIFY APPLICATION USING ALPHABETIZABLE TITLE

About one hundred and fifty years ago, mathematicians developed a knot theory, which describes the way that strings are formed into knots using mathematics. This allowed them to compare different knots to see whether they were similar or not. The mathematics of knot diagrams was worked out in the early 1920s using a new form of algebra. Eventually computer codes were developed to study different knots.

Recently knot theory has been applied to the study of viruses. Viruses are really just pieces of DNA or RNA, which can be thought of as long strands that can be made into knots. The virus inserts itself into the host's DNA causing it to coil, or knot, in ways that are very different from the original piece of DNA. Studying these patterns will, hopefully, bring insight to the fight against diseases that are caused by viruses.

Where to Learn More

Books

Devlin, Keith. *Life By the Numbers.* New York: John Wiley & Sons, Inc., 1998.

Gardner, Robert, and Edward A. Shore. *Math in Science and Nature.* New York: Franklin Watts, 1994.

Web sites

Britton, Jill. "Fibonacci Numbers in Nature." <http://ccins.camosun.bc.ca/~jbritton/fibslide/jbfibslide.htm> (September 25, 2003).

"Fibonacci Numbers and the Golden Section" *University of Surrey* <http://www.mcs.surrey.ac.uk/Personal/R.Knott/Fibonacci/fib.html> (November 1, 2004).

"What is the Fibonacci Sequence" <http://techcenter.davidson.k12.nc.us/Group2/main.htm> (November 5, 2004).

"SeaWiFS Project" *NASA Goddard Space Flight Center* <http://seawifs.gsfc.nasa.gov/SEAWIFS.html> (November 5, 2004).

Negative Numbers

Overview

Negative numbers are like mirror images of positive numbers. Any positive number can be made into a negative number by attaching a negative sign to it: for 5 there is -5, for 0.133 there is -0.133, and so on.

While negative numbers may at first seem intuitively odd, they actually have a prominent role in everyday life.

Fundamental Mathematical Concepts and Terms

When representing physical quantities with numbers, negative numbers allow the "bracketing" of a zero value. If that zero value has some meaning, then the positive and negative numbers that accrue from that point also have meaning.

For example, negative numbers allow benchmarks of physical behavior to be established. A fundamental example is the physical liquid-to-solid change that occurs in water at 32 degrees Fahrenheit (0 degrees Celsius). On several scales, measuring temperature negative numbers because the numbers characterize temperature states (a measure of molecular activity) below the zero point. Absolute zero on temperature scales is characterized by negative numbers on the Fahrenheit, Celsius, and Kelvin temperature scales. Absolute zero—0 Kelvin, $-459.67°$ Fahrenheit, or $-273.15°$ Celsius—is the minimum possible temperature: the state in which all molecular motion of the particles in a substance has ceased.

A Brief History of Negative Numbers

The conceptual roadblock against numbers less than zero dates back thousands of years. The ancient Chinese calculated numerical solutions using colored rods. Their use of red rods for positive quantities and black rods for negative quantities is opposite our present-day accounting standard (where "in the black" means having a *positive* bottom line). Although they accepted that numbers could decrease the value of a quantity, the idea that a final solution could be negative was unacceptable. Instead, they would rearrange the problem so that a positive number answer was obtained.

This line of thinking persists. For example, a section of a 2002 United States government income tax document contained the following section: "If line 61 is more than line 54, subtract line 54 from line 61. This is the positive amount you OVERPAID. If line 54 is more than line 61, subtract line 61 from line 54. This is the positive amount you OWE." In both cases, the solution is presented as a positive, rather than a negative.

The earliest known records of the acceptance of zero as an actual value and the related concept of negative numbers date from India around 600 A.D. Several centuries later, the Indian scholar Brahmagupta routinely used negative numbers and even derived mathematical rules to deal with such numbers.

Yet, even into the 1500s, European mathematicians argued that negative numbers could not exist, since zero signified nothing, and "it is impossible for anything to be less than nothing." In *The Principles of Algebra*, published in 1796, William Fredn stated that "to attempt to take [a number] away from a number less than itself is ridiculous."

By the nineteenth century, however, negative numbers had become accepted as a valid part of numbering systems.

Real-life Applications

TEMPERATURE MEASUREMENT

Whether the thermometer scale is in degrees Fahrenheit (°F) or Celsius (°C) or both, a range of positive and negative numbers brackets the zero mark.

In degrees Celsius, the zero value is arbitrarily assigned to that temperature at which water changes its chemical structure from a liquid to become a solid. Because the temperature of the air can become even colder, it is necessary to have a number scale to relate this degree of coldness in terms that are rational and intuitively understandable.

In a typical thermometer, the expansion and contraction of mercury or alcohol liquid in a column is indicated by the temperature scale. Thus, the mercury column will be shorter (compressed or contracted) on a −10°F-day than the length of the column on a 10°F-day.

The change in temperature that will occur during the course of a day is also described by negative numbers. As the day cools into evening, the temperature will decrease and the mercury column will drop (compress). The temperature decrease is represented by a negative number.

The Kelvin scale developed in the mid-1800s by the British scientist Lord Kelvin. The zero point of the Kelvin scale corresponds to absolute zero. This temperature, which represents the minimum molecular motion, is the equivalent of −273.16°C.

ACCOUNTING PRACTICE

The fundamentally important economic practice of bookkeeping, exemplified in balancing income and expenditures or in the completion of an income tax return, requires negative numbers. Depending on the accounting system used, negative numbers can represent a debt (e.g., −$20 representing a sum owed and to be deducted from an account), an expenditure (e.g., −$20 to be deducted from an account for a purchase, or a remaining balance (e.g., −$20 as a balance when adding together all income and expeditures where expeditures exceed income by $20.

The countdown clock is stopped at the T-minus five minute mark (a negative number) while the Space Shuttle Columbia sits on Launch Pad 39-B on Oct. 15, 1995 at Kennedy Space Center. NASA often uses negative number during countdowns to launch. AP/WIDE WORLD PHOTOS. REPRODUCED BY PERMISSION.

THE MATHEMATICS OF BOOKKEEPING

Bookkeeping describes the process where the amount of money coming in (income) and the amount of money being spent (expenses) are itemized in an arrangement that clarifies which side of the "ledger" is greater.

In a business or a home budget, the ultimate aim is to have a greater income than expense. This remainder can be used for investment or pleasure.

However, a harsh reality can be when expenses exceed income. An annual income of \$75,000 and expenses of \$85,000 will produce a final tally of 75,000 − 85,000 = −10,000. In modern-day bookkeeping parlance, this negative tally is "in the red." This is also known as a deficit.

Similarly, income tax calculations itemize all the sources of income and expense to arrive at either a positive number, which represents the refund payable by the government, or a negative number, which is the amount that the person or business filing the tax return owes the government.

Such negative numbers are not usually a cause for celebration.

However, a business owner can be heartened if the deficit has decreased from that of the previous year. For example, a deficit for one tax (fiscal) year of \$50,000 (−50,000) followed by a deficit of \$10,000 (−10,000) in the following fiscal year would be hailed as a tangible indication that the business is edging toward profitability (a positive number). So, even though the "bottom line" is a negative number, the trend is encouraging.

A bank account is another example of bookkeeping. An algorithm (a set of instructions that denote a method for accomplishment of a task) applied when every deposit or withdraw of funds are made keeps a tally of the money remaining in the account. A negative number, which can be represented by the bracketing of the number, indicates that the account is "overdrawn"; more money has been taken from the account than was actually in the account. The imbalance must be corrected by depositing more money to at least bring the account balance to zero.

SPORTS

In the realm of sports, various statistics and scoring systems are rooted in negative numbers.

Negative numbers are in integral part of a number of sports. In golf, the aim on each hole of the 18 holes that make up a standard course is to hit the ball from the starting area (the tee) into a hole in a determined number of shots. The number of shots representing normal or "par" performance varies from three to five (and, in rare cases, six) depending on the length of the hole. The cumulative score represents what is called "par" for the course.

The vast majority of golfers will never achieve par. Their scores will be greater than the ideal score. For example, on a par 72 score, a competent golfer may routinely shoot 85. This can also be described as shooting 13 over par, or +13.

Professional golfers are able to routinely shoot the par score and even lower, as for example a score of 65 on a par 72 score. This score can be described as a negative number, in this case 7 under par, or −7.

While a negative number is desirable in golf, it is undesirable in football. A team's movement on the football field from their end to the opposition's end is measured in yards. Positive yardage indicates forward movement. Negative yardage is indicative of backwards movement. If the quarterback or running back is tackled behind the place where the play began, the effort results in a negative number of yards. This makes the team's task of reaching the opposition's end of the field all the harder.

Negative numbers are encountered in sporting events where time is measured. Running events on the track or longer distances run on roads are two examples. Often, an athlete will gauge his or her performance by splitting the event into two or more equal distances (called splits) and timing each phase of the race (split times). A desirable goal is to achieve what are called negative splits, where the time to complete the later stage(s) of the race is less than the time spent in the earlier distance. Negative split times can be a way to the winner's podium.

On the track, negative numbers are part of the blindingly quick sprints. A worldclass sprinter can run 100 meters in under ten seconds. Often the field of runners will cross the finish line within a fraction of a second of one another. To sort out their finishing places, cameras positioned at the start and finish lines accurately record the respective times. The winner's time can then become the benchmark, denoted as zero, on which the other times are compared by means of negative numbers. Thus, if the second and third place finishers were 1/100 and 1/10 of a second slower, respectively, their times will be displayed as −0.01 and −0.1. This use of negative numbers permits a rapid assessment of the race's outcome.

FLOOD CONTROL

Land located next to a water body that can rapidly increase in volume can be flooded if the increased volume cannot be accommodated. The flooding potential of the watercourse can be measured by recording the level of the water above or below the flood stage. A positive number, which is above the flood stage, indicates that flooding is occurring. A negative number indicates that the water level has not reached the danger zone.

As with temperature, monitoring the progression of the number change over time reveals the trend, and so can guide subsequent actions. For example, if flood control officials note that the flood measurements are negative numbers and these are increasing with time, they can be confident that the flood danger is past and the river's level is settling back to normal. Flood control procedures can be eased or cancelled.

BUILDINGS

The floors in a multi-story building are denoted by numbers. Sometimes a building has levels below the ground, typically parking space, as well as floors towering overhead. The ground floor forms the demarcation between the above- and below-ground floors, in essentially the same way that zero marks the positive and negative number series on a numerical scale.

The real-life demarcation is readily evident when riding down in an elevator. Beginning on a floor (the fifth, for example), the floor indictor could display 5, 4, 3, 2, 1, 0, −1, −2. The latter two are the underground levels. The practice of which floor to assign zero or ground level often varies from country to country.

Where to Learn More

Books

Pickover, Clifford A. *Wonders of Numbers: Adventures in Mathematics, Mind, and Meaning.* New York: Thomas Dunne Books, 2004.

Schwarz, Alan. *The Numbers Game: Baseball's Lifelong Fascination with Statistics.* New York: PowerKids Press, 2004.

Strazzbosco, John. *Extreme Temperatures: Learning About Positive and Negative Numbers.* New York: PowerKids Press, 2004.

Web sites

Purplemath. "Negative Numbers Review I." <http://www.purplemath.com/modules/negative.htm> (February 11, 2005).

British Broadcasting Corporation. "Skillswise factsheet: negative numbers – practical examples." <http://www.bbc.co.uk/skillswise/numbers/wholenumbers/whatarenumbers/negativenumbers/> (February 11, 2005).

Number Theory

Overview

Number theory is the study of numbers, in particular integers. Integers are the positive and negative whole numbers: . . . -3, -2, 1, 0, 1, 2, 3 . . . Number theory was once considered a branch of pure mathematics, which means that its major focus was to explore the properties of numbers without concern for the real-world application of any of the results. Nonetheless, applications of number theory that are extremely important to the real world have resulted from research in this field. Cryptography, which is the transformation of information into a form that is unintelligible (and the reverse of this process) is commonly used in electronic transactions of all kinds to ensure privacy and security. Error checking codes, which are used in telephone communications, satellite data transfer, and compact discs, ensure that information remains intact. Both of these applications have foundations in number theory.

Fundamental Mathematical Concepts and Terms

Number theory is concerned with the properties of integers. Because of its concern with numbers, some people associate the terms arithmetic and higher arithmetic with number theory. Number theory is subdivided into a number of fields, the major ones being elementary number theory, analytic number theory, algebraic number theory, geometric number theory and Diophantine approximation. Several other fields of study within number theory include probabilistic number theory, combinatorial number theory, elliptic curves and modular forms, arithmetic geometry, number fields, and function fields.

Elementary number theory is one of the major subfields of number theory. The word elementary does not refer to the simplicity of the problems in this subfield, but rather to the fact that the problems studied do not use techniques from any other field of mathematics. Elementary number theory has a certain popular appeal because many of the problems are easily explained, even to people who are not mathematicians. However, finding solutions for these seemingly simple problems is often extremely complex and require great insight.

Some of the important problems involve prime numbers. Prime numbers are numbers greater than 1 that only have two divisors: 1 and the number itself. The prime numbers less than 10 are 2, 3, 5 and 7. As of 2005 the largest known prime number was $2^{25964951} - 1$, which has 2,816,230 digits. It is a special type of prime number called a Mersenne prime. Prime number theory states that there are an infinite number of prime numbers; new ones are being found all the time.

Elementary number theory also investigates perfect numbers. Perfect numbers are numbers that are equal to the sum of all the integers that are its divisors. The number 6 is a perfect number. Its divisors are 1, 2 and 3 and the sum of these three numbers is 6. A second perfect number is 28. Its divisors are 1, 2, 4, 7 and 14, which sum to 28. Ancient Greeks discovered two more perfect numbers: 496 and 8,128. As of 2005, 42 perfect numbers were known and all of them were even. It is unknown if an odd perfect number exists, but if one does number theorists have shown that it will have at least seven different prime factors.

Questions of divisibility and prime factorization are also part of elementary number theory. Divisibility means that a number can be divided by another number without leaving a remainder. For example, both 5 and 6 are divisors of 30. Finding all the divisors that are prime numbers is prime factorization. The prime factors of 30 are 2, 3 and 5.

One of the important operators used in number theory is modulus. Modulus refers to dividing an integer by another integer and calculating the remainder. For example, 10 mod 2 = 0 because 10 is divided evenly by 2 and there is no remainder. In another example, 10 mod 3 = 1 because dividing 10 by 3 is 3 with a remainder of 1. Modulus is used in cryptography as described below.

The Euclidean algorithm is also part of elementary number theory. This algorithm is used to find the greatest common divisor of two integers. Euclid wrote it down in about 300 B.C., making it one of the oldest algorithms known. The greatest common divisor is the largest number that divides two integers without leaving a remainder. For example, the greatest common divisor of 42 and 147 is 21, although 3 and 7 are also common factors.

A second subfield of number theory is analytic number theory. This field involves calculus and complex analysis to understand the properties of integers. Many of these techniques depend on developing functions that describe the behavior of arithmetic phenomenon and then investigating the behavior of the function. This often makes use of the asymptotic nature of certain functions; functions that tend toward certain values called limits at extremely large (or small) values.

A number of statements in elementary number theory are easily described, but require extremely complicated techniques in analytic number theory to solve. For example, the Goldbach conjecture states that every even number greater than 5 is the sum of three primes. This conjecture has never been proven or disproven, but remains a source of much research in analytic number theory. The twin prime conjecture states that there are an infinite number of primes of the form p and $p + 2$. Although most mathematicians argue that this is true, it too has never been proven and remains an active area of research.

The subfield algebraic number theory concerns number that are algebraic numbers, which are numbers that are the solutions to polynomial expressions. All numbers that can be expressed as the ratio of two integers, also called rational numbers, are algebraic numbers. Some irrational numbers are also algebraic.

Some of the important areas of research in algebraic number theory are Galois theory, which studies how different solutions to polynomials are related to each other, and Abelian class field theory and local analysis, which investigate the properties of fields. In mathematics, fields are abstract structures in which all the elements can be subjected to addition, subtractions, multiplication and division (except by zero) and in which the distributive rule, the associative rule and the commutative rule all hold.

Geometric number theory and Diophantine approximation represent another field of study within number theory. Diophantus was an ancient Greek mathematician who lived in Alexandria, Egypt, probably in the third century A.D. He wrote a treatise called *Arithmetica* in which he described many problems concerning number theory. Diophantine equations are attributed to this great thinker and they are equations that have whole numbers as their solutions. Some of the most common Diophantine equations are whole number solutions to the Pythagorean theorem: $x^2 + y^2 = z^2$, which can also represent the length of the sides of a right triangle (a triangle that has one 90° angle). Some solutions include $3^2 + 4^2 = 5^2$ and $5^2 + 12^2 = 13^2$. In fact, Diophantus showed that there are an infinite number of whole number solutions to the Pythagorean equation. Other problems in geometric number theory incorporate the theory of elliptic curves, the theory of lattice points in convex bodies and the packing of spheres in different types of spaces.

Fermat's last theorem is one of the most famous statements in number theory. It claims that there are no solutions to the problem $x^n + y^n = z^n$ for any values of n greater than 2. In the margin of Diophantus's *Arithmetica*, the famous French mathematician Pierre de Fermat claimed "I have a truly marvelous demonstration of this proposition, which this margin is too narrow to contain." In 1665 he died without ever writing down the "marvelous demonstration." The statement was the source of much fascination to mathematicians for more than three centuries. A cash reward was even offered to the person who could provide a proof of the statement. Most mathematician accept that an extremely complex proof using techniques in geometric number theory by mathematician Andrew Wiles finally proved Fermat's last theorem in 1994.

Real-life Applications

Number theory is a pure math discipline, which means it evolved without any attention to developing real-life applications. Nonetheless, number theory has proven to have real-life applications that affect almost everyone. As the Internet and other forms of electronic communication has become a larger part of daily life, the need to keep personal information private and to verify the identity of individuals becomes extremely important. Number theory provides techniques, which can be used to disguise information in order to ensure privacy and security. These techniques form the basis for the field of cryptography and they are used in a broad range of industries from retail stores to finance to government to healthcare. Every time a credit card is swiped, a bank transaction occurs, insurance agencies and hospitals send patient information to each other or the police use a driver's license to verify an identity, techniques from number theory are used to keep the information transferred secure.

While the goal of cryptography is making information harder to decipher, the goal of error correcting codes is to protect information from corruption. Error correcting codes are based in number theory and they are used in everything from the information beamed back to earth from Mars rovers to the compact disks that contain music.

CRYPTOGRAPHY

Cryptography is the set of techniques, usually mathematical, that are used to encrypt and decrypt information. Encryption means converting information from its understandable form to a form that is unintelligible. Most often, a set of mathematical steps called an algorithm is used for this purpose. A second algorithm is then performed to transform the unintelligible version of the message back to it original form. This is called decryption.

A simple example of encryption is the XOR algorithm. It can be used to transform binary codes. Binary codes are strings of 0s and 1s. All information in computers is eventually reduced to binary codes. Binary addition is slightly different from the addition that is commonly used with integers. It has four rules: $0 + 0 = 0$; $0 + 1 = 1$; $1 + 0 = 1$; and $1 + 1 = 0$.

Suppose that a binary message is 1010. A key for encrypting this message could be any string of four 0s and 1s; for example, 1101. Adding the original message to the key (bit by bit, with no carryover from the highest place—also known as the XOR function) results in an encrypted string. If someone were to intercept the encrypted string, they would not know what the original message was without the key. With symmetric keys, the same key is used for encryption and decryption.

There are several symmetric key systems in common use. The data encryption standard (DES) is one of the most popular, though it is not considered particularly secure because more than one person knows the key. The Diffie Hellman key agreement algorithm provides a higher degree of security because the parties involved in the exchange of information negotiate the key that they want to use as they exchange information. Because the key is developed as it is used, the chances that it will be intercepted by a third party decreases. In addition, the algorithm relies on the fact that the people involved in the negotiation will only have to do simple calculations to establish the key, but an eavesdropper would have to do very difficult calculations to steal it.

Encryption techniques that employ asymmetric keys, also called public keys, require that different keys be used for encryption and decryption. One of the most commonly used public key systems is the RSA Public-Key System. It is named for the last names of its developers, Ron Rivest, Adi Shamir and Leonard Adleman, who first developed the algorithm in the 1970s at MIT. These mathematicians build a company around the algorithm called RSA Data Security, headquartered in Redwood City, CA. RSA technology has been incorporated into a broad range of computer software including Microsoft Windows, Netscape Navigator, Intuit's Quicken, Lotus Notes, as well as operating systems for Apple, Sun and Novell computers. It is part of the Society of Worldwide Interbank Financial Telecommunications standards for financial transfers as well as standards used by the United States banking industry.

The RSA algorithm makes use of two important features of number theory: prime numbers and the modulus function. Its security depends on the fact that it is very hard to factor very large numbers. For example, the algorithm usually uses a modulus that is somewhere near 2^{800}; in order to discover the private key, an eavesdropper would need to find a way to factor the modulus.

Several types of factoring algorithms have been developed and they can be used to estimate the difficulty of

The RSA Public-Key Algorithm

The RSA algorithm is one of the most popular public key algorithms. It is probably best understood by example. Assume that a customer wants to make an electronic deposit of $3 using an automatic teller. The number 3 is the original message, *M*, which must be encoded for transfer and then decoded when the bank receives it. The automatic teller acts as the keymaker, generating numbers that act as keys for encryption and decryption.

In the first step of the RSA algorithm the keymaker generates two prime numbers: say $p1 = 11$ and $x2 = 2$. Next the product of the two numbers is calculated: $n = (p1)(p2) = (11)(2) = 22$. This number *n* is part of both the encryption key and decryption key. It is the modulus that is used later in the algorithm.

Next a number is calculated using Euler's totient function. This number is referred to as *t* and it is equal to $(p1-1)(p2-1) = (11-1)(2-1) = (10)(1) = 10$. The keymaker then selects a number, let's call it *e*, such that *e* is less than *t* and the greatest common divisor of *e* and *t* is 1. In this case the number chosen is 3, because it is less than 10 and the greatest common denominator of 3 and 10 is 1.

The next calculation requires finding a number *d* such that when the product of *e* and *d* is divided by *t* the remainder is 1. Another way to write this is $ed = 1 \bmod t$. In this case, *d* is 7 because $3 \times 7 = 21$ and $21/10 = 2$ with a remainder of 1.

The public key, used to encrypt the message is *e* and *n*, in this example 3 and 22. The keymaker may make this key known to everyone. The private key is *d* and *n* or 7 and 22, and only the keymaker knows it.

The keymaker now transforms the message by raising the message *M* to the power *e*, dividing by *n*, and calculating the remainder. This calculation can also be written $M^e \bmod n$. In this example $M^e = 3^3 = 27$. The number 27 is divided by $n = 22$ and the remainder is 5. The encrypted message, *E*, is 5.

The automatic teller then sends the encoded message ($E = 5$) to the bank, along with the private key, which in this example is *d* and *n* or 7 and 22.

The encrypted message and the private key is received by the bank, they decrypt it using the calculation, $E^d \bmod n$. In this example, $E^d = 5^7 = 78{,}125$. The number 78,125 is divided by $n = 22$ and the remainder is 3, which was the original message.

Although very small numbers were chosen for *p*1 and *p*2 in this example, in practice they are usually on the order of 2^{400}, which makes *n* extremely large, somewhere near 2^{800}. The difficulty in factoring such big numbers is crucial to the security of the algorithm. If the factors of *n* were easy to find, then discovering the private key would not be that hard to do.

factoring very large numbers. According to one of these algorithms, in order to factor a value number that is close to 2^{800} would require about 2^{77} steps. In 2005, the average computer could do about 100 million instructions per second. This corresponds to 100,000,000 instructions/second $\times$ 60 seconds/minute $\times$ 60 minutes/hour $\times$ 24 hours/day $\times$ 365 days/year $= 3 \times 10^{15}$ instructions/year, which can also be written approximately as 2^{51} instructions per year. As a result it would take about $2^{77} / 2^{51} = 2^{(77-51)} = 2^{26}$ or roughly 70 million years to factor the modulus.

In addition to RSA, other groups of public key algorithms have been developed. One is called ElGamal and it relies on similar mathematics as the RSA algorithm. ElGamal can also be used to verify that information sent has not been compromised during transmission. It does this by means of a digital signature and special mathematical functions called Hash functions. Digital signal algorithm (DSA) can also be used for digital signatures. Another public key algorithm relies on functions called elliptic curves, which are studied in number theory and have become increasingly popular for use with cryptography.

ERROR CORRECTING CODES

As binary information (information coded as strings of 0s and 1s) is transmitted, errors can occur in the string, which make the information unintelligible. Error correcting codes are algorithms that ensure that information is transmitted error-free and many of these algorithms depend on results from number theory.

Claude Shannon and Richard Hamming working at Bell Laboratories in the late 1940s developed a method of repeating strings to ensure that the information sent was received. They worked out theories, which optimized the number of repetitions necessary to ensure that the information received was correct. Another researcher,

Key Terms

Algorithm: A set of mathematical steps used as a group to solve a problem.

Binary code: A string of zeros and ones used to represent most information in computers.

Decryption: The process of using a mathematical algorithm to return an encrypted message to its original form.

Divisibility: The ability to divide a number by another number without leaving a remainder.

Encryption: Using a mathematical algorithm to code a message or make it unintelligible.

Greatest common divisor: The largest number that is a divisor of two numbers.

Integer: The positive and negative whole numbers.. –4, –3, –2, –1, 0, 1, 2, . . . The name "integer" comes directly from the Latin word for "whole." The set of integers can be generated from the set of natural numbers by adding zero and the negatives of the natural numbers. To do this, one defines zero to be a number which, added to any number, equals the same number.

Key: A number or set of numbers used for encryption or decryption of a message.

Modulus: An operator that divides a number by another number and returns the remainder.

Perfect number: A number that is equal to the sum of its divisors.

Prime factorization: The process of finding all the divisors of a number that are prime numbers.

Prime number: Any number greater than 1 that can only be divided by 1 and itself.

Public key system: A cryptographic algorithm that uses one key for encryption and a second key for decryption.

Symmetric key system: A cryptographic algorithm that uses the same key for encryption and decryption.

John Leech, also developed theories related to error correcting codes. His work included some abstract mathematical in number theory such as groups and lattices.

Error correcting codes were put to immediate use at NASA, where satellites were equipped with powerful error checking codes. A typical algorithm of this type is capable of correcting seven errors in every 32 bits sent back to earth. The redundancy in the data sent is immense; in every 32 bits only 6 are data. The rest are for error checking.

When they were initially developed, compact discs were highly sensitive to scratching and cracking. But by incorporating two redundant codes that are interleaved, CD players can recover up to 4,000 consecutive errors. Additional error checking algorithms are built into CD players to further correct problems with the signal.

Where to Learn More

Books

Gardner, Martin. *The Colossal Book of Mathematics.* New York: W.W. Norton & Company, 2001.

Gullberg, Jan *Mathematics: From the Birth of Numbers.* New York: W.W. Norton & Company, 1997.

Hoffman, Paul. *The Man Who Loved Only Numbers.* New York: Hyperion, 1998.

Paulos, John Allen. *Beyond Numberacy: Ruminations of a Numbers Man.* New York: Alfred A. Knopf, 1991.

Periodicals

Schroeder, Manfred R. "Number theory and the real world," Math. *Intelligencer 7* (1985), no. 4, 18–26.

Web sites

Department of Mathematics University of Illinois at Champaign Urbana. "Guide to Graduate Study in Number Theory." January 23, 2002. <http://www.math.uiuc.edu/Research Areas/numbertheory/guide.html> (April 27, 2005).

Grabbe, J. Orlin. "Cryptography and Number Theory for Digital Cash" October 10, 1997 <http://www.aci.net/kalliste/cryptnum.htm;> (May 2, 2005).

The Mathematical Atlas. "Number Theory." January 2, 2004. <http://www.math.niu.edu/~rusin/known-math/index/11-XX.html> (April 27, 2005).

Pinch, Richard. "Coding theory: the first 50 years" Plus Magazine. September 1997. <http://pass.maths.org.uk/issue3/codes/> (May 4, 2005).

The Prime Pages "The Largest Known Primes." <http://primes.utm.edu/largest.html> (May 9, 2005).

RSA Security. "Crypto FAQ" 2004. <http://www.rsasecurity.com/rsalabs/node.asp?id=2152> (May 4, 2005).

Weisstein, Eric W., et al. "Number Theory." MathWorld—A Wolfram Web Resource. <http://mathworld.wolfram.com/NumberTheory.html> (April 24, 2005).

Odds

Overview

Probability is a form of statistics used to predict how often specific events will occur, and is used in fields as varied as meteorology, criminal justice, and insurance underwriting. When probabilities are calculated, they are frequently expressed in terms of odds. Odds provide a simple, shorthand language for communicating probabilities, regardless of the specific situation being assessed. Odds can be expressed using differing terminology and notations, but the basic principles remain constant, regardless of the application.

Fundamental Mathematical Concepts and Terms

Several systems of terminology can be used to express the odds of a particular event occurring. Consider the case of a standard deck of 52 playing cards, which consists of four suits of 13 cards apiece. A dealer takes this deck, shuffles it thoroughly, then draws a single card; what are the odds that he will draw the single Ace of Hearts? The odds of drawing this particular card out of the fifty-two in the deck are denoted 1:52. This probability can also be described as one chance in 52 of successfully drawing the desired card, or expressed as a fraction: 1/52.

Odds are also expressed in reverse, giving the odds against an event happening. In the previous example, the odds of drawing the desired card from the deck could also be expressed as 51:1, meaning that of the 52 possible outcomes, 51 would be undesirable while only 1 would be the hoped-for outcome. In some cases, the odds for and the odds against are used interchangeably: odds of 1 in a million and odds of a million to 1 are both used to describe extremely unlikely events, and are almost identical mathematically. However, this same relationship does not hold true for smaller values, with odds of 1 in 3 (33%) being significantly better than odds of 3 to 1 (25%).

A Brief History of Discovery and Development

Because odds are simply the language of probability, the history of odds runs parallel with the history of probability, and is discussed extensively in the entry on that subject. However, as the language of odds has been applied to an expanding array of applications, a unique vocabulary has developed around the use of odds. Unfavorable odds, such as odds of one in a million, have come to be called long odds, meaning the event they describe is highly improbable. Long odds are also sometimes referred to as a long shot; slow race horses and unknown political candidates are often described as long shots, suggesting that their odds of winning are remarkably small. Odds are sometimes expressed in terms of a percentage, or on a base of 100. A salesman who claims to be 90% sure he can deliver his product on time is offering odds of 9 in 10 that he will succeed. A wildcatter drilling a new oil well might give odds of 60/40 that the new well will be a gusher, placing the odds at 6:4, which can be reduced to 3:2 and then reduced further to 1.5 to 1 that he will succeed. A stock analyst who gives a stock a 50:50 chance of rising is giving it a 1 in 2 chance, equal to the chance of flipping heads on a single toss of a coin.

In a few cases, odds are used to imply that an event is absolutely certain to occur; theoretically these odds would 1:1, or 100%, though the certainty of any future event is always less than 100%. However in these cases, an event is often referred to as a lock, or a sure thing, suggesting that it will certainly occur. However the history of gambling and athletics is replete with sure things which failed to materialize, suggesting that the sure thing and the lock are more a result of wishful thinking than of rigorous statistical analysis.

Real-life Applications

SPORTS AND ENTERTAINMENT ODDS

Poker is a popular card game in which odds are used to develop strategy, and successful poker players often possess an innate sense of the odds associated with certain hands. In the course of a typical poker game, players are often forced to make quick decisions on whether a hand is winnable and should be played, or is unwinnable and should be folded. For example, a player holding a 5, 6, 7, 9, 9 must decide whether to keep the 9s and hope to be dealt a third 9, or to discard one of the nines in the hope of drawing an 8 to complete a straight. Using a basic understanding of probability and the rules of the game, an experienced player will probably keep the 9's, knowing that the odds of ending up with a winning hand are significantly better using this strategy.

Gambling in any form has been a popular pastime for most of recorded history. During the past century, gambling, or gaming, as it is sometimes called, grown from a casual pastime into a multi-billion dollar industry. As the gambling industry has grown, casino owners have increasingly turned to fields such as psychology and marketing in order to increase their earnings. The very existence of these extravagant entertainment centers, some costing more than $1 billion to build, simply confirms the efficiency with which casino owners separate players from their money.

Modern casinos are scientifically designed to lure players in and keep them playing as long as possible. A typical casino contains a variety of games, offering a wide array of playing styles and varying odds of winning though one fact remains: in every form of casino gambling, the odds of the game favor the casino, or as it's known in the industry, the house. In most cases, the tilt in favor of the house is slight, allowing some players to beat the house over the short-run and leading to impressive tales of huge jackpots. But the ultimate result is the same as in any other activity governed by the laws of probability: over the long-run, the house will always win.

The house edge, or how strongly the odds of a particular game favor the casino, vary from game to game. The game of roulette, in which a ball is dropped onto a spinning number wheel, offers a house edge of 5.6%, meaning that in practice, for each $100 wagered, a player will lose an average of $5.60. If a roulette player spends two hours playing, betting $25 per spin and averaging 30 spins per hour, the casino will expect to make about $75.00 in that time, and the customer will have paid roughly $37.50 an hour for the privilege of watching a small marble drop onto a shiny spinning wheel. Of course two other outcomes are also possible. A player could actually win several times in a row and walk away with his winnings, taking the house for a loss; this possibility is what keeps die-hard gamblers coming back for more action. The other possibility is that the player hits a run of tough luck and loses his entire stake sometime during the session. In this case, the casino's edge has simply been felt earlier than expected, and the player goes home empty-handed.

Roulette offers some of the lowest odds of any casino table game, meaning the house edge is larger in this game than in most others. Blackjack, a card game in which a player tries to collect cards totaling 21, offers a theoretical house edge of only 0.80%, though in practice few players play the game with such computer-like precision, making the actual house edge higher. Assuming a gambler can follow the optimal betting strategy without error, he should be able to wager $100 during his session and lose only 80 cents to the casino. Similar odds accompany the game of craps, in which dice are rolled and games are won and lost based on the outcome of the roll.

Some of the worst odds in the casino are offered by one of the most popular games, the slot machine. Aptly nicknamed the "one armed bandit," these flashing, beeping machines involve no skill whatsoever, requiring players only to insert a coin and pull a lever or push a button. Based on the outcome of a set of spinning wheels, prizes are paid according to a table on the front of the machine. The house edge for slot machines is difficult to calculate, because machines can be programmed to return a higher or lower amount of player money. As a general rule, slot machine payouts vary depending on the amount required to play, with higher play values receiving better odds. The typical house edge for a nickel slot machine would be somewhere near 8%, meaning that for each $100 bet, a player would typically lose about $8.00. Odds are much better on higher-value slot machines, meaning that a player willing to spend $5.00 per play will face a less severe house edge. Unfortunately, he will also burn through his funds much more quickly. Slot machines offer high efficiencies to casino operators. By combining low operating costs, a high house edge, and the potential for players to bet several times per minute, one armed bandits are among the casino's best money-makers, probably explaining why most Las Vegas casino entrances are lined with a sizeable collection of the shiny machines.

Why do people gamble? Few other activities offer a guaranteed chance to lose money, yet gambling today is more widespread in both the physical and the virtual world than ever before. For some players, gambling is simply a form of entertainment. These players typically allot a set amount to spend on an outing, wager and enjoy the experience and the excitement, then leave, having paid relatively little for their entertainment. For other players, gambling is perceived as a chance to improve their lot in life by offering a fast route to large amounts of cash.

Some percentage of gamblers behave irresponsibly, wagering far more than they can afford to lose and creating serious problems for themselves and their families. Compulsive gamblers are similar to compulsive drinkers in that they are unable to moderate their behavior; in some cases, compulsive gamblers spend entire paychecks or close out bank accounts attempting to recoup previous gambling losses. Gamblers Anonymous, an organization created to help compulsive gamblers recover, provides a list of questions to help gamblers determine whether they have a problem. Questions such as, "Did you ever gamble to get money to pay debts," "Have you ever sold anything to finance gambling," "Did you ever gamble down to your last dollar," and "Did you ever have an urge to celebrate good fortune with a few hours of gambling?" are intended to help gamblers assess their situation. According to the National Council on Problem Gambling, in 2005 between 3,000,000 and 12,000,000 Americans had gambling problems of varying degrees.

The game of roulette offers a house edge of 5.6%, meaning that in practice, for each $100 wagered, a player will lose an average of $5.60. ROYALTY-FREE/CORBIS.

ODDS IN EVERYDAY LIFE

While the question "paper or plastic?" is probably the most common dilemma faced by shoppers, one shopper's quandry (perplexing question) is as old as the supermarket: how can a shopper tell which check-out line will move fastest? Obviously if one line has fewer shoppers in it, that is probably the line to choose, though smart shoppers also know that if the light above that line's cashier is flashing it's probably a danger sign. But what if all the lines have the same number of customers waiting? What chance is there of choosing the fastest line? All things being equal, a shopper's chance of choosing the fastest line is actually pretty small, a statistical reality born out by many shoppers' frustrating experiences.

Consider this quandary as a probability question. Assume that the shopper has no information about which checkers currently working are faster or slower, meaning that in this case, his decision comes down a random selection. Once he has made his choice, in this case choosing lane three, he will be forced to stand and watch the other lanes to learn whether or not he chose wisely. From lane three he is able to see lanes one and two on his left and lanes four and five on his right, and chances are good that he will soon find out he did not choose the fastest lane. A simple odds calculation explains why.

The total group of possible outcomes consists of the five cashiers from which the shopper can choose, while the outcome of interest is the particular lane the shopper ultimately selects. All other factors being equal, the odds of choosing the fastest line are 1 in 5; put more pessimistically, the odds of choosing incorrectly are 4 in 5, meaning that most days, most shoppers will watch at least one other line move faster than the one they have chosen. Given these poor odds, one might instead opt for a new development, self check-out, which in most cases is markedly slower than being checked out by a professional, but which eliminates the annoying wait in line, as well as providing a reasonable distraction from the process.

Human beings have an innate fascination with events or objects which seem to defy the laws of probability. Chinese basketball star Yao Ming fascinates most Westerners, not just because he is a basketball star or even because he is an unlikely 7'6" tall. Most Americans are stunned by Yao's towering height because he beat the odds by growing so tall in a nation where the average man is around three inches shorter than the average American man. While Yao seems like the ultimate example of a long shot, two factors make his height far more understandable. First, his parents are exceptionally tall, even by U.S. standards, with his father standing 6'7" tall and his mother measuring 6'3". Second, regardless of the average national height, China is more likely to produce another giant player simply by virtue of its enormous population: with more than 1.3 billion residents, mainland China is more than four times as populace as the U.S., dramatically boosting its chances of producing another seven-footer or two. In addition, both the U.S. and China are demographically diverse, and there are areas of China where the average height exceeds the average world and U.S. height for males.

ODDS IN STATE LOTTERIES

Many states and countries now generate revenue by operating lotteries. In a typical lottery, players are encouraged to buy a ticket with a set of numbers on it, such as six numbers between 1 and 45. Tickets are sold for a set period of time, then a drawing is held in which 6 numbers are randomly selected. Matching some of the selected numbers is rewarded with a cash prize determined by the number of numbers guessed correctly. In a typical lottery, the largest prize, the jackpot, is won by correctly guessing all the numbers selected, with the prize normally being more than one million dollars. Because most lotteries are run by government agencies, they are required to publicize details of the games, such as the odds of winning at each level and the actual use of the lottery's earnings.

State lottery managers must make several decisions in order to maximize the number of players and, by extension, the total number of dollars earned for the state; for this reason, lottery rules change frequently in order to keep players interested. One state lottery in 2005 used the following formula: two sets of numbers are used, each running from one to 44. From the first set of numbers, five values are randomly selected, and from the second set of numbers, a single bonus number is chosen. A player lucky enough to match all five of the initial numbers has managed to beat odds of 1.1 million to one and will collect a sizeable prize. But a player who manages to combine this feat with a correct pick of the bonus number wins the top prize, frequently in the tens of millions of dollars. Jackpots often go unclaimed for several weeks, since the odds of picking all six values correctly are 1 in 47 million. A player's odds of winning any prize (prizes start at $3.00) in a single play are 1 in 57.

Where does lottery money go? Lottery proponents are quick to point out that the net proceeds of a lottery are typically spent on education and other popular projects. However, the actual education income from lottery tickets is typically less than one-third of the money spent playing the game. The Texas State Lottery in 2004 published a breakdown of how its income was spent. For each dollar wagered, the program returned fifty-two cents to players in the form of prizes, making lotteries among the worst bargains in gambling when compared with almost any casino game. Seven cents of each lottery dollar also went to administrative costs associated with running the lottery itself, including salaries for administrators and advertising costs, while another five cents was paid to retailers in return for their work selling the tickets.

Once all these costs are removed, this particular lottery program contributes the remaining thirty cents of each dollar to the state's education agency for use in local school programs. Is this level of return high or low? The answer depends on whether the lottery is analyzed as a business or as a non-profit fund-raising agency. For most business CEOs, managing to pass 30% of their gross revenues along to their owners would make them among the most successful and admired business leaders in the hemisphere. But a non-profit fund-raising organization which consumes 70% of its revenues paying administrative and other costs before passing less than one-third of contributions along to its beneficiary is generally considered either unethical or grossly incompetent. Because state lotteries fail to cleanly fit either model, the appropriateness of this 30% pass-through rate remains controversial.

Lotteries have risen from relative obscurity in America during the early twentieth century to a point where most states operate the programs. Several facts have contributed to this rise in popularity. One trend which helped lotteries flourish was a general resistance to additional taxes, beginning with the Reagan presidency in the 1980s and continuing through the turn of the century. In an atmosphere where even proposing a tax hike could be politically fatal, lotteries provided a sizeable revenue boost without the political costs of a tax hike. A second argument has gained steam as lotteries have spread to cover most of the country. Economic studies have looked at out-of-state revenue garnered by lotteries, concluding that in many cases, residents of non-lottery states drive across the border to buy tickets when jackpots grow large. In response, some non-lottery states eventually conclude that initiating their own lottery is the lesser of two evils when compared with continuing to watch residents take their dollars to neighboring states.

A final argument in favor of starting lotteries is the voluntary nature of participation. For many voters, the choice between a hike in property taxes, which impacts most citizens, and a state lottery which produces the same amount of revenue but is purely voluntary, may seem straightforward. However, given the lottery's common nickname of "the stupid tax," (a perspective on the fact that the "tax" is regressive in that it is generated even if voluntarily) from people with lower levels of education and income as opposed to the general population.

While the long-term impact and success of state lotteries remains to be proven, various challenges have already faced lottery administrators. For example, in one case a larger than expected number of players of one particular game won jackpots during the first quarter of the year. Because lottery profits are tied directly to the number of tickets sold, this statistical fluke did not endanger the lottery's solvency. However, because of the psychological appeal of larger prizes, this run of smaller jackpot winners did substantially reduce the number of ticket sold, slashing income during the first months of the year. By mid-year, directors were evaluating a change in the game rules to reduce the number (and increase the size) of jackpots won.

OTHER APPLICATIONS OF ODDS

Large numbers are used for a variety of business purposes, including security. A typical credit card uses a sixteen digit account number. For a thief trying to make an online purchase, how hard would it be to simply guess random numbers until he chose one that was valid? The total number of credit card account numbers possible using all sixteen digits is 10^{16}, meaning that the odds of guessing a particular number on a single try are 1 in 10,000,000,000,000,000. Since most credit cards start with the same few sets of four digits, those digits are not available for creating account numbers, reducing the number of possible account numbers to 1 in 10^{12}, or 1 in 1,000,000,000,000. If United States consumers held one billion credit cards, the odds of guessing a correct number would improve dramatically, to roughly 1 in 1,000. Using modern software, a thief could easily try 1,000 numbers in order to find one that would work; for this reason, most credit card companies now also require a three to four digit code from the back of card, as well as complete personal information including the billing address, making the guessing tactic relatively useless. In response, twenty-first century thieves are far more likely to focus on hacking into massive credit databases where they can steal millions of valid card numbers and billing addresses at one time, rather than spending time guessing card numbers.

While providing a set of odds lends an air of credibility to a claim, odds are sometimes assigned to an event based on little evidence. Surgeons and other care-givers are frequently asked to give worried family members the odds of a patient's recovery from a serious illness. How can a concerned doctor provide these odds? An experienced surgeon can probably scan his memory for similar cases, or refer to medical reference volumes that estimate the likelihood of recovery. Unfortunately, given the wide variations in actual patient conditions, these methods are subject to wild swings in accuracy, and in many cases, a doctor is probably forced to provide an estimate based on little more than his instincts.

Ironically, in the case of a doctor giving odds for a patient's survival, some incentive exists for the doctor to actually overstate the danger and give lower odds than he might otherwise. For example, imagine that a doctor assigns a serious case 1 chance in 10 of surviving; if the patient dies, the concerned family and friends will likely be unsurprised, since the odds provided were not favorable. On the other hand, if this long shot patient manages to pull through, the family will be elated, potentially concluding that doctor is a medical genius. In this case, with the doctor providing unfavorable odds, he is ultimately perceived more favorably whether the patient lives or dies.

Conversely, consider the situation in which a doctor provides an overly optimistic assessments. If a doctor gives the patient survival odds of 9 in 10, the patient's recovery will occur only as an expected event. However, if the patient takes a turn for the worse and ultimately dies, the stunned family will have been emotionally unprepared, and may in fact blame the doctor or file lawsuit based on their expectations rather than the merits of the case. Subconsciously, the doctor who consistently gives his patients good odds only to watch them die may begin to question his own performance. While most doctors undoubtedly try to give accurate assessments of a patient's prognosis, little incentive exists for a doctor to give optimistic assessments. Subconsciously, this may lead doctors to paint a grim picture while hoping for a positive outcome.

Key Terms

80/20 rule: A general statement summing up the tendency for a few items to consume a disproportionate share of resources, such as cases in which 20% of a store's customers lodge 80% of the total complaints.

Long odds: Poor odds, or odds which suggest an event is highly unlikely to occur.

Odds: A shorthand method for expressing probabilities of particular events. The probability of one particular event occurring out of six possible events would be 1 in 6, also expressed as 1:6 or in fractional form as 1/6.

Probability: The likelihood that a particular event will occur within a specified period of time. A branch of mathematics used to predict future events.

Doctors are not alone in struggling to understand the odds of common events. Despite numerous news stories on the topic, many people still believe that flying is a more dangerous form of travel than driving. In truth, the odds of dying while driving to the airport are frequently higher than the odds of dying in the ensuing plane trip. People's irrational fear of flying is perhaps because hundreds of automobile accidents occur each day without notice, while every passenger plane crash receives extensive media coverage.

A similar situation exists in the energy industry. Many people believe that nuclear energy is the most dangerous form of power generation, and that the odds of being killed or injured in a nuclear power accident are far higher than the odds of being injured by processes related to coal or other low-tech energy sources. Ironically, far more people have died as a result of working in coal mines or in the coal industry than from the U.S.'s single nuclear power accident at Three Mile Island. Each year, a person's odds of being killed by a simple electric shock or by falling down in the bathtub are far higher than her odds of being exposed to dangerous levels of radiation.

Because most people have such a hard time assessing the odds of an event occurring, some companies market products designed to allay people's fears of these events occurring. Insurance companies, well-aware of both the slim odds of an airplane crash and the public's irrational fear of flying, have long offered flight insurance policies at airports, comfortable that these policies will provide both peace of mind to their buyers and steady income to their sellers.

In some situations, people take elaborate precautions to protect against hazards with relatively low odds, while ignoring other events with much higher odds of occurring. Most responsible drivers recognize the hazards of driving while intoxicated and avoid this behavior because of its potential for disaster. However, the number of drivers who talk on their wireless phones while driving remains high, despite advertising campaigns informing drivers that the chance of being in an accident are roughly the same in both situations. While the odds may be the same in both cases, most chatting drivers apparently do not recognize the hazard they are creating.

Because the language of odds provides a shorthand way of discussing possible outcomes, informal odds are frequently assigned to events as part of a discussion. One example of this informal use of odds is the well-known 80/20 rule. While not a strict mathematical set of odds, this rule is commonly used to explain a variety of situations in which some small portion of the whole has a larger than expected influence on the outcome. For example, managers sometimes use this rule in describing employees who require exceptionally large amounts of time, implying that the trouble-prone 20% of their employees are responsible for 80% of the manager's total problems. Teachers sometimes use this same rule to describe certain students who require constant assistance. Fund-raisers who spend their careers soliciting donors for causes such as education, disease research, or religious work, are well aware that the 80/20 rule accurately approximates the distribution of their donors, with the most generous 20% contributing 80% of the total funds, while the other 80% collectively give only 20%. Inventory managers frequently observe a similar pattern, with a relatively small number of products making up a significant majority of total order volume received.

The 80/20 principle can guide decision-making in a variety of scenarios. A fund-raiser who recognizes the pivotal role played by his larger contributors will go to extraordinary lengths to remain in favor with these donors, since he knows that a large donor is both far more painful to lose and far more difficult to replace. In optimizing inventory management, a warehouse supervisor can observe that 80% of orders received will include at least one of the top 20% of products, meaning he will keep a sizeable supply of these items on hand for immediate shipment. On the other side of this equation, this same manager knows that he can safely inventory smaller quantities of the less popular items, since the odds of any single one being ordered on a given day is quite small.

In a few cases, people can improve the odds of favorable outcomes. A male college applicant might consider the ratio of men to women at a particular campus, assuming that a larger number of single women on a campus would raise his odds of finding an appropriate date or mate. While the ratio of men to women is roughly 1:1 in the entire world, specific locations feature some surprisingly large variations in this mix.

Potential Applications

Because odds are simply the language of probability, future developments in the use of probability will be reflected in future uses of odds. With the growing popularity of poker as both a participant and a spectator sport, odds may become a more common topic of discussion, and fans of the game may commit the odds of completing particular hands to memory. In addition, recent advances in computer technology will likely lead to more accurate assessments of odds for numerous events, such as predictions of the future state of the U.S. and world economies.

Where to Learn More

Books

Epstein, Richard A. *The Theory of Gambling and Statistical Logic.* New York: Academic Press, 1977.

Glassner, Barry. *The Culture of Fear; Why Americans are Afraid of the Wrong Things.* New York: Basic Books, 1999.

Orkin, Mike. *What Are the Odds? Chance in Everyday Life.* New York: W.H. Freeman and Co., 2000.

Web sites

BBC News World Edition. "Clumsy Horse Beats all the Odds." <http://news.bbc.co.uk/2/hi/uk_news/wales/4377413.stm/> (March 30, 2005).

CNN Law Center. "Chinese Height Discrimination Case." <http://www.cnn.com/2004/LAW/05/31/dorf.height.discrimination/> (March 27, 2005).

National Council on Problem Gambling. "What is the Biggest Problem Facing the Gambling Industry?" < http://www.ncpgambling.org/media/pdf/g2e_flyer.pdf> (March 29, 2005).

Percentages

Overview

A percentage is a fraction with a denominator of 100. A percentage may be expressed using the term itself, such as 25 percent, or using the % symbol, as in 25%.

The calculation of various kinds of rates by way of percentages is the backbone of a wide range of mathematical applications, including taxes, restaurant tips, bank interest, academic grades, population growth, and sports statistics.

Fundamental Mathematical Concepts and Terms

Percentages are the natural mathematical extension of three other familiar concepts: fractions, ratios, and proportions. A fraction is a number expressed as one whole number divided by another; for example, one half is expressed as ½. A ratio is the relationship between two similar magnitudes. For example, as of 2005, the relationship between the population of Canada, estimated at 31 million people, and that of the United States, measured at approximately 310 million people, is a ratio of 1 to 10.

A proportion is a pair of ratios expressed as a mathematical equation. For example, if in a city of 100,000 residents, 1,000 people had red hair, the proportion of the population with red hair will be expressed as 1,000/100,000, or 1/100. The equation 1,000/100,000 = 1/100 is a proportion.

All percentages are an expression of a relationship based on 100. Every fraction, ratio, and proportion may be expressed as a percentage. Percentages may also be expressed where decimals are required, as in the figure 66.92%.

An important application of the concepts concerning percentages is that of percentiles. A percentile, which is one of the 99 points at which a range of data is divided to make 100 groups of equal size, is an important tool used in a vast number of statistical areas. For example, students in a class or across a larger population are given percentile rankings on a national test. The determination of the percentile ranking is a way of measuring relative standing to every other person in the class or larger group.

DEFINITIONS AND BASIC APPLICATIONS

A percentage is a fraction with a denominator of 100. A percentage may be expressed in any of the following ways: 38 percent, 38%, 38/100, or 0.38 as a decimal notation.

- To convert a decimal to a percentage, move the decimal point two places to the right and tack on a % sign. For example, the decimal 0.09 equals 9%.
- To convert a percentage to a fraction, remember that x% will always mean x/100; for example, 40% = 40/100. The simplest form of this fraction is 4/10, or 2/5.
- To convert a fraction to a percentage, find the decimal equivalent of the fraction and convert the decimal to a percentage as described above. For example, 3/4 = .75 = 75%.
- Finding a percentage of a quantity is common. For example, 15% of a shipment of 350 books is stated to be damaged by water. To find the actual number of damaged books in the shipment, proceed as follows: the word "of," as used here, means multiply, or, (15/100) × 350 = 52.5.

RATIOS, PROPORTIONS, AND PERCENTAGES

To properly understand the many ways that percentages can be applied in modern life, it is important to understand the relationship between ratios, proportions, and percentages. These terms are commonly applied, and each has a separate meaning and a distinct mathematical purpose.

A ratio is defined as the expression of the relative values of numbers or quantities, using one of three forms:

- use of the word "to," as in "a ratio of 8 to 5"
- use of a colon, as in 8:5
- use of a fraction, as in 8/5

A ratio may also be expressed where different quantities are related. For example, the relationship of 20 minutes to one hour is the same relationship as 20 minutes to 60 minutes, or a ratio of 20:60, or 2:6, and ultimately, a ratio of 1:3.

Other examples of ratio conversion include 5 tons to 500 pounds, 10,000 pounds to 500 pounds, 10,000/500, and a ratio of 20:1.

Ratios are a common form of expression in certain forms of sports wagering and games of chance. When a certain horse is favored to win at a racetrack, the probability of that horse winning its race, referred to as the odds of winning, is expressed in ratio form, for example, 3 to 2. In this context, the ratio means that for every two dollars agreed, the bettor will win three dollars if the horse wins.

The calculation of odds finds itself in other aspects of daily living. If in a particular place, over the course of an average year, 35 young drivers (under the age of 21) out of a sample of 100 young drivers were involved in motor vehicle accidents, and 10 older drivers (over the age of 50) were involved in accidents, what are the odds of a young driver being involved in accident versus those of an older driver? The odds are calculated as follows: 35/65 × 10/90 = 4.85. Therefore, the odds of the young driver being involved in an accident might be said to be almost 5 (rounding up the 4.85 figure).

Proportions result when two ratios are set equal to one another. For example, 6:9 = 12:18: a/b = c/d.

A Brief History of Discovery and Development

The term percent is derived from two Latin words: *per*, meaning by, and *cent*, meaning one hundred. The use of multiples of 10 as the basis for arithmetic, the forerunner to the modern decimal system, first gained acceptance with the Pythagorean school of mathematicians based in Greece in approximately 400 B.C.

However, the percentage is a relative latecomer as mathematical developments are gauged. The decimal had been developed as an effective way to easily distinguish between fractions with different denominators (for example, on first observation, the fractions 4/13 and 5/17 have similar values, but the corresponding decimal conversions for each, 0.307 and 0.294, are clearly different values). The decimal point became standard throughout the European scientific community in the early 1600s.

The introduction of the decimal fraction was one of the great advances of mathematics. This occurred because the decimal simplified numerical calculations, thus engineers, surveyors, and scientists could express their work to any desired degree of accuracy. The decimal fraction eliminated the potential for errors when fractions were compared with one another or converted in the course of measuring or other mathematical calculations.

The percentile concept was first developed in 1885 by English physician and mathematician Sir Francis Galton (1822–1911). His motto, "Whenever you can, count" is as appropriate today as when Galton coined the expression, given the role of the percentile in the modern world's obsession with measuring and ranking an infinite range of activities, from business to government to sport.

The percentage is now used as both a general descriptive term (in phrases such as "play the percentages," "there is no percentage in that"), as well as a mathematical tool of comparison and analysis.

The understanding of the various ways that percentage calculations may be used is crucial to the successful navigation of commercial, academic, and social worlds. Because society is now so accustomed to percentages being advanced in support of a particular viewpoint or concept, percentages can sometimes convey a superficial or misleading sense of certainty about a topic. Broad statements made in the media by business leaders, government officials, and others speaking on public issues often incorporate the expression of percentages. An example, "The economy will grow by 2% this year," has the ring of authority because a specific figure, 2%, is stated. However, an understanding as to how a particular percentage figure was arrived at is more important than the figure itself.

Similarly, an NBA basketball player may take pride in making 65% of his shots in the course of a playing season. If he only takes five shots per game, when his team is regularly scoring 100 points or more per game, the superficial impression and the impact of the high shooting percentage is much less, and the 65% figure is deceptive.

Real-life Applications

IMPORTANT PERCENTAGE APPLICATIONS

Percentages are calculated in a multitude of real-life situations. The understanding and proper applications of various percentage calculations are critical to daily living. The most relevant of these applications are set out below:

- The calculation of any type of rate: bank interest rate, a student loan rate, tax rate, mortgage rate.
- Education: determination of student grades. The ranking of students will often be determined by their grades, usually expressed as a percentage, as well as determined by the calculation of a related application to that of the percentage, the determination of a percentile.
- Science: in fields such as chemistry, pharmacology, or medicine, it is essential to be able to calculate the concentration of a particular substance in a mixture or solution.
- Food industry: percentages are used to determine the relative amount of the contents of food and beverage products, including the amount of certain fats, the amount of alcohol by volume in liquors, and the amount of a recommended vitamin or mineral.
- Retail sales: pricing increases and sales discounts are almost always expressed as a percentage. It would be difficult for businesses and customers alike if a price reduction was expressed as 2/7 of every dollar off, as opposed to a percentage figure.
- Social studies: any analysis of population growth, income, spending, inflation, or unemployment is expressed as a percentage.
- Meteorology: weather forecasts express the possibility of certain changes in the weather as a percentage; for example, a 20% possibility of precipitation.
- Sports: percentages are used to make comparisons in all types of competition. The shooting percentage in basketball, a quarterback's pass completion percentage in football, or baseball's batting average have become essential to the manner in which these sports are understood.
- Business: a company's current performance, the prospects for future growth, and measures of profitability and returns on investment will all be measured by percentage applications.
- Government and public service: trends in government spending, the increases or decreases in all aspects of the size, nature, and extent of public service, and future projections of every kind.

EXAMPLES OF COMMON PERCENTAGE APPLICATIONS

The 1% method This method of calculation is often useful for quickly determining small percentages. Determine 1% of the given number, and then compute the value of the desired percent.

To calculate 3% of 1,800: if 1% of 1,800 = 18, then 3% of 1,800 = 3 × 18 = 54.

To calculate 2.5% of 1,250: if 1% of 1,250 = 12.5, and 2% = 25.0 and 0.5% = 6.25, then the total = 31.25.

FINDING THE RATE PERCENT

Rate is the comparison between two numbers expressed as a ratio, written as a common fraction. For example, to express what percent of x is y: y:x or y/x, what percent of 20 is 8: rate = 8/20 = 0.40 = 40%.

FINDING THE BASE RATE

The determination of the base rate is often a feature of real-life calculations in business. Business finances will often involve determining a number of rates, from how quickly inventory is being distributed, to comparing spending from month to month or year to year, to salary and benefits increases or decreases. All of these determinations require an understanding of base rate calculations.

Base rates are calculated by creating an equation. For example, to determine what number is 25% of 88: x = 25% of 88, the percentage is changed to a fraction, creating x = 1/4 × 88, then x = 22. If it is desired to determine

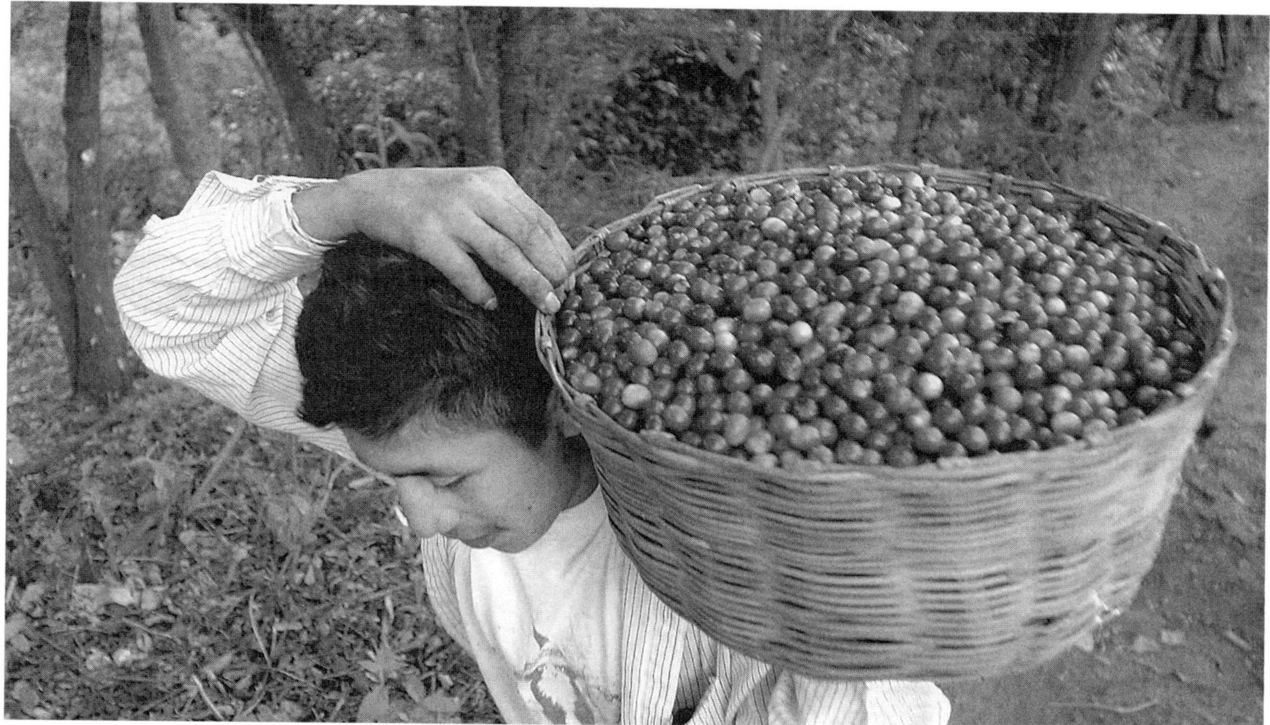

A man carries a basketful of organic coffee beans harvested throughout the morning at a coffee plantation in Guatemala City. Guatemala is ranked number one in percentage of its crop rated as highest quality. AP/WIDE WORLD PHOTOS. REPRODUCED BY PERMISSION.

60% of what number is equal to 42: 60% of x = 42, the percentage is changed to a decimal, 0.6x = 42, therefore x = 42/0.6, x = 70.

To apply this method of base rate calculation to a business example: a company keeps track of its sales on a monthly basis. What were the sales for the month of October if the November sales of $14,352 were 115% of October sales? October sales = x; 14, 352 = 115% of x; 14,352 = 1.15x; x = 14,352/1.15 = 12,480.

PERCENTAGE CHANGE: INCREASE OR DECREASE

Whenever a problem is expressed by words such as "is 15% more than," or "is 60% less than," or "has increased by 180%," or "has decreased by 38%," the problem requires a calculation of the percentage change, as either an increase or a decrease.

For example, 44 increased by 25% is what number? The new number will be represented by x, whereby x = 44 + 25% of 44; x = 44 + 1/4 (44); x = 44 + 11; x = 55.

Another example: 90 decreased by 40% is what number? The formula is x = 90 − 40% of 90; x = 90 − 0.4 (90); x = 90 − 36; x = 54.

FINDING THE RATE OF INCREASE OR DECREASE

The rate of change may be expressed as the following equation: rate of change = amount of change / original number. For example, if 40 increases to 46, rate of change = (46 − 40)/40 = 15%.

FINDING THE ORIGINAL AMOUNT

If the quantity after the change in a circumstance is known, the original quantity may be found as follows: 96 is 60% more than what number? In this example, 96 is the number after the increase. Let the original number be x: x + 60% of x = 96; x + 0.6x = 96; 1.6x = 96; x = 96/1.6; therefore, x = 60.

CALCULATING A TIP

A number of studies in recent years have determined that in North America, between 30% and 50% of a typical family's yearly budget for food is spent outside of the home, in restaurants ranging from the typical fast food emporiums of all types to more formal restaurant dining. Tips and tipping are the common terms used to describe the gratuity typically paid to a server in a restaurant for their assistance to a diner during the course of a meal. The money spent on the tip, which is in addition to the cost of the food and the taxes that may apply to the purchase of a meal, is therefore an important factor in the measurement of the total cost of food spending outside the home.

From the perspective of a server working in a restaurant, the correct calculation of the tip is important because it has a direct impact upon their personal income, as typically the tips earned by a server for their work will constitute an important part of their earnings.

The calculation of a tip involves a percentage-based application, usually related to the total amount of the bill, not including sales tax. It is generally accepted that a 15% tip recognizes good service, while a 20% tip tells the server that the service was outstanding. Tips of less than 10% are treated as an expression of the diner's dissatisfaction with the server and the establishment about the meal.

Assume a 15% tip in the following examples: 15% = 15/100 = 0.15. Where a restaurant bill totals $28.56, without tax being added to the total, to calculate the tip: .15 × 28.55 = 4.28. Thus, the 15% tip on this bill is $4.28.

It is unusual to leave a precise amount such as this for the tip, especially if the bill is being paid by cash. Custom may dictate that if the patron is paying by cash, a rounded figure that approximates the 15% will be left for the server, perhaps $4.25 or $4.50 in this example.

Note that when using the 1% method, this tip could be also calculated as follows: 10% of 28.55 = 2.85; 5% is ½ of 10% = 1.43. Thus, the total is 4.28.

COMPOUND INTEREST

Bank interest is expressed as a percentage. If funds are left in a bank account as a savings, they will attract what is referred to as compound interest, which is interest calculated both on the principal amount as well as the accumulated interest over time.

For example, in Year 1, $10,000 is deposited to a bank account that will pay the depositor 4% per year. The interest earned in Year 1 will equal $400. The interest to be calculated in Year 2 will be calculated on the original $10,000 as well as the Year 1 interest of $400, for a total of $10,400. 4% = 0.04, so Year 2 interest = 10,400 × 0.04 = $416.

Total monies in the account at the end of Year 2 will be $10,816. The 4% rate will apply indefinitely until the money is withdrawn in this example.

RETAIL SALES: PRICE DISCOUNTS AND MARKUPS AND SALES TAX

Many aspects of retail sales advertising are expressed in percentage terms. Sale prices, discounts, markups on merchandise, and all sales tax calculations depend on percentages. The various methods set out below assist in determining the various ways that retail sales are dependant upon percentages.

Discounts and markups: a discount is any sale where the seller claims that the goods are being sold at less than the regular or listed price. In some cases, the original price of the item is known, as is the percentage discount. The sale price is not known and it must be calculated, as follows: A refrigerator was said to have a list or regular price of $625. In the appliance showroom, there is a tag placed on the refrigerator advertising the item as on sale at 40% off its regular price. To find the sale price, 40% = 0.40; 40% of 625 = 0.40 × 625 = 250; 625 − 250 = 375.

In this example, the discount of 40% is $250, and the sale price is therefore $375. As an alternative method for calculating the sale price, 100% − 40% = 60%; 60% = 0.60; 0.60 × 625 = 375.

The next type of discount application commonly required in retail sales is the computing of the percentage discount advertised in any given situation. A used motor vehicle is advertised by its owner as being for sale at a price of $8,500. The advertisement states that the vehicle is worth $12,000 and that it is being sacrificed at the $8,500 price because the owner is relocating to another country to take a new job. The percentage by which the vehicle price is being discounted is calculated as follows:

Percentage discount = original price − sale price / original price × 100%; percentage discount = 12,000 − 8,500 / 12,000 × 100% = 3,500 / 12,000 × 100% = 29.17%.

The opposite concept in retail sales is the notion of the markup. While discounts are typically a part of the retail process that is advertised to the public, the markup is primarily an internal mechanism within a particular retailer.

Items that are sold in retail stores are often manufactured or assembled elsewhere, and they are purchased by the retailer on what is known as a wholesale basis. The ultimate sale price offered by the retail store to a purchaser will be the price paid by the retailer to obtain the item, plus an amount reflecting the relationship between what the retailer paid for an item themselves and what it will be sold for to the public. This amount is the markup. It is also referred to in some businesses as a margin, as in a business operating on a small margin, or the markup may also be described as the gross profit (the profit before costs and overhead is deducted). The relationship between cost, markup, and the retail or selling price for any item may be expressed in this simple equation: cost + markup = selling price.

Markups will be quoted as either a percentage of the cost price or of the selling price of an item, depending upon what is customary in that particular business. To compute selling price, the following example sets out the process: A hardware store buys drills a cost of $145 per drill. The store marks up the cost 65% based on its cost. The selling price is determined by 65% = 0.65; markup = 0.65 × 145 = $94.25; selling price = 94.25 + 145 = $239.25.

Alternatively, the known markup can be added to 100%, creating a total percentage figure, to perform the calculation: 100% + 65% = 165% = 1.65; selling price = 1.65 × 145 = $239.25.

SALES TAX CALCULATIONS

In most jurisdictions in the world, anyone purchasing consumer goods, ranging from bubble gum to motor vehicles, will be faced with the imposition of a sales tax. Such taxes, depending upon the location, may be imposed by city, state or province, or national governments. Tax rates vary from place to place; it is common to find 5% sales taxes. In some countries what are referred to as goods and services taxes, when combined with existing local taxes, can have a combined impact of 15% or more on a consumer purchase.

When assessing the price of an item offered for sale by a retailer, the total cost of the item must be assessed with the applicable taxes taken into account. For example, a new vehicle dealer is selling a pickup truck for $21,595, plus taxes. If the applicable tax rate is 4.5%, the total cost of the item is 4.5% = .045; tax = .045 × 21,595 = $971.76; total cost = 971.76 + 21,595 = $22,566.78.

Another factor in relation to the calculation of costs is the fact that the retailer may also have paid taxes on their purchase, which are being passed along. For this reason, the actual savings on a discounted item that is purchased has two components: the available discount on the price of the goods in question, and a reduction in the sales tax otherwise applicable to the price.

For example, a television is listed at a regular price of $649. It is then the subject of a "one third discount." The total savings available to the consumer are as follows: Price discount is ⅓ discount = 33.3%; discount = 0.333; discount = 0.333 × 649 = 214.17; discount price = 649 − 214.17 = $432.88.

If the applicable sales tax was 5% the sales tax payable on the discounted price would be tax rate 5% = 0.05; tax on discount price = 0.05 × 432.88 = 21.64; total cost of discounted item = tax + discount price = 432.88 + 21.64 = $454.52.

Had the television been purchased at the regular price, the sales tax would have been taxed at regular price = 0.05 × 649 = 32.45. The total cost of the television at its regular price is 649 + 32.45 = $681.45; total savings on the discounted television purchase is regular price total cost − discounted price total cost: 681.45 − 454.52 = $226.93.

REBATES

A variation on the notion of discounts is that of the rebate. A rebate occurs where a retail business sets a particular advertised or published sale price, and then will offer to refund or discount to the customer a fixed amount or percentage of the sale price. Rebates are frequently advertised in retail sales, and they are most common in the automotive sector, and they are also employed in the sale of various kinds of electronic devices and computer hardware.

For most circumstances, a rebate will have the same effect on a transaction as does a discount: a price that is the subject of a 10% rebate will have the same effect on a transaction as a 10% discount. However, there is one distinction between the impact of a discount and that of a rebate when the rebate is not offered at the retailer, but by way of the format known as a mail-in rebate.

For example, at a computer store that offers various types and brands of computers for sale, a particular computer manufacturer is offering a new computer monitor for sale at a price of $399, less a $50 mail-in rebate. The computer is purchased in accordance with the following transaction: sale price = 399; sales tax rate = 5% = 0.05; sales tax = 0.05 × 399 = $19.95; total cost = 399 + 19.95 = $418.95.

The purchaser is provided with a mail-in rebate card, which sets out the terms of the rebate, namely that upon receipt of the card, the manufacturer will send the sum of $50 payable to the purchaser within 60 days. Therefore, after 60 days, plus the time it takes to deliver the rebate to the manufacturer, the net cost to the purchaser shall be $368.95.

Two percentage-based calculations come into play in this mail-in rebate example. First, the difference is sales tax payable between the mail-in rebate and an identical discount; second, the 60 days or greater that the customer's $50 is out of the customer's control.

SALES TAX CALCULATION: IN-STORE DISCOUNT VERSUS MAIL-IN REBATE

If a $50 discount had been applied to the computer monitor purchase at the time of the transaction, the sale price would have been reduced to \$349, resulting in a total cost to the purchaser of sales tax = 0.05 × 349 = \$17.45; total cost = 349 + 17.45 = \$366.45.

The difference between the rebate being obtained by the mail-in method and the discount being calculated at the time of purchase at the store is \$2.50. To calculate the percentage difference between the total cost of the in store discount purchase and that of the 60-day rebate purchase: rebate cost / discount cost × 100% = percentage difference, or 368.95 / 366.45 × 100% = 1.006%.

To express the cost difference between the in-store discount and the mail-in rebate, the mail-in rebate process is 1.006% more expensive. This calculation as set out here does not place a value on other likely costs, including the time the purchaser would take to complete the rebate form, mail the rebate, and other associated steps required to have the rebate processed.

IMPACT OF THE 60-DAY REBATE PERIOD ON THE COST OF THE PRODUCT

As was noted in the examples dealing with the calculation of percentages, an interest rate measures the value of money over a period of time. Interest rate calculations are useful not only to calculate an increase in the value of money (such as the rate on interest being compounded on money being held in a bank account), but as is illustrated by the 60-day rebate, the interest rate percentage calculation can be used to confirm a loss of value over a period of time.

The calculation of the difference in the total cost of the refrigerator confirmed that the in-store discount total price of \$366.45 was \$2.50, or 1.006% less than the mail-in rebate total price of \$368.95. The next calculation will illustrate what happens to the \$50 rebate during the 60-day rebate period.

Assume that if the \$50 were placed in a bank account, it would earn interest at a rate of 4% per year. Had the customer purchased the refrigerator by way of an in-store discount, the \$50 discount would have been an immediate benefit to the purchaser, deducted at that point from the price paid to the retailer.

By waiting 60 days to receive the rebate (the minimum period, given that as a mail-in rebate there are additional days of mail and processing by the manufacturer), the purchaser lost an opportunity to use that \$50 sum. The percentage interest calculation will place a value on that loss of opportunity: loss = value of rebate × number of days rebate not available / length of the year × interest rate; value of rebate = \$50; mail-in period = 60 days; year = 365 days; interest rate = 4% = 0.04; loss = \$50 × 60 / 365 × 0.04; loss = 50 × 0.164 × 0.04; loss = 0.205.

In this example, the loss of opportunity for the purchaser on the \$50 rebate paid to the purchaser after 60 days is a small figure, 20.5 cents. The total difference in cost between the in-store discount purchase and the rebated purchase is the difference in total cost, \$2.50, and the loss of opportunity on the \$50 rebate, \$0.205, for a total of \$2.705.

However, as with most retail sales examples using relatively small numbers, it is easy to understand the importance of these percentage calculations where the retail price is 10 or 100 times greater. The percentages do not change, but where the percentages are applied to larger numbers, the potential impact on a purchaser is considerable.

UNDERSTANDING PERCENTAGES IN THE MEDIA

It is virtually impossible to read a news article, whether in paper format or by way of Internet service, that does not make at least one reference to a statistic that is described by way of a percentage. Sports, television ratings, employment, stock prices: all are commonly described in terms a percentage. In the media, it is common for percentage figures to be stated as a conclusion. For example, the income tax rate will be increased by 2.5% next year, for all persons earning more than \$75,000 per year.

To properly understand how things such as the consumer price index, the inflation rate, the unemployment rate, and similar issues are reported in the media, it is important to keep in mind the mathematical rules concerning percentages and how they are calculated.

The Consumer Price Indexes (CPI) program produces monthly data on changes in the prices paid by urban consumers for a representative basket of goods and services. Comparisons between prices on a month-by-month basis are useful in determining whether living costs are going up or down. To put it another way, the CPI tells how much money must be spent each month to maintain the same standard of living month to month, as the CPI values the same items to be purchased each period.

The CPI is based upon a sample of actual prices of goods that are grouped together under a number of categories such as food and beverages, clothing, transportation, and housing. Each individual item is priced, and the entire costs of the categories are compared with a selected base period. There are a number of adjustments that are also factored into the calculations, to take into account seasonal buying patterns at holidays and well-known sale periods.

The CPI calculations are made as follows: the base period, representing the time against which the current comparison will be made, is equal to 100, based upon 1990 reference data. Assume that the period to be compared is in November 2005: 1990 base price = \$100.00; November 2005 price = \$189.50.

The increase in the CPI index from 1990 to November 2005 is 89.5% or (189.50 − 100.00)/100. If the December cost of the consumer basket is 191.10, the increase from the base period of 1990 is 91.10% or (191.10 − 100.00)/100. To calculate the percentage increase between November and December, the following process must be carried out: the November value of 189.50 must be subtracted from the December value of 191.10, for an increase of 1.60% when compared to the 1990 rate. To calculate the percentage change between November and December: 1.6%/ 189.5% × 100% = 0.0084 × 100 = 0.84%. Therefore, there was a 0.84% increase in the consumer price index in this example between November and December.

PUBLIC OPINION POLLS

From time to time, specialist organizations, known as polling companies, will be commissioned to gather data from a segment of the population concerning particular issues. The question asked of the people polled may involve a large national issue, such as whether capital punishment ought to be permitted, or whether the maximum speed limits on national highways should be increased or decreased. In some instances, the polling organization may be hired to obtain the opinions of the public in relation to issues that pertain to a local concern, such as whether a town should permit a casino to be constructed within its boundaries.

The manner in which public opinion polls are carried out is a branch of social science. The methods used by the pollsters in the asking of the questions, the number of people who form the sample upon which calculations are made, and the age and the background of the responders are all factors that may impact upon the answers given to the polling company.

From the perspective of percentages, it is important to appreciate that virtually all such public opinion polls are translated, and reported in the media, as a percentage figure. The meaning to be attached to the percentage quoted as the result of the poll must be examined carefully.

For example, a sample of 4,000 people were asked the following questions: Should cigarette sales in their city be banned completely? Should smoking be banned in every public place in their city? For the first question, the following results were noted: 1,900 said, "yes"; 1,800 said, "no"; 250 were "not sure", and 50 "refused to answer." For the second question, the following results were noted: 2,100 said, "yes"; 1,550 said, "no"; 300 were "not sure"; and 50 "refused to answer."

What are the different ways that the results of each of these questions can be expressed as a percentage? Depending upon how the percentage calculation is used in each case, what answers may be given to each of the questions? The percentage calculation for each answer to question 1 on the ban of cigarette sales is "yes" = 1,900/4,000 = 47.5%; "no" = 1,800/4,000 = 40%; "not sure" = 250/4,000 = 6.25%; "refused" = 50/4,000 = 1.25%.

If the poll was to exclude those who refused to answer the question, and only calculate the responses from people who did answer, the percentages for each answer are "yes" = 1,900/3,950 = 48.1%; "no" = 1,800/3,950 = 345.6%; "not sure" = 250/3,950 = 6.3%.

If the poll were further defined as all respondents who had made up their minds and therefore had a positive opinion on the issue, the formula is "yes" = 1,900/3,700 = 51.35%; "no" = 1,800/3,700 = 48.65%. By taking these steps, the polling company might choose to state this result as "more than 50% of respondents to the poll who had formed an opinion on the question were in favor of a ban on the sale of cigarettes in the city."

If the poll is defined by who is in favor of the question, the formula is "yes" = 1,900/4,000 = 47.5%; "all other responses" = 2,100/4,000 = 52.5%. The polling company might state this result as "less than 50% of all respondents to the poll stated that they were in favor of a ban on cigarette sales in the city."

The result to the question 2 to ban cigarette smoking in public places generates the following percentage calculations: "yes" = 1,650/4,000 = 41.25%; "no" = 1,550/4,000 = 38.75%; "not sure" = 700/4,000 = 17.5%; "refused" = 100/4,000 = 2.5%.

Using the same analysis as carried out with question 1, if the persons who refused to answer the question are also eliminated from the sample: "yes" = 1,650/3,900 = 42.3%; "no" = 1,550/3,900 = 39.8%; "not sure" = 700/3,900 = 17.9%.

If the persons who were not sure in their answers to the question are removed from the sample: "yes" = 1,650/3,200 = 51.5%; "no" = 1,550/3,200 = 48.5%.

In the same manner as is set out in the question 1 analysis, the manner in which the percentages are calculated in each case can support different conclusions. With the question 2 calculations, when the whole sample of 4,000 answers is examined, only 41.25% of those questioned supported the ban on smoking in public places. By restricting the sample to those with a definitive opinion, a majority of those questioned may be said to support the proposed ban.

Miami Heat's Dwayne Wade goes up and scores against the Atlanta Hawks in the game in Miami. Players are often rated by percentages, such as their field goal percentage. AP/WIDE WORLD PHOTOS. REPRODUCED BY PERMISSION.

USING PERCENTAGES TO MAKE COMPARISONS

It is common in media reports to compare different results in related topics. For example, government spending may be reported in a particular year as having increased 5% over the previous year. The population of a particular state may be stated as having increased by 3% over the past decade.

These calculations are relatively straightforward, because the comparison is being made between single entities, namely a government budget, which would be calculated and measured to be reflected as a total figure, or population, which would have been measured by way of a population count, known as a census.

Percentages are more difficult to put into perspective when they are employed to compare less certain items. For example, if the two public opinion questions and the various answers are compared by way of percentage calculations, the results are not always certain.

In question 1, when only the respondents who had either a yes or a no opinion were calculated, the number of those in favor of the ban on cigarette sales was 51.35%, and those opposed to such a ban was 48.65%. In the question 2 analysis, when only the respondents with a yes or no opinion were counted, the number of those in favor of banning smoking in all public places was 51.5%, those opposed totaled 48.5%.

Based upon the determination of percentage figures that are virtually identical (51.35% and 51.5%) in each question, it would be possible to state the following as a conclusion from the two sets of polling questions, namely a majority of people in the city are in favor of both a ban on cigarette sales and a ban on smoking in all public places.

However, having worked through the calculation to each of the percentages that form the basis of this statement, it is also clear that the use of those percentages in the manner contemplated by this conclusion is not the entire picture. If other parts of the calculation are used to determine a conclusion, it could also be stated that as 47.5% of all respondents were in favor of the ban on cigarette sales, and then a further 41.25% were in favor of the public places ban, the following conclusions are valid: less than 50% of persons polled were in favor of any restriction upon cigarette purchase or usage in the city; a little over 2 out of 5 people polled were in favor of these restrictions.

Percentages and statistics of all types are often stated as a definitive answer or conclusion to an issue. As illustrated in the questions posed above, it is important that the method employed in calculating the percentage be understood if one is to truly understand the significance of the percentage figure that is stated as a conclusion. Where the methodology behind a particular percentage is not stated in a particular media report, the percentage must be regarded with caution.

SPORTS MATH

Another common media report in which percentages are employed in a variety of ways is that of the sports commentary. There are a seemingly limitless number of ways that sport and athletic competition commentaries are enhanced by the use of statistics, many of which are dependent upon percentage calculations.

In the media, there is a recognition that certain statistics go beyond analysis of an individual performance, but are descriptors that convey a definition of enduring excellence. The "300 hitter" is a description applied to a solid offensive professional baseball player, while a "400 hitter" is in an ethereal world inhabited by legends like Ted Williams and Ty Cobb. A 90% free-throw shooter in basketball has a similar instantaneous public recognition.

The American humorist Samuel Langhorne Clemens, better known as Mark Twain (1835–1910), once said that there are three kinds of lies: % lies, damn lies, and statistics. Whenever a percentage is referenced in a sports article, as with any other media usage of percentages, care must be taken to determine whether the percentage figure being quoted is an accurate indicator of performance, or whether at best it is a lesser or insignificant fact adding only color, and not necessarily insight, concerning the sporting event.

Sports examples of percentage calculation usage are based on daily examples found in the media around the world. For instance, in basketball, an example would be Amanda and Claire as members of their girls' high school basketball team. The coach of the team has been asked to select a most valuable player for the season. While the coach has a personal view of each player based on his assessment of their play through practice and games all season, he decides to do an analysis of their respective offensive statistics. Each player had the following statistics after the completion of the 20-game high school season: Amanda scored 160 total points; 108 2-point shots attempted; 62 2-point shots made; 10 3-point shots attempted; 6 3-point shots made; 21 free throws attempted; 18 free throws made; 17.5 minutes played per 32-minute game. Claire scored 322 total points; 341 2-point shots attempted; 125 2-point shots made; 22 3-point shots attempted; 5 3-point shots made; 81 free throws attempted; 57 free throws made; 28.8 minutes played per 32-minute game. The team scored 887 points on the season.

How can percentages be used to help determine who is having the better season? Conversely, do percentage calculations distort any elements of the performance of these players?

If the 2-point shooting of each player is compared, by calculating the percentage accuracy of each player through the entire season, the following comparison can be made: Amanda = 62 shots made/108 shots attempted = 57.4%. Claire = 125 shots made/341 shots attempted = 36.66%.

The 3-point shooting percentage calculation is as follows: Amanda = 6 shots made/10 shots attempted = 60%. Claire = 5 shots made/22 shots attempted = 22.7%.

The players' free-throw shooting percentages are calculated as follows: Amanda = 18/21 = 85.7%. Claire = 57/81 = 70.4%.

If a newspaper report was written setting out the coach's analysis of the respective play of Amanda and Claire, it is quite possible that such a report might describe Amanda as a better shooter than Claire because her shooting percentages in every area of comparison (2-point shooting, 3-point shooting, and free-throw shooting) are better than Claire's. Conversely, Claire has scored the most points and she has played more minutes per game than Amanda. When those statistics are assessed, the following percentage calculations can be determined: For Amanda, 160 points scored/887 team points scored × 100% = 18% of the team offense. For Claire, 322 points scored/887 team points scored × 100% = 36.3% of the team offense.

Further, Amanda generated her 18% of the team offense while playing 17.5 minutes per game. Claire produced her 36.3% of the team offense while playing 28.8 minutes per game.

There are certain hard conclusions that the coach in this scenario may have reached based upon the percentage calculations that pertain to Amanda and Claire. Amanda is a more accurate shooter in every aspect of the shooting game. It is likely that based upon these percentages, the coach will create opportunities for Amanda to shoot more often next season.

However, as with many applications of the percentage calculation in a sports context, it is important to have more information about the team and the players to give the percentage statistics more context, and to put the percentages into a better perspective. If Amanda is a weak defensive player, her offensive percentages are placed in a different light. If Claire had performed all season known to all rivals as the team's best player, and thus attracted extra attention from opponents, her shooting percentages would be weighed differently.

Baseball statistics may be the most identifiable percentage in sport, usually expressed as a decimal. For example, a strong hitter in the North American professional leagues will be referred to as a "300 hitter," meaning a batsman with an average of 0.300, or a 30%, success rate. This percentage is calculated by the following formula: Number of hits/Number of at bats × 100% = Batting average.

However, as befits a sport that has been played professionally in North America since the 1870s, statistics have grown out of the game, some clear to even the average fan, and some very obscure. A key percentage used to calculate offensive contributions is that of "on base percentage," which measures how often a batter advances to first base by any of the means available in baseball, namely hit, walk, hit by pitched ball, etc. The percentage is calculated by the following formula: Total number of times on base / Total number of at bats or plate appearances × 100% = On base percentage.

A very intricate set of percentages has made its way into the analytical end of baseball through the work of Bill James. His approach, which he termed sabermetics, is an attempt to use scientific data collection and interpretation methods that employ various types of percentages in different aspects of baseball to conclude why certain teams succeed and others fail.

North American football is also riddled with statistics. One of those measurements is that concerning the most prominent player on the field, the quarterback. How often the quarterback may successfully throw the ball down field is an important statistic, referred to as passing completion. This percentage is calculated by: Number of passes completed/Number of passes thrown × 100%.

However, much like the basketball examples set out above, this percentage on its own is potentially deceiving. A quarterback who throws 80% of his passes for completions, but never throws a pass for a score, is unlikely to be as successful as the 50% passer who throws for 20 touchdowns in a season.

TOURNAMENTS AND CHAMPIONSHIPS

With the rise in the popularity and the sophistication of college sports in the United States, coupled with the impossibility of having hundreds of teams in a given season playing one another head to head, statistical tools were developed to weight the relative abilities of teams that would not necessarily meet in a season, but each of whom would seek selection to an elite end-of-season tournament or championship.

In American college basketball, hockey, and football, tournament selection is made using what is known as the RPI, or ratings percentage index. This interesting and much debated tool is defined in college basketball as follows: RPI = Team winning percentage/25% + Opponents winning percentage/50% + Opponents' opponents winning percentage/25%.

If a team had a record of 16 wins and 12 losses in a season, they would therefore have a team winning percentage of 16 of 57.14%. The team played opponents whose total record was 400 wins and 354 losses. The opponents' winning percentage is 53.05%. These opponents played teams whose winning percentage was 49.1%, the opponents' opponents' winning percentage: RPI = 57.14/25 + 53.05/50 + 49.1/25, which is RPI = 2.28 + 1.06 + 1.96 = 5.304.

A team will typically have a bigger and better RPI if the team combines its own success with an ability to beat strong opponents that have themselves played a strong schedule. Therefore, a team at the end of a particular season that has a lesser record than a rival, but that has played what the RPI determines to be a demonstrably more difficult schedule, may be selected to compete over the team with the better win/loss record. The RPI has a number of nuances that are not the subject of this text, but it is important to understand that the percentage calculation is at the root of any RPI determination.

Percentiles

The percentile is a ranking and performance tool that is closely related to the concept of percentages. A percentile represents a place on a scale or a field of data, providing a rank relative to the other points on the scale. Percentiles are calculated by dividing a data set into 100 groups of values, with at most 1% of the data values in each group.

Percentages can be expressed in any number from 0 to virtual infinity, with either a positive or negative value as circumstance may determine. However, it is commonly accepted that in many applications where a percentage calculation determines a grade or a score in a particular activity, the percentage is expected to be between 0% and 100%. For example, where a school assignment was graded at 17/20, the assignment has a percentage grade of 85%.

In situations where there is a large class of students, it is often desirable to rank them in order of performance. Ranking provides a measure of how a particular student has performed relative to every other comparable student. For example, hundreds of thousands of potential university students in the United States, with many thousands more worldwide, test for the standard Scholastic Aptitude Test (SAT) every year. The SAT is tested at a multitude of test sites, at various times. Each test in a given year is similar, but the exact questions asked on each of the tests will vary. The SAT has a complicated scoring system generating scores from 0 to 1600, and the administrators of the test recognize that assessing students who have taken different versions of the SAT is very difficult. For this reason the percentile ranking becomes important, as it measures where every student stands relative to every other student who took the test.

Determining where an individual students stands relative to everyone else who took the test is a terrific tool with which to assess relative performance. This determination is done by calculating the percentile.

SAT SCORES OR OTHER ACADEMIC TESTING

The percentile grew from the concept of percentages; for that reason, founded upon the concept of 100, and if the data comprising the test results is regarded as a unit of 100, percentile ranking proceeds in bands from 0 to 99, with the 99th band being that that includes the highest score or scores in the sample.

Each percentile in the sample may have more than one score within it. Further, percentiles are not subdivided. For example, there may be as many as 20,000 test scores produced from one round of SAT testing. If eight students scored a perfect 1600 on the SAT, they would each be described as having a result in the 99th percentile even if, say, 10 students with slightly lower scores were also in the 99th percentile. Similarly, if the 55th percentile, representing 1% of all scores from that test, was determined to be all of the scores between 1010 and 1040, all scores within that percentile band would be described as in the 55th percentile.

One formula to calculate the percentile for a given data value is: Percentile = (number of values below x + 0.5)/number of values in the data set × 100%.

As an example, the following is a sample of the shoe sizes for a 12-member high school boys basketball team: Sample: 14, 12, 10, 10, 13, 11, 10, 9, 9, 10, 11, 9. How is the percentile rank of shoe size 12 determined? First, the shoes sizes must be arranged in values smallest to largest, which create this set: 9, 9, 9, 10, 10, 10, 11, 11, 12, 13, 14. The number of values below 12 is eight, and the total number of values in the data set is 12. The formula to express the percentile rank of the value 12 is $(8 + 0.5)/12 \times 100\% = 70.83\%$. The percentile ranking of the value of the size 12 can therefore be expressed as the 1st percentile.

To calculate the percentile ranking of the size 10, there are three identical sizes in the data set. There are three values in the set below 10. The formula would be $(3 + 0.5)/12 \times 100\% = 29.1\%$. The percentile ranking of the value of all three of the size 10s is expressed as the 29th percentile.

It is also common to express a ranking using a broader term. For example, a student may be described as being in the top 20% of their class, or in the top quarter. These expressions are a paraphrasing of the percentiles known as deciles (groups of 10 percentiles) and quartiles (groups of 25 percentiles). Deciles divide the data set into 10 equal parts, and quartiles divide the set into four equal parts.

The 50th percentile, the 5th decile, the 2nd quartile, and the median are all equal to one another.

Final grades in academic courses are typically expressed as a percentage. Even where alternate methods are used to express performance (as with alpha grades A through F), or as a grade point average, each alternative has an equivalency expressed as a percentage. The percentages are then matched to a particular letter grade that has a range of percentages within it. For example, A+ is the equivalent of 90–100%; A is the equivalent of 80–89%; B is the equivalent of 70–79%; C is the equivalent of 60–69%; D is the equivalent of 50–59%; and F is the equivalent of below 50%.

Letter grades function in a similar way as percentiles, in that each grade includes a potential range of percentage scores, and like the percentile, the percentage scores are not ranked within the assigned grade.

Any area of human performance that is subject to ranking will likely employ percentiles as a measuring stick. Topics can be as diverse as the relative rate of obesity in children, ranking increases or decreases in funding rates for hospitals and schools, and comparing the relative safety rates in relation to speed on highways. These are three of the almost limitless ways that percentiles can be used to assist in a ranking of performance.

Potential Applications

The better understanding of a multitude of everyday concepts and activities will be determined, directly or indirectly, by an appreciation of the ability to perform the percentage calculation.

As further examples, percentages play a key role in the following areas:

- Voting patterns and election results: Percentages are used to take the large numbers of persons who may vote in an election, and reduce the figures to a result that is often easier to understand.
- Automobile performance: Octane is a term that is familiar to everyone who has ever used gasoline as a fuel for a vehicle. In general terms, the octane rating refers to how much the fuel can be compressed before spontaneously igniting, an important factor in optimizing the performance of the internal combustion engine. While the public generally associates high octane requirements as required for certain motor vehicle models with more powerful engines and vehicle performance, the octane rating represents the percentage between the hydrocarbon octane (or similar composition) in relation to the hydrocarbon heptane. For example, an 87 octane rating (a common minimum in the United States) represents an 87 percent octane, 13 percent heptane mixture in the fuel.

Key Terms

Fraction: The quotient of two quantities, such as 1/4.

Percentage: From Latin *per centum*, meaning per hundred, a special type of ratio in which the second value is 100; used to represent the amount present with respect to the whole. Expressed as a percentage, the ratio times 100 (e.g., 78/100 = .78 and so .78 × 100 = 78%).

Ratio: The ratio of a to b is a way to convey the idea of relative magnitude of two amounts. Thus if the number a is always twice the number b, we can say that the ratio of a to b is "2 to 1." This ratio is sometimes written 2:1. Today, however, it is more common to write a ratio as a fraction, in this case 2/1.

- Clothing composition and manufacture: Most clothing is sold with a tag or other indication as to its material composition. For example, it is common to see a label on a shirt indicating 65% cotton, 35% polyester, or a sweater marked as 100% wool.
- Vacancy rates: The availability of vacant apartment space in a particular city is of great importance to prospective residents and existing apartment dwellers alike. The vacancy rate is expressed as a percentage to provide interested persons with an indicator as to the relative ease or difficulty to obtain particular types of rental accommodation. Vacancy rates can be viewed as of a particular period (for example, the vacancy rate in Spokane was 1.8% in April), or as a calculation increase or decrease from period to period (for example, the vacancy rate in Toronto fell 0.7% last month).

Where to Learn More

Books

Boyer, Carl B. *A History of Mathematics.* New York: Wiley and Sons, 1991.

Upton, Graham, and Ian Cook. *Oxford Dictionary of Statistics.* London: Oxford University Press, 2000.

Web sites

College Board. "Scholastic Aptitude Test." (March 29, 2005.) <http://www.collegeboard.com>.

NCAA Tournament Selection, 2005. (March 29, 2005.) <http://www.ncaa.com>.

Perimeter

Overview

A perimeter is the boundary of an area or shape. Its measurement is the total length along the border or outer boundary of a closed two-dimensional plane or curve. The origin of the word perimeter comes from the Greek words *peri* (around) and *metron* (to measure).

The application of perimeters in everyday life is widespread when determining a wide range of mathematical problems such as the amount of fencing needed to encompass a homeowner's property; the number of miles of beach property along a lake; and the distance around the equator of Earth.

Fundamental Mathematical Concepts and Terms

One of the simplest equations for solving a perimeter is that of a square or rectangle, which is the sum of its four sides. The general equation for determining the perimeter of a rectangle is $p = 2W + 2L$, where $W =$ width of the rectangle and L is the rectangle's length. Knowing that a rectangle always has four sides with opposite, equal widths and lengths, a rectangle (for example) with length of 4.3 meters (about 14.1 feet) and width of 6.4 meters (21 feet) has a total perimeter length of $p = 2$ (6.4 meters) $+ 2$ (4.3 meters) $= 12.8$ meters $+$ 8.6 meters $= 21.4$ meters (about 70.2 feet).

The equation that determines a perimeter of a circle (also known as its circumference) is $p = 2\pi r$ or $p = \pi d$ (where $\pi =$ approximately equal to 3.14159, $r =$ radius of the circle, and $d =$ circle's diameter and $d = 2r$). As a specific example, a circle with a diameter of 7.5 meters (about 24.6 feet) has an approximate perimeter of $p = \pi$ (7.5 meters) $= 3.14159$ (7.5 meters) $= 23.6$ meters (about 77.4 feet). By knowing the shape of a simple figure, such as a triangle, hexagon, square, or pentagon, its perimeter can be easily calculated. More complicated figures, such as an ellipse, need the tools of calculus in order to calculate its perimeter.

A Brief History of Discovery and Development

Archimedes is known to have found the approximate ratio of the circumference to diameter of a circle with circumscribed and inscribed regular hexagons. He computed the perimeters of polygons obtained by repeatedly doubling the number of sides until he reached ninety-six sides. His method for finding perimeters with the use of circumscribed and inscribed hexagons was similar to that used by the Babylonians (whose civilization endured from the eighteenth to the sixth century B.C. in Mesopotamia, the modern lands of Iraq and eastern Syria).

Real-life Applications

SECURITY SYSTEMS

A physical barrier around the perimeter of a building may stop or at least delay potential intruders from penetrating inside. Such physical barriers include fences, brick or concrete walls, and metal fencing. A well-known outer perimeter barrier surrounds the White House complex in Washington, D.C., which includes very substantial physical fencing, Secret Service agents, and an assortment of television cameras and high-tech sensors. An effective perimeter security system, especially for critically important properties, may include a combination of several physical barriers, an electronic detection system, and numerous manual procedures. A single barrier completely around the perimeter of a protected property could take only a few seconds to penetrate, while multiple barriers will typically take longer to penetrate. Taller and stronger perimeter barriers will further increase the time it takes an intruder to gain entry to a site.

In all cases, in order to effectively secure a property, the physical barrier must completely surround the property's perimeter. As a result, the installers of a perimeter barrier must first measure the number of feet (or meters) in the perimeter. Because of this measurement, these professionals must know the appropriate equations to calculate the perimeter of a square, rectangle, circle, and other shapes. In many instances, numerous equations will need to be combined due to irregular-shaped perimeters around a facility or property. Because of increased risks of terrorism and criminal activities around the world, security that involves total perimeter protection is becoming more popular at governmental, industrial, and commercial facilities such as airports, correctional centers, court houses, entertainment complexes, military bases, and police stations, along with residential homes.

LANDSCAPING

The use of perimeters in landscaping is a common way to design for particular purposes. For instance, commercial properties may use certain plants and shrubs along the perimeter of their facility for the following reasons: to completely isolate the facility from the public; to create a visual separation between the facility and the public; to soften the appearance of streets, parking areas, and other exterior buildings and structures; and to provide summer shade on parking areas.

Defining a landscape's outer boundaries (its perimeter) with respect to the interior buildings, gardens, and other structures and materials often help to create a better visual effect for the entire property. Homeowners with small urban properties, where neighbors live in close proximity to each other, naturally lean toward defining their perimeters with the use of fencing, hedges, shrubs, trees, and other similar structures. These materials are used for such reasons as identification of property lines, privacy, and overall aesthetic beauty. When larger properties are involved, perimeter framing is less used because of fewer concerns for privacy and other such considerations. However, large properties without visible exterior boundaries will often allow such an open area to look more exposed and unfinished—thus detracting from the overall beauty. Simple placement of plantings along the perimeter will make the entire area look more organized and cohesive. Unless privacy, unattractive outside views, or intrusion of wildlife are a concern, most perimeter plantings only need a light planting of trees and shrubs of various densities, sizes, and textures. In all cases, accurate calculations with respect to the total length of the perimeter is essential.

Perimeters are not only used to define the boundary line of a property. Landscaping within a property can also use perimeter-planting when planting around the boundary of a perennial flower gardens, houses, swimming pools, or other such structures. In each instance, the measurement of perimeters is important when designing an outside landscape.

SPORTING EVENTS

Knowledge of the perimeter of various sport fields is important with respect to the watching, playing, and discussing of the games. For example, the perimeter of an American football field (excluding the end zones) is 920 feet (280 m): two lengths of each 300 feet (91 m) and two widths each of 160 feet (49 m). Since each end zone is 30 feet (9 m) long, the perimeter of each end zone is $(30 + 30 + 160 + 160)$ feet $= 380$ feet. Thus, the total perimeter of a football field including the two end zones is 1,680 feet (about 512 m). Playing strategies by coaches and players depend on knowing the exact measurements of a field's perimeter in such sports as football, soccer, tennis, baseball (which can vary depending on the size of the stadium), basketball, and hockey.

BODIES OF WATER

The calculation of perimeters of bodies of water such as lakes and swimming pools is important for many

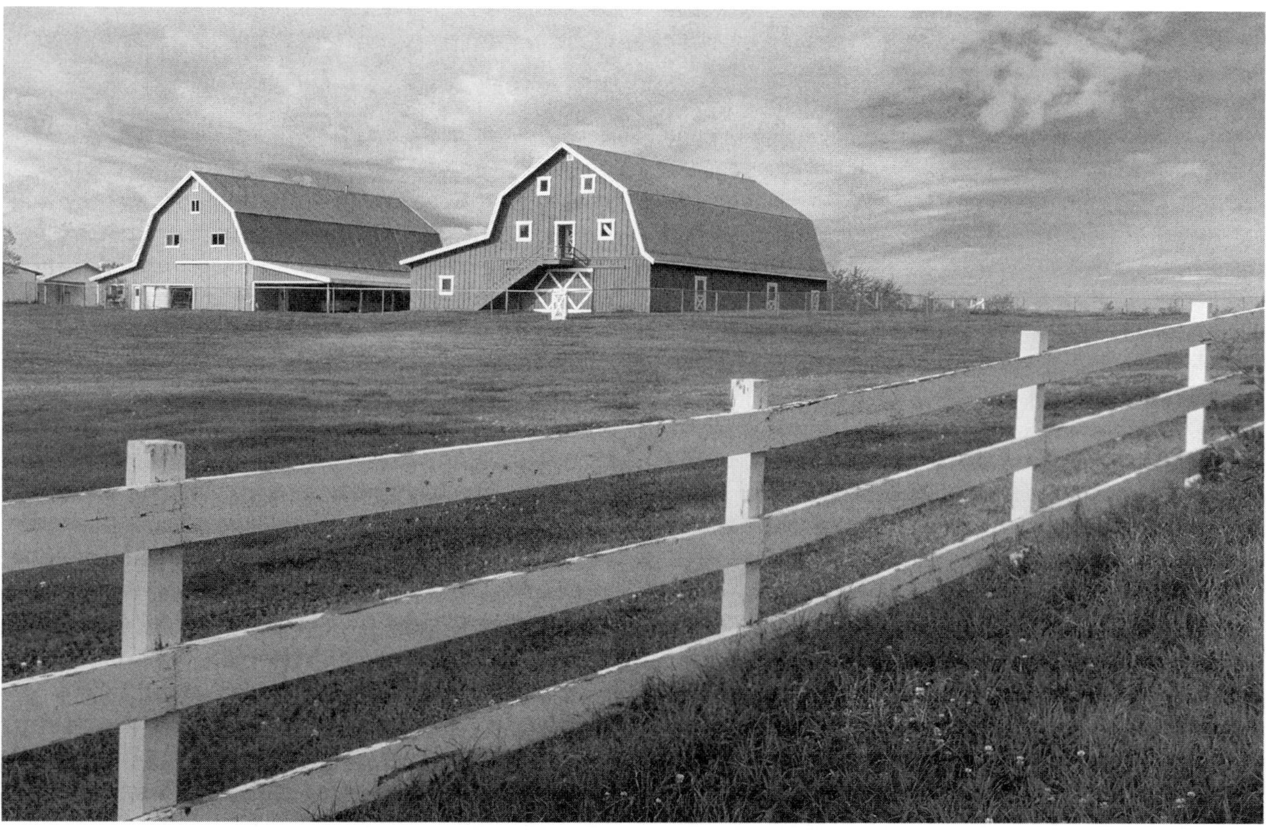

One side of the perimeter of a farm is marked with a fence. TERRY W. EGGERS/CORBIS.

reasons. Because shorelines are very valuable property with regards to investments, people like to build expensive houses along lakes. Therefore, it is important to accurately measure the perimeter around a lake so, by knowing the length of each house lot, the possible number of total houses built can be figured. This information is very important, for instance, when surveyors and building contractors are first plotting out new lakeside developments.

When first building swimming pools that are to be used for competitions, it is important to know the perimeter of the pool so that the proper number of lanes can be built. For example, the world swimming organization FINA (International Amateur Swimming Federation or, in French, Fédération Internationale de Natation Amateur) states that the official dimensions for pools used for Olympic Games and World Championships are to be of a total length of 50 meters (164 ft) and a total width of 25 meters (82 ft), with two empty widths of 2.5 meters (8 ft) at each side of the pool. With this information, it is easily calculated that an Olympic-sized pool must have a perimeter of 150 meters (about 492 ft) and contain eight lanes, each with a width of 2.5 meters. That is, the total of 25 meters of width consists of 20 meters (66 ft) of lanes (8 lanes × 2.5 meters per lane = 20 meters) and 5 meters (16 ft) of empty lanes (2 empty lanes × 2.5 meters per lane = 5 meters).

MILITARY

The United States military has an important need for physical security barrier walls and systems that can protect its ground forces, military fighting assets such as airplanes and tanks, and critical infrastructure assets from hostile actions. These materials are set up around the perimeter of critical structures, soldiers, and materials in order to assure that enemy forces do not penetrate, attack, and destroy such critical personnel and hardware. These perimeter security devices can be simple, portable coaxial cables laid around the perimeter of buildings, properties, or assets, which emit multiple radio-frequency signals. Strategically placed receivers monitor the signals and trigger an alarm when there is a disturbance along the protected perimeter. Other more complex perimeter security devices can be high-technology corrugated metal barriers that can withstand the blast of high-order detonations and anti-ram barriers that can withstand the repeated assault by enemy tanks and other motorized vehicles.

Potential Applications

PLANETARY EXPLORATION

Perimeter is such a general term within mathematics that its use will always be important for new applications. For example, as mankind ventures further into the solar system, unmanned rovers with portable power supplies, such as rechargeable batteries, may depend on supplementary power generated on stationary landers. As the rover explores a pre-determined area of a celestial body, such as the moons of Saturn and Jupiter, it would return to the central lander to recharge its power supply. This method is very similar to how motorists check their fuel gauge to make sure they are not too far away from a gas station when the arrow points near empty. In such a scenario, aerospace scientists would calculate the straight-line perimeter of maximum exploration for the rover in order to assure that the rover would never venture too far from its power supply. Knowing this maximum number of kilometers, the scientists then keep track of the actual mileage of the rover, most likely within an internal sensor of the rover, to accurately predict when to return to base camp.

ROBOTIC PERIMETER DETECTION SYSTEMS

The U.S. Department of Defense's Defense Advanced Research Projects Agency (DARPA) and Sandia National Laboratories' Intelligent Systems & Robotics Center (ISRC) are developing and testing a perimeter detection system that uses robotic vehicles to investigate alarms from detection sensors placed around the perimeter of protected territories and buildings. Such advanced technologies that involve the use of perimeters allow humans to perform other, more important tasks, and eliminate the loss of human lives from investigating possible intrusions.

Where to Learn More

Books

Bourbaki, Nicolas (translated from French by John Meldrum). *Elements of the History of Mathematics.* Berlin, Germany: Springer-Verlag, 1994.

Boyer, Carl B. *A History of Mathematics.* Princeton, NJ: Princeton University Press, 1985.

Bunt, Lucas N.H., Phillip S. Jones, and Jack D. Bedient. *The Historical Roots of Elementary Mathematics.* Englewood Cliffs, NJ: Prentice-Hall, Inc., 1976.

Web sites

Rores, Chris Rorres. Drexel University. "Archimedes." Infinite Secrets. October 1995. <http://www.mcs.drexel.edu/~crorres/Archimedes/contents.html> (March 15, 2005).

Sandia National Laboratories. "Perimeter Detection." November 4, 2003. <http://www.sandia.gov/isrc/perimeter detection.html> The Intelligent Systems & Robotics Center. (March 15, 2005).

Thordarson, Olafur, Dingaling Studio, Inc. "Project for an Olympic Swimming Pool, 1998." October 1995. <http://www.thordarson.com/thordarson/architecture/laugar dalslaug.htm> (March 15, 2005).

Perspective

Overview

Perspective is the geometric method of illustrating objects or landscapes on a flat medium so that they appear to be three dimensional, while considering distance and the way in which objects seem smaller and less vibrant when they are farther away. The items must be portrayed in precise proportion to each other and at specific angles in order for the effect to be realistic. In art, perspective applies whether the painting or drawing depicts a landscape, people, or objects.

Fundamental Mathematical Concepts and Terms

Basically, perspective works when a series of parallel lines are drawn in such a way that they all seem to head for, and then disappear at, a single point on the horizon called the vanishing point (see Figure 1). The parallel lines running toward the vanishing point are referred to as orthogonals. The vanishing point itself is considered the place that naturally draws the eye in relation to the other objects in the composition, regardless of the size or subject of the work of art, and the horizon is a straight line that splits the image, placed according to the artist's point of view. The higher the artist's vantage point, the lower the horizon appears in the rendering, and vice versa. More than one vanishing point can be applied to a work of art, giving the illusion that the picture bends around corners or has several points of focus. These compositions are referred to as having two-point, three-point, or four-point perspective.

Perspective is based upon the assumption that one is viewing the image from a single point, and is therefore, sometimes referred to as centric or natural perspective. It is also possible to examine three-dimensional space from two points, the study of which is known as bicentric perspective.

A Brief History of Discovery and Development

Early paintings and drawings, prior to the invention of perspective, tended to appear flat and out of proportion. They lacked a sense of realism. Linear perspective, the first method of creating art that was more precise in its portrayal of its subjects, was invented by Filippo Di Ser Brunellesci (1377–1446), a sculptor, architect, and engineer in Florence, Italy. Brunellesci was responsible for building several of Florence's most famous structures including the Duomo (dome of the main cathedral) and

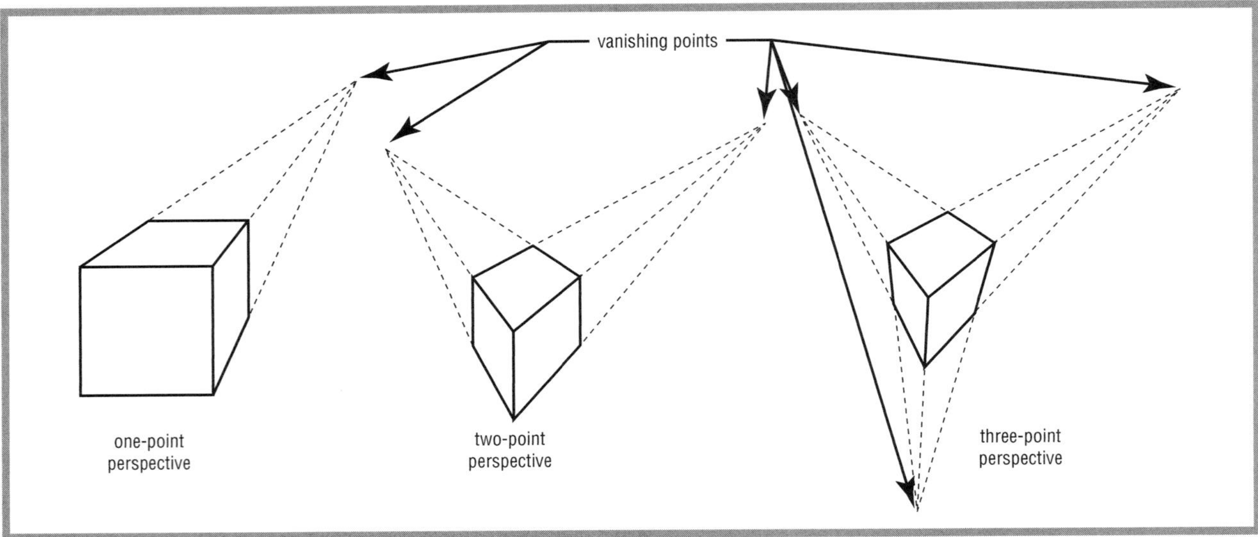

Figure 1.

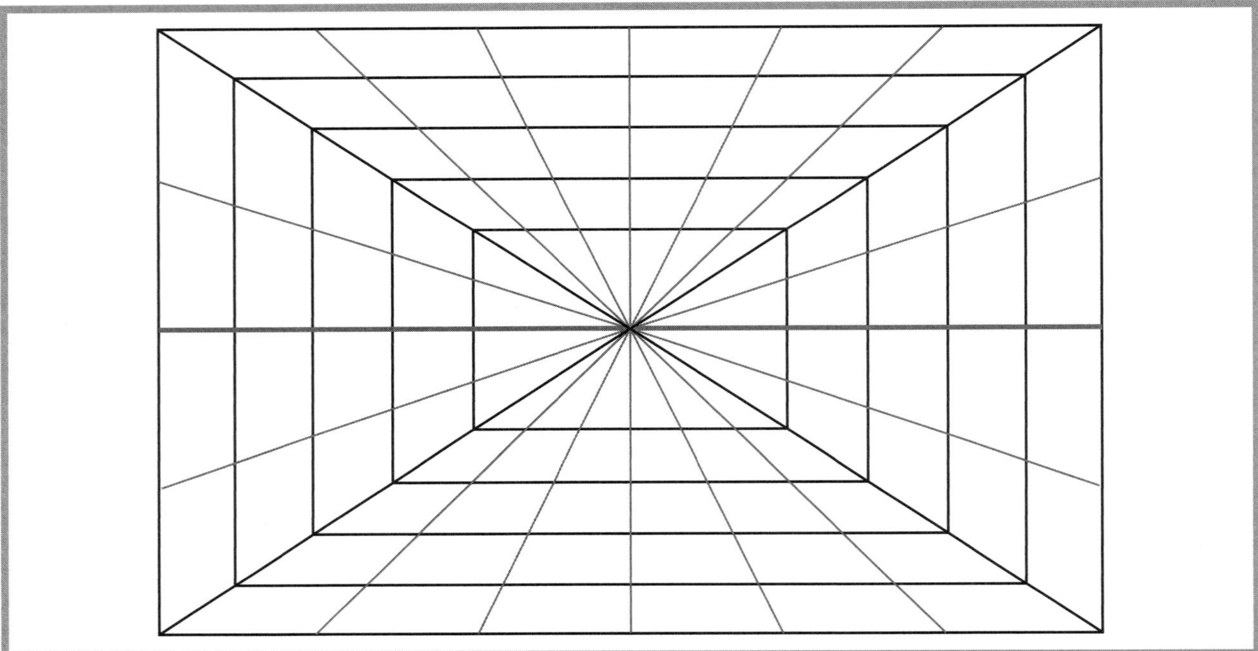

Figure 2.

church of San Lorenzo. Brunellesci experimented with creating a single line of sight, toward a vanishing point, by viewing a reflection of a picture or image through a peep hole in a sheet of paper and thereby focusing his vision on a single line (see Figure 2).

Brunellesci never recorded his findings, but may have passed them on to other artists and architects through demonstrations or word of mouth. The first written account of the use of perspective was recorded by the Italian architect Leon Battista Alberti (1404–1474), who initiated the use of a glass grid through which the artist would look at the subject while painting in order to assist in creating the proper perspective. Alberti determined that he could use a geometric technique in order to mimic what the eye saw, and also that the distance from the artist to the scene being painted had an effect on the rate at which the image appeared to recede. Alberti said that the artist created a sort of visual pyramid, turned on its side, between himself and the painting, where his line of sight connected to the vanishing point on the work of art. The surface of the painting itself was the base of the pyramid and the painter's eye formed the summit. Alberti considered it necessary to maintain that position in order for the artist to accurately capture the perspective of his subject on the canvas.

The first surviving example of the use of perspective in art is credited to Donato di Niccolò di Betto Bardi (1386–1466), more commonly known as Donatello, an Italian sculptor during the early part of the Renaissance. Of his surviving work, most prominent are sculptures he created for the exterior of the Florentine cathedral, including St. Mark and St. George. The latter is a marble relief that depicts Saint George killing the dragon, and the work shows some indication that Donatello attempted to use perspective within the scene. Some of the lines used to create the illusion were most likely inaccurate, as the perspective is less than perfect, so it cannot be said for certain that he was applying this then-new methodology.

However, in later works, it becomes more obvious that Donatello was aware of the principles of perspective. In a bronze relief panel he designed for the font at the Siena cathedral, titled *Feast of Herod*, Donatello clearly utilized a vanishing point and orthogonals. While there is a slight imperfection in the panel, in that the orthogonals do not meet precisely at the same point, it is likely this defect was not part of the original sketches, but instead resulted at some point during the execution in bronze.

Masaccio (c. 1401–1428), considered with Donatello and Brunellesci to be among the founding artists of the Italian Renaissance, showed no signs of attempting to use perspective in his first known painting, *Madonna and Child with Saints*. However, his three most famous works painted near the end of his life all use linear perspective. One of these, *Trinity*, which was done for Saint Maria Novella in Florence, is thought the oldest perspective painting to still survive today. It depicts the crucifixion of Jesus Christ, with key figures such as John the Baptist and the Madonna framing him in a pyramid fashion, and God hovering above. Masaccio supposedly discussed "Trinity" with Brunellesci. The work itself was painted based on a strict grid that was applied to the surface before any painting began. Every detail is in precise perspective, down to the nails holding Christ to the cross. In another perspective painting, "Tribute Money," Masaccio used linear perspective not just to create a realistic portrayal of the scene from the lives of St. Peter and St. Paul, but also to direct the viewer's eye in such a way that the painting becomes a narrative. Christ stands in a group of his followers, and it is his head that is the vanishing point on which the viewer focuses.

The advent of the camera obscura in the mid-fifteenth century offered another way to examine perspective. Based on similar techniques as the peephole experiments, the camera obscura allowed light into a darkened room through a small hole. An image was then projected onto a wall and the artist attached paper to the surface in order to trace it. The act of tracing guaranteed the artist would achieve the proper angles and proportions of perspective.

Other artists went on to do additional experiments in perspective, and to perfect the technique. Leonardo da Vinci, noted as an artist, inventor, and mathematician, did much to further the understanding of how perspective applied to distance, shape, shadows, and proportion in art. He was the first artist to work with atmospheric perspective, where the illusion of distance was created through using fainter or duller colors for objects meant to be farther from the viewer. By combining this knowledge with other mathematical references, such as the standard proportions of the parts of the human figure, he was able to create compositions that appeared realistic and natural. Albrecht Dürer, a noted German Renaissance artist and print maker, experimented with using tools to assist in attaining proper perspective, and kept detailed records of his discoveries. In 1525, he wrote a book in order to teach artists how to represent the most difficult shapes using perspective.

During the seventeenth century, Dutch artists were particularly known for their exemplary use of perspective in their paintings. Pieter de Hooch and Johannes Vermeer were two such painters renown for including such details as floor tiles, elaborate doorways, and multiple walls incorporating perspective in order to achieve the most realistic effect.

Real-life Applications

ART

Artists display the most obvious need for a clear understanding of perspective in their work. In order to

Figure 3.

fashion any realistic depiction of a scene, whether in a simple sketch or a detailed painting, an artist must use the rules of perspective to guarantee that the proportions and angles of the images appear three-dimensional. Landscapes particularly require exact application of perspective in order to give the illusion of depth and distance. A common illustration of this technique (see Figure 3) depicts a train track heading toward the horizon, the parallel lines of its rails appearing to become closer together as they grow farther away, until they eventually converge at the vanishing point. The picture becomes more complex if the artist wishes to add something along the side of the train tracks, such as trees or telephone poles. Although the artist knows the phone poles must appear smaller as they grow more distant, he needs to determine at what rate their size decreases. By applying the rules of perspective, the artist may sketch in the orthogonals, the diagonal lines that stretch from the vanishing point to the edge of the paper, in order to provide a guideline for the heights of the poles as they gradually shrink into the distance.

This method can be applied to any number of subjects that may appear in a painting, such as a row of buildings that reaches to the skyline or clusters of people scattered across a large room for a party. Orthogonal lines can be placed at any height in relation to other subjects so that smaller objects remain in proportion to larger ones, regardless of their placement in the scene. If a man who is six feet tall stands next to a child who is only three feet tall, the child will appear half the height of the man if they are sketched at the front of the painting or back near the horizon, even though the actual size of each will be adjusted to represent their placement in the composition.

Perspective helps artists render drawings that include buildings much more accurately, as well. If an artist wishes to paint a landscape that includes a house and a barn that are situated at an angle, with the corners of the buildings facing the viewer, perspective allows him to draw the edges of the buildings and their roofs at the correct angles. The horizontal lines that form the top and bottom edges of the buildings, as well as the horizontal lines for the door and windows—if extended straight out to the side—should eventually intersect at a vanishing point. The slanted lines that form the side edges of a pitched roof will also intersect in the same way. If the painting includes a split-rail fence around the farmland, the rails must all angle so that the lines would extend to a vanishing point. In these types of landscapes, the artist will frequently use two-point or three-point perspective in order to set the angles for the different sides of the buildings.

Artists often use the vanishing point as a focal point when composing the layout of a painting. If several people are depicted, it is common for an artist to have their attention directed toward the vanishing point. A person gesturing with an arm might likewise be indicating something at the vanishing point.

ILLUSTRATION

One specific application of artistic talent, illustration, provides books and other publications with artwork to accompany the text. Children's books are a prime example of this, and the simplicity of many of the pictures that illustrate children's stories does not preclude the need to apply perspective to the composition. A child will notice if a picture seems out of proportion, just as an adult will, and as the illustrations carry much of the weight of the storytelling for pre-readers, it is important that everything is rendered correctly and in proportion.

Comic books or graphic novels are other examples of illustration as an art form. As with picture books for children, comic books rely heavily on the pictures to tell the story, with only a small amount of narrative and dialogue to move the plot forward. Each panel of a comic is drawn in perspective, with the occasional pane drawn in such a way as to indicate the action happens in the foreground and is therefore, more important. Using perspective for emphasis allows comics to convey heightened emotion and action in a relatively small space.

ANIMATION

Animation, an art form unto itself, would not be possible without perspective, as the figures would appear flat and lifeless on the screen despite their ability to move. Early animated films were hand drawn a single frame at a

Art and Mathematics—Perspective

Perspective provides flat, two-dimensional works of art with the means to appear three dimensional and realistic. No painting, sculpture, or frieze can seem to have depth or illustrate distance from the viewer if the artist fails to apply the rules of perspective to the composition. In reality, the curvature of the planet combines with the eye's ability to look into the distance and creates the visual effect of perspective where lines appear to converge upon a single point, even when the lines never actually meet, as is the case with the two rails of a train track. This trick of the eye, or perspective, must be replicated as an optical illusion on a flat canvas in a painting in order for it to considered a precise representation of the three-dimensional view seen in real life.

A student of art must learn to apply perspective to whatever he is attempting to create. This holds true of paintings done from life and those created solely from the imagination. While it is possible to sit at an easel and recreate the landscape just beyond the top of the canvas, it is more difficult to create an accurate rendering when the subject is not visible. For this reason, art students learn the principles behind the illusion of perspective. An artist can sketch a horizontal line onto a canvas and create both horizon and vanishing point, then add orthogonal lines to assist in creating an accurate, realistic landscape, even in a room without a view.

time, and the precise measurements required to achieve perfect perspective made it easier for the artist to recreate the background of the film over and over, while limiting variances that might have made the finished film appear inconsistent or fake.

As animation has grown more technical and the art has shifted from paper to computers, it has become more important that the angles and lines required to give the illusion of a three-dimensional setting remain constant. Animators can now feed mathematical calculations into a computer where a graphics program will plot the coordinates for the horizon and the vanishing point. Once this information is computerized, it is saved in the machine's memory and applied whenever that particular background is needed for the film. The computer software allows the animators to program shifts in shadow based on time of day or night for the story, to alter the camera angles, or even to add in new background structures such as a new building or taller trees due to the passage of time. The changes are made automatically within the parameters of the perspective already programmed into the computer.

One modern example of the use of this technology is the Walt Disney Company's film *Beauty and the Beast.* This animated movie applied new technology to centuries-old theories of perspective to create a scene where the Beast and Belle dance in an animated virtual reality ballroom. The scene consists of a large ballroom with rounded walls and a tiled floor, and the film gives the illusion of a living couple twirling around the dance floor as the camera pans around them. The animators programmed the computer to maintain the proportions of the room, with the apparently rounded backdrop, and the tiles on the floor decreasing in size as they grew more distant from the camera. As the animated couple dances and the camera follows them, the vanishing point is required to shift with each movement so that it will remain steady in relation to the eye of the audience and the illusion of depth may be maintained.

FILM

Animated films are not the only ones concerned with perspective. As live action films include more and more special effects that require actors to perform in front of green screens or blue screens, perspective becomes the concern of special effects artists. Obviously the effects artists need to apply perspective when generating the background, as they would with an animated film, but in addition they must maintain the size ratio between the live actors who will be part of the finished scene and any computer graphics components, including scenery and creature effects. The actors must also perform in relation to special effects that are not present while they are filming. While stand-ins are sometimes utilized, it is also helpful to apply the same lines of perspective that an artist would use when composing a painting. An actor might address himself toward what will end up being the vanishing point of the scene, allowing the special effects artists to fill in the graphics around the same point, creating the illusion that all of the components of the film actually took place at the same time.

An example of combining live action with digital backgrounds is the film version of the Frank Miller graphic novels, *Sin City.* In this film, the actors performed their scenes against a green screen, often without even the benefit of another actor to whom they could address their lines. The background, a heightened noir-style city in stark black and white, was created on the computer using a three-dimensional digital program. Using the graphic

Study for perspective with animals and figures by Leonardo da Vinci. BETTMANN/CORBIS.

novel as a template, the director recreated the look and feel of each panel of the comic by mimicking the perspective of each shot. The background maintained the perspective and all of the angles from the original source material, and the actors were placed in relation to that background to make it seem as if the graphic novel itself had come to life.

Another optical illusion popular in film—particularly fantasy or science fiction films—is making actors of similar heights appear vastly different in size. *The Lord of the Rings* trilogy faced this challenge when the filmmakers attempted to create a world shared by several species of varying heights. When an actor playing a short Hobbit filmed a scene with an actor playing a normal sized person, it was not only necessary to have the actors appear to be different heights. The sets around them also had to be altered so that items that appeared average size for the man would be oversized for the Hobbit. Props, such as a ring or a mug of ale, could be duplicated in varying sizes and then substituted for each actor according to their character's size, but the background and furnishings were more complicated. The set designers used perspective to determine the precise proportions for each item and then used forced perspective filming in order to create the optical illusion that the two actors were actually using the same items. For example, in a scene where the wizard, Gandalf, and the Hobbit, Bilbo, are seated at a long table, the front of the table was cut down to be smaller than normal, so that Gandalf would appear to be cramped. The back half of the table was sized normally so it would appear to fit Bilbo. Items placed on the table at the joining point helped disguise that the table was not all one size, and the camera was placed at an angle to shoot down the table's length, taking advantage of the fact that perspective would help make it seem to grow smaller at a distance. The actors themselves stood several feet apart, but staring straight ahead, and were filmed in profile to give the illusion of their facing each other. Perspective made the more distant actor playing Bilbo appear smaller than the actor closer to the camera.

INTERIOR DESIGN

Interior designers and decorators are responsible for the layout and design inside a house, and frequently use perspective as a tool to maximize the potential of a living space. An architectural detail such as exposed beams—which were originally solely a functional aspect of a house, used to brace walls and support the roof—can make a room appear to be longer than it really is. Looking carefully at the beams running parallel to each other, they seem to grow closer together as they move toward

Leonardo da Vinci's "Window" for Recording Proper Linear Perspective in Art

Italian artist, inventor, and mathematician Leonardo da Vinci (1452–1519) understood that linear perspective was necessary in order for a painting to appear realistic. In order to practice transposing the exact lines and angles of the world as he saw them, Leonardo began to use a window as a framework. When he looked out the window, whatever he saw became the subject of his painting, as if the edges of the window were the edges of a canvas. He would then attach a piece of paper to the window so that the natural light shone in from outside and he was able to see the outline of the scene through the paper. It was necessary for him to cover one eye when working, so that he would, in effect, be looking at the three-dimensional world from a two-dimensional viewpoint. He would then go on to trace what he saw through the window onto the paper. Leonardo da Vinci accurately captured all of the lines of perspective as they appeared in nature. This exercise enabled him to learn how perspective affected the composition. He discovered that his own distance to the window, as well as the distance of the objects outside to the window, changed the perspective of the scene. If he shifted to the left or the right, the vanishing point on the horizon also shifted on his paper. It was also possible for Leonardo to sketch in guiding lines, orthogonals, to help him maintain the size ratio between various items in the composition, regardless of where they appeared in relation to the vanishing point. Leonardo proceeded to apply what he learned to his painting. Early sketches of his work illustrate how he composed his work to include a vanishing point that was logical in relation to the subject of the painting.

The famous painting, *The Last Supper*, clearly illustrates Leonardo da Vinci's use of perspective. While the scene itself shows only minimal depth, concentrating more on the length of the dining table as it stretches the width of the painting, with Christ and his disciples positioned along the back, Leonardo applied his knowledge of perspective to create the rear walls of the room. Jesus himself, seated at the center point between his followers, provides the focus of the painting, and his head serves as the vanishing point on the horizon for the composition.

the opposite end of the room from the viewer, just as train tracks seem to converge toward a vanishing point when viewed from a distance. In a house, the beams reach the supporting wall before they appear to meet each other, but the vanishing point still exists. If one could see through the wall and extend the beams indefinitely, they would illustrate a textbook example of perspective. As it stands, the optical illusion they create gives a home a more spacious feel. Anything that adds horizontal lines to the overall look of a room—tiles or hardwood flooring, a chair railing or molding, decorative detail on a ceiling, built in bookshelves that run the length of a wall—gives the impression that a room is longer and more spacious.

A similar illusion that also uses perspective to make a room seem larger is adding a large mirror to a wall. If an entire wall contains a mirrored surface, it will seem to double the size of the room by reflecting it back upon itself. By staring into the mirror, a viewer will notice that the reflected walls seem to angle inward, just like the train tracks in a perspective painting. The illusion of additional space suddenly looks more like the view out a window than an addition to the room. The mirror effect is particularly popular when a designer can place it opposite a window, thereby reflecting not only additional space from the room, but the light and the view from outside as well, creating an open effect.

Another decorating effect that makes use of perspective is the artistic treatment known as *trompe d'oeil*. Literally meaning "trick of the eye," this painting technique involves rendering a highly realistic looking painting or mural directly onto the wall of a room in an attempt to make it appear completely authentic to the viewer. In some cases, the painting is something simple, such as a statue on a pedestal standing in an alcove. Someone looking at the painting from a distance will be tricked into believing that the wall really does curve back at that point, and that the piece of art in question is actually a three-dimensional statue. Only when they draw nearer will they realize that the statue is painted on the wall. The artist uses lines of perspective to create the illusion, perhaps giving the alcove portion of the painting a tiled pattern or gradually lightening the tone of the paint used since colors fade at a distance, all in order to make the wall seem to curve.

Key Terms

Bicentric perspective: Perspective illustrated from two separate viewing points.

Centric perspective: Perspective illustrated from a single viewing point.

Orthogonals: In art, the diagonal lines that run from the edges of the composition to the vanishing point.

Vanishing point: In art, the place on the horizon toward which all other lines converge; a focus point.

Other examples of the use of *trompe d'oeil* may include a painted window or doorway, including the view through that opening. Perspective is applied as it would be in any landscape, so that the view through the painted window or door mimics what one might see through an actual hole at that point, or else the artist might create an entirely imaginary landscape, giving a city apartment the luxury of a view of the beach or the countryside.

Trompe d'oeil may also be applied to an entire wall, as in a mural. This sort of effect can involve multiple illusions, depending on the images chosen for the composition. Some of the wall might be painted as if it were still part of the house, with the rest providing some sort of outdoor view. Examples might include a painting of a balcony that overlooks the garden, with the majority of the perspective applied to the images that are meant to be more distant, and other, more subtle techniques used for the supports of the balcony that are meant to be much closer. However, the lines of the balcony must remain in harmony with the lines of the view, maintaining the same vanishing point, in order to maintain the overall effect.

LANDSCAPING

Landscapers and landscape architects do for the outdoors what interior designers do for the inside of a building. By applying the rules of perspective when laying out a garden, park, or other property, landscapers can make a small piece of land seem larger or grander than it might otherwise appear. A building with a straight driveway can be made to appear farther from the road by planting a series of trees along each side of the drive. The effect is similar to that of a painting of a road with trees lining it, the road converging on the vanishing point and the trees shrinking into the distance. Likewise, details such as long, narrow reflecting pools, hedges, stone walls, flower beds, and flagstone or brick pathways help draw the eye in a particular direction and direct the visual focus of the landscape in whatever way the designer sees fit.

Potential Applications

COMPUTER GRAPHICS

Any work done with computer graphics can make use of the rules of perspective. Programs that allow images to appear on the computer screen in three dimensions apply to a range of work, including architecture, city planning, or entertainment.

Architects and engineers can use preprogrammed angles of perspective to create virtual images of buildings or bridges or other large-scale projects, enabling them to test the effect of the new construction in its intended setting without having to build detailed models. City planners can in turn use perspective to get an accurate idea of the layout of a town from the comfort of a desk. Streets and traffic flow, how roads converge, where traffic lights might be most effective, entrances to major thoroughfares, and placement of shopping or public facilities, all may be programmed into a computer and illustrated in a realistic, three-dimensional layout.

Computer game designers can apply perspective to their creations, enabling enthusiasts to enjoy the most realistic experiences possible when playing their games. Accurate perspective can enhance a variety of games, such as those where the participant drives a racecar, pilots an airplane, or maneuvers a space ship through an asteroid field in a faraway galaxy. Likewise, games that involve role play or character simulation can provide realistic settings, such as towns or the interiors of buildings.

Where to Learn More

Books

Atalay, Bulent. *Math and the Mona Lisa: The Art and Science of Leonardo da Vinci.* Washington, D.C. Smithsonian Books, 2004.

Parker, Stanley Brampton. *Linear Perspective Without Vanishing Points.* Cambridge, MA: Harvard University Press, 1961.

Woods, Michael. *Perspective in Art.* Cincinnati, OH: North Light Books, 1984.

Periodicals

Ashcroft, Brian. "The Man Who Shot Sin City." *Wired.* April, 2005.

Web sites

Dartmouth College Web site. "Geometry in Art and Architecture." <http://www.dartmouth.edu/~matc/math5.geometry/unit11/unit11.html/> (April 8, 2005).

Disney's Beauty and the Beast: Unofficial Pages. "Gallery of Key Scenes." <http://www.ilhawaii.net/~beast/pages/bb_gllry.htm> (April 8, 2005).

Drawing in One Point Perspective. Harold Olejarz. <http://www.olejarz.com/arted/perspective/> (April 8, 2005).

Leonardo's Perspective. <http://www.mos.org/sln/Leonardo/LeonardosPerspective.html> (April 8, 2005).

Perspective from MathWorld. <http://mathworld.wolfram.com/Perspective.html> (April 8, 2005).

Wired News. "Sin City Expands Digital Frontier." Jason Silverman. April 1, 2005. <http://www.wired.com/news/digiwood/0,1412,67084,00.html> (April 8, 2005).

Other

The Lord of the Rings: The Fellowship of the Ring. Extended Edition DVD special features. New Line Home Entertainment, 2002.

Photography Math

Overview

Photography, literally writing with light, is full of mathematics even though modern auto-exposure and auto-focus cameras may seem to think for themselves. Lens design requires an intimate knowledge of optics and applied mathematics, as does the calculation of correct exposure. When mastered, the mathematics of basic photography allow artists, journals, and scientists to create more compelling and insightful images whether they are using film or digital cameras.

Fundamental Mathematical Concepts and Terms

THE CAMERA

In its simplest form, the camera is a light-tight box containing light sensitive material, either in the form of photographic film or a digital sensor. A lens is used to focus light rays entering the camera and produce a sharp image. The amount of light striking the film and sensor is controlled by shutter, or curtain that quickly opens and exposes the film or sensor to light, and the size of the lens opening, or aperture, through which light can pass.

FILM SPEED

The speed of photographic film is a measure of its sensitivity to light, with high speed films being more sensitive to light than low speed films. Film speed is most commonly specified using an arithmetic ISO number that is based on a carefully specified test procedure put forth by the International Organization for Standardization (ISO), for example ISO 200 or ISO 400. Each doubling or halving of the speed represents a doubling or halving of the sensitivity to light. Thus, ISO 400 speed film can be used in light that is half as bright as ISO 200 speed film without otherwise changing camera settings. Some films, particularly those intended professional photographers or scientific applications, also specify speed using a logarithmic scale that is denoted with a degree symbol (°). Each logarithmic increment represents an increase or decrease of three units corresponds to a doubling or halving of film speed. ISO 400 film as a logarithmic speed of 27° but ISO 200 film, which is half as fast, has a logarithmic speed of 24°.

Photographic films are coated with grains of light-sensitive silver compounds that form a latent image when exposed to light. Film speed is increased by increasing the size of the silver grains, and the grains in high speed films can be so large that they produce a visible texture, or graininess, in photographs that many people find distracting. Therefore, photographers generally try to use the slowest possible film for a given situation. In some cases, however, photographers will deliberately choose a high-speed film or use developing methods that increase grain in order to produce an artistic effect. The choice of film speed is also affected by factors such as the desired shutter speed and aperture.

LENS FOCAL LENGTH

The focal length of a simple lens is the distance from the lens to the film when the lens is focused on an object a long distance away (sometimes referred to as infinity, although the distance is always finite), and is related to the size of the image recorded on the film. Given two lenses, the lens with the longer focal length will produce a larger image than the lens with the shorter focal length. Most camera lens focal lengths are given in millimeters. A lens with a focal length of 100 mm (3.9 in) is in theory 100 mm (3.9 in) long, but camera lenses consist of many individual lens elements designed to act together. Therefore, the physical length of a camera lens will not be the same as the focal length of a simple lens. Zoom lenses have variable focal lengths, for example 80–200 mm (3.1–7.9 in), and also variable physical lengths. The physical lens length will also change as the distance to the object being photographed changes.

Lenses are often described as telephoto, normal, and wide angle. Normal lenses cover a range of vision similar to that of the human eye. Wide angle lenses have shorter focal lengths and cover a broader range of vision whereas telephoto lenses have longer focal lengths and cover a narrower range of vision. All of these terms are relative to the physical size of the film being used. A normal lens has a focal length that is about the same as the diagonal size of the film frame. For example, 35 mm (1.4 in) film is 35 mm (1.4 in) wide and each image in a standard 35 mm (1.4 in) camera is 24 mm (0.9 in) by 36 mm (1.4 in) in size. The Pythagorean theorem can be used to calculate that the diagonal size of a standard 35 mm (1.4 in) frame is 43 mm (1.7 in). Lenses are usually designed using focal length increments that are multiples of 5 mm (0.2 in) or 10 mm (0.4 in) and 40 mm (1.6 in) lenses are not common so, in practice, the so-called normal lens for a 35 mm (1.4 in) camera is a 35 mm (1.4 mm) or 50 mm (2.0 in) lens. Manufacturers of cameras with film sizes or digital sensors of different sizes will sometimes describe their lenses using a 35 mm (1.4 in) equivalent focal length. This means that the photographic effect (wide angle, normal, telephoto) will be the same as that focal length of lens used on a 35 mm (1.4 in) camera.

Camera lens. UNDERWOOD & UNDERWOOD/CORBIS.

SHUTTER SPEED

The amount of light striking the film is controlled by two things: the length of time that the shutter is open (shutter speed) and the lens aperture. Shutter speed is typically expressed as some fraction of a second, for example 1/2 s or 1/500 s, and not as a decimal. Manual cameras allow photographers to choose from a fixed set of mechanically controlled shutter speeds that differ from each other by factors of approximately 2, and the shutter is opened and closed by a series of springs and levers. For example, 1/2, 1/4, 1/8, 1/15, 1/30, 1/60, 1/125, 1/250, and so forth. Note that the factor changes slightly between 1/8 and 1/15, and then again between 1/60 and 1/125. In order to make the best use of limited space on small cameras, film speed dials or indicators in many cases use only their denominator the shutter speed. Thus, a camera dial showing a shutter speed of 250 means that the film will be exposed to light for 1/250 s. Electronic cameras, whether film or digital, contain microprocessors and can offer a continuous range of shutter speeds. The shutter speeds can be set by the photographer or automatically selected by the camera.

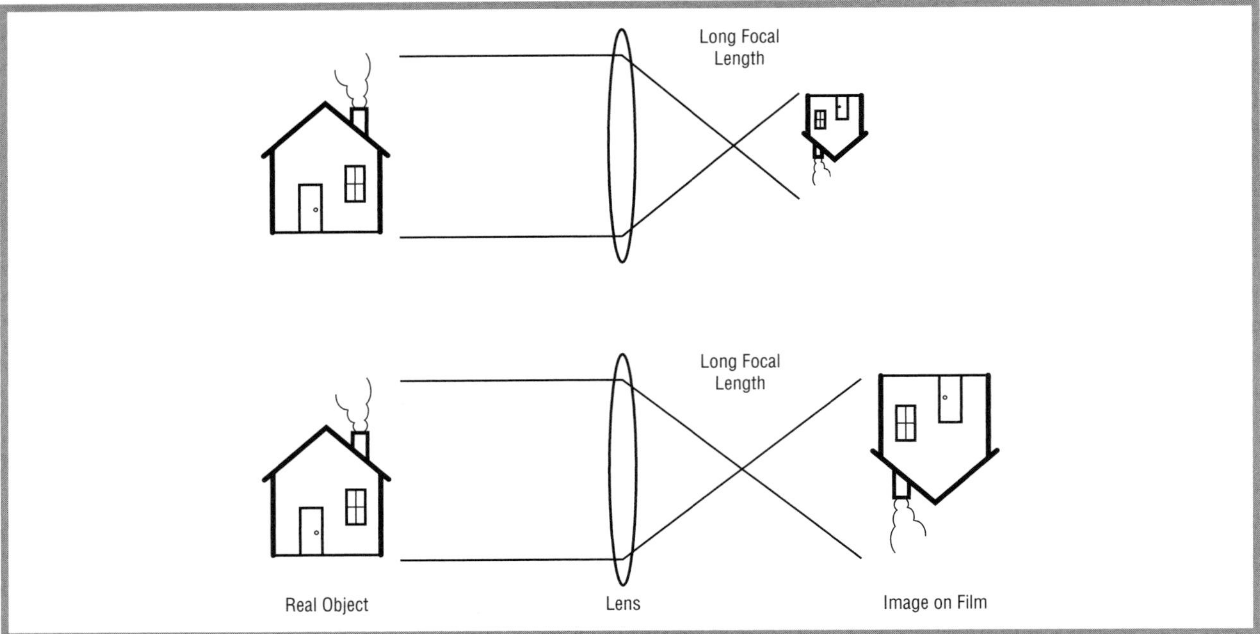

Figure 1.

LENS APERTURE

Lens aperture is the diameter of the opening through which light passes on its way to the film. The larger the aperture, the wider the opening and the more light that will pass through the lens to the film. Aperture is expressed as a so-called f-stop or f-number that is the quotient of the lens focal length divided by the diameter of the aperture, and is controlled by a diaphragm consisting of moving metal blades within the lens. A lens with a focal length of 100 mm (3.9 in) and an opening 50 mm (2.0 in) in diameter is said to have an aperture of 100/50 = f/2, but a 400 mm (15.7 in) lens with the same opening would have an aperture of 400/50 = f/8. Therefore, the size of the opening must increase proportionately with focal length in order for lenses of two different focal lengths to have the same aperture. This is why the large telephoto lenses used by sports and nature photographers are so long and wide. They must have both long focal lengths and wide openings to transmit enough light to properly expose the film.

The term f-stop refers to the fact that photographers have traditionally adjusted the aperture of their lenses by rotating a ring on the lens to choose among several pre-set apertures. Each pre-set aperture is marked by a sensible and audible click, or stop, hence the name f-stop. The pre-set apertures were chosen so that each stop halved or doubled the radius of the opening, using the same logic as pre-set shutter speeds, thus halving or doubling the amount of light passing through. The result was this progression of f-stops: f/1, f/1.4, f/2, f/2.8, f/4, f/5.6, f/8, f/11, f/16, f/22, and f/32. Although many modern lenses have continuously adjustable apertures, and some are electronically controlled with no aperture rings at all, the f-stop terminology and progression of f-stops marked on lenses persists. Physical constraints make it difficult to design lenses with large apertures, so the range of most lenses begins above f/1, typically in the range of f/2 or f/2.8. The additional difficulty of designing zoom lenses, especially if they are to be affordable to large numbers of people, sometimes motivates lens designers to use maximum apertures that change according to the focal length. An 18–70 mm (0.7–2.8 in) f/3.5-4.5 zoom lens would have a maximum aperture that ranges from f/3.5 at 18 mm (0.7 in) focal length to f/4.5 at 70 mm (2.8 in) focal length.

DEPTH OF FIELD

Depth of field refers to the range of distance from the lens, or depth, throughout which objects appear to be in focus. A lens can be focused on objects at only one distance, and objects closer to or farther away from the lens will be out of focus on the plane of the film. In the case of a point of light that is out of focus, the result is a fuzzy circle known as a circle of confusion. Depth of field is increased by decreasing the size of the circles of confusion in an image, which is accomplished by reducing the aperture of the lens, until objects over a wide range of distances appear to be in focus to the human eye.

Although a small aperture reduces the sizes of the circles of confusion in an image, it also increases the relative importance of diffraction around the edges of aperture. As light passes through the movable metal blades that control aperture, some of it is scattered or diffracted. When the aperture is large, the effects of diffraction generally go unnoticed. As the aperture decreases, diffracted light becomes an increasingly large proportion of all the light passing through the aperture and image sharpness can decrease. Therefore, setting a lens to its smallest aperture will not generally produce the sharpest possible image. The sharpest images will generally be obtained by setting the lens to an aperture in the middle of its range.

For a specified aperture, lenses with long focal lengths will always have shallower depths of field than lenses with short focal lengths. This is because the longer lenses must have physically larger openings than the shorter lenses, even if the aperture (f-stop) is the same. A larger opening transmits more light, which in turn produces larger circles of confusion.

RECIPROCITY

Reciprocity is a mathematical relationship between shutter speed and lens aperture. If a photographer increases the light passing through the lens by opening the aperture one f-stop and then doubles the shutter speed (which will reduce the length of time the shutter is open by one-half), the amount of light reaching the film will not change. An aperture of f/4 and a shutter speed of 1/500 s, for example, will deliver the same amount of light as an aperture of f/5.6 and a shutter speed of 1/250 s. In other words, aperture and shutter speed share a reciprocal relationship and many different combinations of shutter speed and lens aperture will provide the same amount of light. The reciprocal relationship also extends to film speed. If film speed is doubled, either the shutter speed can be increased (producing a shorter exposure) or aperture can be decreased by the same factor without changing the amount of light that reaches the film.

In practice, there are some limitations to reciprocity. Photographs with very slow shutter speeds, for example minutes or hours instead of fractions of a second, can be appear too dark (underexposed) because the reciprocity relationship does not extend to such long exposures. This is known as reciprocity failure and can pose a problem for photographers working at night in situations where artificial lights cannot be used, for example when astronomers are attempting to take photos of the night sky using their very sensitive equipment. Film manufacturers publish tables that allow photographers to compensate for reciprocity failure in different kinds of film.

DIGITAL PHOTOGRAPHY

Virtually everything written in this article applies to digital photography as well as film photography. The primary difference is that a digital camera uses an electronic sensor instead of a piece of plastic film coated with silver compounds. In place of the film used in a conventional camera, a digital camera uses an electronic sensor. Two sensor types are commonly used: CCDs, or charged-coupled device sensors, and CMOSs, or complementary metal oxide semiconductor sensors. Both kinds are composed of rows and columns of photosites that convert light into an electronic signal. Each photosite is covered with a filter so that it is sensitive to only one of the three components of visible light (red, blue, or green). One widely used configuration, the Bayer array, consists of rows containing red and green filtered photosites alternating with rows containing green and blue photosites. When the image is being processed by the camera, values for the two missing colors are estimated using the mathematical technique of interpolation.

Two primary measures are used to characterize digital images: resolution and size. Resolution refers to the ability of a sensor to represent details, and is generally specified in terms of pixels per inch (ppi). Image size refers to the total number of pixels comprising an image, and is typically given in terms of megapixels. A pixel is the smallest possible discrete component of an image, typically a small square or dot, and one megapixel consists of one million pixels. As of early 2005, the best commercially available digital cameras had resolutions of approximately 20 megapixels and many professional quality digital cameras had resolutions of 5 or 6 megapixels.

Digital photographers can adjust the sensitivity of the sensor to light just as film photographers can use films with different ISO speeds. In digital cameras, however, there is done with a switch or button on the camera and the sensor is not physically removed. Although digital cameras commonly have ISO settings, they vary from manufacturer to manufacturer and do not follow the consistent ISO standard. Instead, they are an approximate gauge of the sensitivity. The digital equivalent of film grain is electronic noise, which can appear in images as visual static or randomly colored pixels, and is most often a problem using high digital ISO settings. The size of the sensor can also contribute to the amount of noise in a digital image, because the photosites on a small sensor are closer to each other than those on a larger sensor and can interfere with each other.

In order to capture action photos, photographers must use math to set shutter and film speeds properly. AP/WIDE WORLD PHOTOS. REPRODUCED BY PERMISSION.

A Brief History of Discovery and Development

The basic concept of using a device to project an image onto a flat surface dates from the camera obscura of ancient times, in which light passed through a small hole that focused the image and projected it in a darkened room. The modern day descendent of the camera obscura is the pinhole camera, which uses a hole without a lens to project an image onto a piece of photographic film. The quality of camera obscura images increased as lenses were developed in the sixteenth century. Still, there as no way to preserve the image except by drawing or painting on the projection screen. The discovery of photosensitive chemicals in the nineteenth century was a major step forward because it allowed images to be preserved without drawing or painting, and many different techniques were invented for creating photographs on paper, glass, and metal sheets. In 1861, Scottish physicist James Clerk-Maxwell invented a system of color photography using black and white images taken through red, green, and blue filters and then combined. George Eastman started his photographic company in 1880, and the first Kodak camera was introduced in 1888. This surge in technology gave rise to an explosion in the technical, journalistic, and artistic use of photography as mechanical cameras and lenses were continually refined throughout the first half of the twentieth century. The advent of computer-aided design in the 1960s and 1970s represented another major step forward, allowing much more sophisticated camera and lens designs, and auto-focus and auto-exposure cameras arrived on the scene shortly thereafter.

Real-life Applications

SPORTS AND WILDLIFE PHOTOGRAPHY

Sports and wildlife photographers often share the same goals. They want to produce photographs of fast moving subjects from a distance. Therefore, they prefer long telephoto lenses with large maximum apertures. By

Key Terms

Aperture, lens: The size of the opening through which light passes in a photographic lens.

Reciprocity: The mathematical reciprocal relationship between shutter speed and aperture, which states that there are many combinations of lens aperture and shutter speed that will supply the same amount of light to the film or digital sensor in a camera.

virtue of reciprocity, these large aperture lenses can be used with higher shutter speeds that freeze action, whether it be a gazelle or a linebacker. Long lenses with large maximum apertures also add an artistic element, helping to blur the background and focus the viewer's eyes on the subject of the photograph. For the same reason, portrait photographers will often used moderately long telephoto lenses that set their subjects apart from the background. Photographers describe the aesthetic quality of the blurred areas with the Japanese word *bokeh*, and a lens that produces pleasingly out-of-focus areas is said to have good *bokeh*.

DIGITAL IMAGE PROCESSING

The ability to create high-resolution digital images, either using a digital camera or by scanning a film negative or transparency, allows photographers to adjust the details of their photographs without entering a darkroom. Each pixel contains a red, green, and blue value that can be brightened or darkened. The overall range of tones, known as contrast, can also be easily adjusted and unwanted tints can be removed. A photographer, for example, can remove the cool bluish cast in shadowy light by adding more red and green to the image. Images can also be sharpened to some degree, although it is impossible to sharpen an image that is truly out of focus. This is done using a technique called unsharp masking, which derives its name from a technique developed by astrophotographers using film many years ago. In order to sharpen a slightly fuzzy image, the photographer would make a deliberately blurred copy of the film negative. The two images would be carefully aligned and a sharpened print made.

PHOTOMICROGRAPHY

Photomicrography uses an optical microscope, rather than a traditional lens, to produce photographs of objects such as microorganisms and mineral grains. In the case of geological photomicrography, small slices of rock are glued to microsope slides and then ground down to a thickness of 30 microns (0.001 in). The slice of rock is nearly transparent at that thickness, allowing it to be examined under the microscope. Digital image processing techniques can also be applied to photomicrographs in order to enhance edges or increase the visibility of subtle details.

Potential Applications

Computer designed lenses and cameras, both film and digital, continue to increase in sophistication each year. Current commercial activity emphasizes the development of improved digital sensors with increased resolution and decreased noise, vibration resistant camera bodies and lenses that compensate for the photographers moving hands, and zoom lenses that cover focal length ranges from wide angle to telephoto.

Where to Learn More

Books

Enfield, Jill. *Photo-Imaging: A Complete Guide to Alternative Processes.* New York: Amphoto, 2002.

Jacobson, Ralph, Sidney Ray, G.G. Attridge, and Norman Axford. *Manual of Photography: Photographic and Digital Imaging,* 9th ed. New York: Bantam, 1998.

Web sites

Greenspun, Philip. "History of Photography Timeline." 2005. <http://www.photo.net/history/timeline> (February 15, 2005).

Kodak. "Photography in Your Science Fair Project: Photomicrography." No date. <http://www.kodak.com/global/en/service/scienceFair/photomicrography.shtml> (February 15, 2005).

Plots and Diagrams

Overview

The use of plots and diagrams is an integral part of everyday life. Plots and diagrams can be found in many applications in scientific study and in real life.

Effective graphs can significantly increase a reader's understanding of complex data sets. The basis of scientific procedure is data collection. Scientists are required to examine and analyze the data they collect. The most efficient way to do this is graphically. A graph is a visual representation of two variables relative to each other. Graphs are one- or two-dimensional figures. Three-dimensional graphs also exist, however, these are often more complex and more difficult to understand than basic two-dimensional graphs. A graph usually has two axes, the x-axis and the y-axis. There is also an origin, which is the point (0, 0). This is where the two axes cross each other. Each point on a scatter graph is represented by a pair of coordinates. These are written in the form (x, y). The number x represents how far along the x-axis the point is, and y represents how far along the y-axis the point is. If a point lies on the y-axis, its co-ordinates would be (0, y), because it is at the 0 point along the y-axis (remember the axes cross each other at 0). Accordingly, if a point is on the x-axis, then its co-ordinates are (x, 0).

PROPERTIES OF GRAPHS

A graph should have at least a title and a scale that is numbered in specific and constant intervals and labeled. This allows the reader to know what the graph is about and what the graph is measuring or showing. The more information that is included on the graph, the easier it is to understand and interpret the data it shows. However, too much information must not be included, as the graph may become cluttered. Some graphs require a legend or key. This helps the reader understand different shading and colors that have been used. A legend is useful if the graph becomes too cluttered with all the labeling. The purpose of a graph is to provide clear, concise information. This is difficult to accomplish if there are large numbers of labels covering the data.

DIAGRAMS

Diagrams can come in many shapes and forms, depending on the application for which they are being used. Most graphs are about numbers; in other words, they are number oriented. However, with diagrams this need not be the case. Some diagrams do present quantitative (number oriented) data, but most diagrams present qualitative (non-numerical) data. They are widely used in both science and everyday life. The type of diagram directly depends on the subject data. Diagrams are usually pictures. Around or on this picture is usually written extra information. This information could be providing details about the diagram, such as a diagram of the body, or the diagram could be there for easier comprehension of details, such as a weather map.

Fundamental Mathematical Concepts and Terms

STEM AND LEAF PLOTS

Stem and leaf plots are similar to histograms (vertical graphs with touching bars) in the way they represent information. However, they usually contain a little more information. Stem and leaf plots show the distribution (or the shape of the data) as well as individual data. These types of plots are useful in organizing large groups of data. In a set of data containing numbers from 1 to 100, the digits in the largest place, the tens, are referred to as the stem. The digits in the smallest place, the units or ones, are referred to as the leaf. When there is a large amount of data, sometimes the stem needs to be represented twice. The first time it is associated with the leaves 0 to 4, and the second time it is associated with the leaves 5 to 9. If a stem is shown five times, then similar rules apply as when it is represented twice. The first stem is associated with 0 to 1, the second with 2 to 3, and so on. This is to make the plot easier to read.

BOX PLOT

A box plot (also known as a box and whisker plot) is a diagram of the measure of spread. It is a graph of the 5-number summary. Data can be divided into four even sections called quartiles. The number of values in each quartile is the same. The middle number is called the median. The value between the median and the minimum value is the first quartile and the value between the median and the maximum value is the third quartile. The 5-number summary is the minimum, the first quartile, the median, the third quartile and the maximum. The inter-quartile range is the distance from quartile 1 to quartile 3. A quartile is 25% of the numbers of the entire set of data. A box plot shows the spread of a set of values. This is an important factor in some statistical analyses.

SCATTER GRAPH

Scientists most often utilize scatter graphs. They are useful for fast and easy analysis of data. These types of graphs are usually a series of points on a grid. Each of the axes is used to represent a value data. The value of the variable along the y-axis (the vertical axis) is dependent on the value of the variable along the x-axis, which is the independent variable.

Scatter graphs are usually used to determine a relationship between two variables. Once two sets of data have been plotted against each other (such as distance against time), a line of best fit can be drawn through the points to determine whether there is a relationship between the two variables. Scatter graphs are most commonly used for scientific purposes. This is because they do not negate individual data. Every single piece of data is included in a scatter graph. However, scatter graphs can also show two sets of data that had the same variables measured, but one was changed.

Three mathematical concepts that are unique and integral to scatter graphs are the line of best fit, the correlation coefficient, and the coefficient of determination. These three tools are important in helping scientists analyze the data that they gather. In real-life applications, the interpretation and understanding of data is the most important part of scientific process. Without interpretation, and thus tools of interpretation, data would just be a meaningless set of numbers.

A line of best fit, also known as a line of regression, is a line that is drawn to represent the trend of the data. A regression line always exists, whether there is correlation, a relationship between two variables, or not. The easiest way to draw this line is to draw a straight line through as many points of data as possible. However, this is usually impossible, especially when scientific errors are taken into account. Then the best method to draw this line is to have an equal number of points above the line and below the line. This averages out the line. There are complicated methods of determining the exact line of best fit that involve long and laborious calculations. A line of best fit is where the vertical deviations (the up or down distances) from the observed point (the ones determined experimentally) and the calculated points (the ones taken from the regression line) are as small as possible. In other words, the line of best fit is a refined line of regression, although the two terms are usually used to represent the same line. It would take a long time to determine the line of best fit if drawing the line of regression by hand. Computer programs for data analysis exist now that can compute and draw the line of best fit automatically. The computer does all the calculations much faster than a person would be able to do it.

The correlation coefficient is an important concept to understand when interpreting graphs and their lines of best fit. The correlation coefficient is a way to measure how close the points are to a regression line. The correlation coefficient is commonly known as r and lies between -1 and 1. When $r = \pm 1$, then there is perfect correlation between the two variables and all the points lie on the line. Then $r = 0$, there is no correlation between two variables and they are all independent of each other and the line of regression. A correlation coefficient between 0.0 and ± 0.3 is considered a weak, a correlation coefficient between ± 0.3 and ± 0.7 is considered a moderate, and a correlation coefficient between ± 0.7 and ± 1.0 is considered a high. Mathematically, the correlation coefficient is the sum of the squares of the individual errors, which are the vertical deviations, to measure how well a function, usually the line of regression, predicts y from x.

The coefficient of determination, R^2, is another measure of how well two variables are related and how well a regression line fits to the set of points. R^2 describes how much of the deviance in the y values can be explained by the fact that they are related to an x value. In simple linear regression, R^2 is the square of the correlation coefficient, in other words, $R^2 = r^2$. Both of these coefficients can help people determine whether data is credible or not, especially in a scientific context. Usually a scientist will have a thesis or aim that he or she wants to prove or disprove. Here the correlation between height and arm span will be used and the aim is to show that they are related to each other. The scientist will take an ample amount of data and then analyze this data, probably using a scatter graph. If the correlation coefficient or R^2 value is below standard to prove the aim correct, then the scientist may have to revise the data or gather more data. This process, especially the R^2 value, is integral to the process of scientific information, especially if a scientist is looking to present credible data.

AREA CHART

A variation of the line chart is an area chart. Line charts look like various line graphs together with the sections between them colored in. They are used where there is one independent series and several dependent series. The independent series together have a constant sum.

PIE GRAPH

Another type of graph is a pie graph. These graphs are aptly named, as they have a circular shape and sections are cut separated by a line making the whole graph look like an unevenly cut pie. The idea is effective because it takes advantage of the everyday principles people use when, say, they are cutting a cake into portions. This makes the pie chart something people can relate to and thus more easily understand. Although these graphs are not often seen, they are the most useful in expressing discrete data in specific categories. They are used to show how one piece fits into the entirety; in other words, pie graphs are used when the values have a constant sum, such as a population or when using percentages.

Pie graphs are best utilized when there is significant variation between the portions. In other words, having five equal areas is quite useless, unless that is the point being made. Pie graphs often have the sections labeled directly on the diagram instead of having a separate table informing the reader of which section is which. However, a pie graph does have its limitations. The number of categories (or portions of the pie) needs to be small or the graph may become cluttered. Generally, the number of categories should be between 3 and 10, although this may vary slightly.

BAR GRAPHS

Bar graphs are another popular style of graph. Bar graphs are a versatile type of graph and can show many different types of data. A bar graph, also called a column graph, is easy to recognize because several long or short rectangles represent data categories. These rectangular bars can be vertically or horizontally orientated. A bar graph has the bars orientated horizontally and a column graph has them orientated vertically. Sometimes this distinction is not made, however, it is important to know the difference. On a bar graph, the bars are usually the same width. They are used as a comparative type of graph and usually compare several things, people, objects, cities or departments, units or entities, in a data series. On a column graph, these categories are along the horizontal axis whereas on a bar graph they are on the vertical axis.

FISHBONE DIAGRAM

Fishbone diagrams have a strange name that resembles the style of graph. Fishbone graphs are primarily used in problem solving, especially in quality improvement programs in business. The problem is written inside a box at the head of the fish. This provides the aim for the diagram. The words and ideas that extend from the backbone are the possible causes of the problem. They allow people to organize thoughts about problems and what may be causing them. It is a systematic way to organize and analyze data that is related to solving a quality problem. Fishbone diagrams may also be called cause-and-effect diagrams or Ishikawa diagrams.

POLAR CHART

There are several specialized types of graphs that have been developed to deal with unusual and different types of data. Even though scatter graphs, pie charts, bar charts, and line graphs deal with a wide variety of data, sometimes even they cannot handle specific data. One such graph that has been developed is the polar chart. A polar chart is used with discrete data where each point has a direction value from a source, in other words, a direction, usually expressed in degrees. The data also has a quantity, or a specific distance it is away from the source. Essentially, this graph represents polar data. Polar graphs are used in the study of polar equations and also in vector studies.

TRIANGULAR GRAPH

Another specific type of graph is a triangular graph. These graphs are most commonly used in geographic applications. They are used to plot discrete data in which each point has three values. These values have a constant sum that is usually expressed as a percentage. It is triangular in shape, hence its name, the triangular graph.

THREE-DIMENSIONAL GRAPH

A more complex type of graph is a three-dimensional graph. These are used when three interdependent variables need to be plotted. If data are grouped, then a three-dimensional column graph can be used. This allows the reader to associate certain things relative to others in one graph instead of having to use many different graphs. When data is displayed in this manner, it may initially look quite complex. However, with further understanding of what the graph represents, it is a suitably useful way of displaying data. If data are from a continuous distribution then a surface plot is used. This is another type of three-dimensional plot. These are more difficult to interpret, as the data is continuous. Three-dimensional graphs can be shown as a continuous surface or as a series of contour lines.

A Brief History of Discovery and Development

The origins of plots and diagrams date to prehistoric ages when people made cave drawings. Although these may not be the sophisticated plots and diagrams that exist in the twenty-first century, they were, nonetheless, diagrams. The earliest map was a tenth century map of China. Diagrams of planets and planetary motions were some of the earliest, more complex diagrams that existed. Graphs made their first appearance around 1770, and became accepted and widely used around 1820. In 1795, graphical scales were used to help convert old measurements to the new, metric measurements. The French mathematician Johann Heinrich Lambert (1728–1777) used graphs extensively in the eighteenth century. He was one of the only scientists of the time to do so. He applied many of the principles now applied to graphs, such as a line of best fit. From this time, graphs and diagrams developed to aid people in determining angles, analyzing data, and providing information. Bar graphs emerged for data that could not be sorted. General x-y graphs did not appear in publications until the twentieth century. From simple things, such as pictorial instructions, to complex graphs, everyday life has been greatly affected by plots and diagrams.

Stem and leaf display

3	2337
2	001112223889
1	2244456888899
0	69

Figure 1: A stem and leaf plot display.

Real-life Applications

STEM AND LEAF PLOTS

Stem and leaf plots can be used for series of scores on sports teams, series of temperatures or rainfall in a month, or series of classroom test scores. In a stem and leaf plot, data is arranged in place value. (See Figure 1.)

BOX PLOT

Box plots are usually drawn as composite box plots. (See Figure 2.) Two different box plots displaying, for example, the heights of boys in a class and the heights of girls in a class, can provide more statistically useful data when compared than a single box plot. It is a simple and clear graphical representation of information that may be difficult to decipher as just a series of numbers on a page. The two graphs together allow the reader to easily interpret the ranges of girls' height with respect to the boys, and vice versa. Box plots also show whether there is a data point that is an outlier, that is, it does not fit within the specified set.

SCATTER GRAPH

Scatter graphs are used to plot much of the experimental data that scientists collect. For example, if a scientist

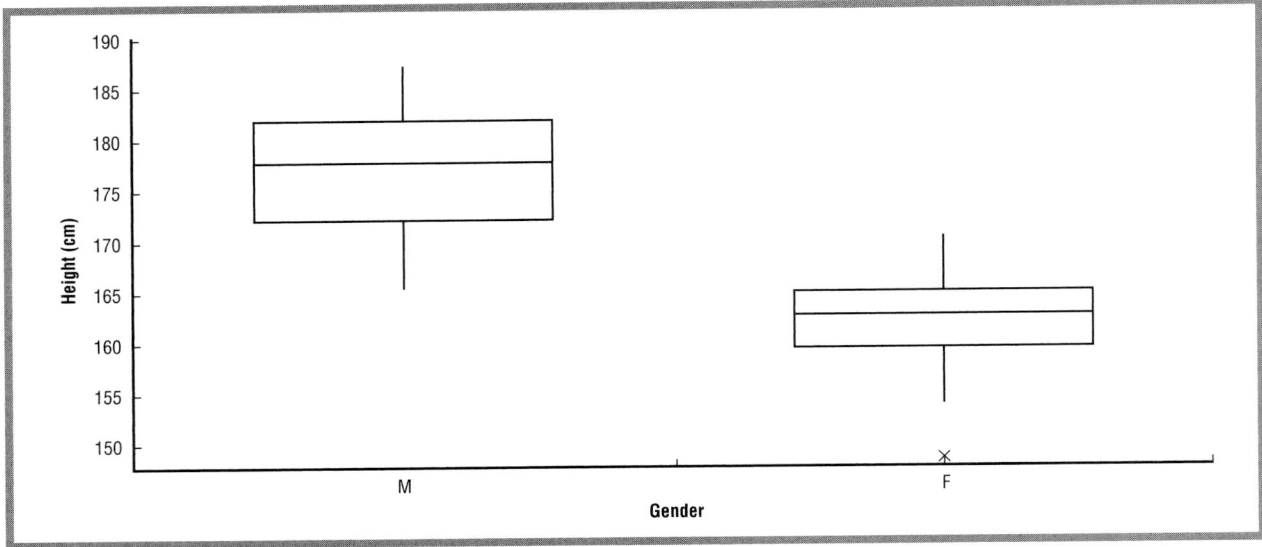

Figure 2: A composite box plot.

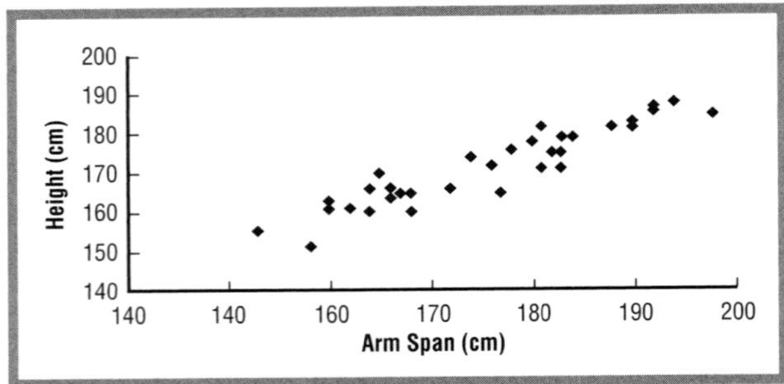

Figure 3: A scatter plot on a scatter graph.

measured the distance a car, traveling at a constant speed, traveled over a period of time, a scatter graph would measure the time and the speed. In this case, the time is the independent variable because the distance the car travels depends on the amount of time the car is traveling. When comparing two sets of data, a scientist could compare the braking distances of cars traveling at different speeds. The speed the car is traveling at would be the independent variable, and the braking distance would be the dependent variable. Different brands and types of cars can be shown on a scatter graph to compare and contrast them to each other. This can help a person wishing to purchase a car to determine which car they want to buy if safety is one of their primary buying criteria.

A scatter graph can also be used to determine if there is a correlation between a person's arm span and their height. The initial scatter graph would show the grouping of the data and can show whether there is evidence of a trend or not. (See Figure 3.)

Once a set of data is plotted, a line of best fit can be drawn to show whether the relationship between two variables is worth investigating. An example would be a scatter graph illustrating the growth of a rabbit population. The population was counted at various intervals and the data was plotted on the graph. A line of best fit was drawn and the correlation coefficient was determined. The use of scatter graphs can help scientists determine important statistical data. People can then use this information when they are studying the growth of populations for assignment or more in depth studies. These graphs help people to understand the way things work without having to be scientists. (See Figure 4.)

LINE GRAPH

Line graphs represent changes in numbers over time. It is one of the most widely used types of graphs in real life. Stock market graphs are an example of line graphs. They provide stockbrokers and the general public with long-term information about a particular stock. It is easiest to read how much a particular stock has gone up in one day. However, to determine whether the stock would be a good investment in the long-term, it is best to look back at years of data. However, poring over page after page of numbers is difficult and laborious, so this data is best represented in a scatter-plot that has each point joined together. This enables people to tell whether its value has fallen or risen over a period of time, all with just

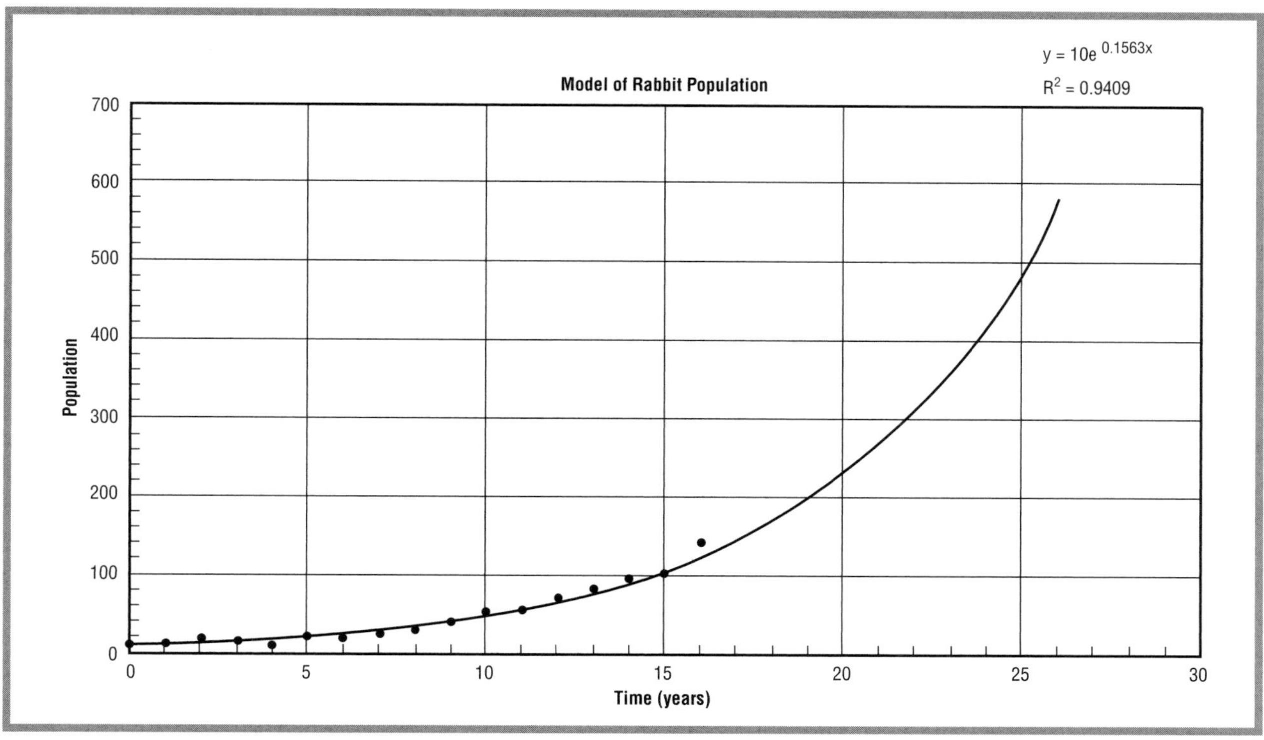

Figure 4: A scatter graph with best line averaging points and transitions between data point that illustrates the growth of a rabbit population.

one look at a single image. Stock market graphs are found in newspapers and can be seen on television. Although these may be difficult to understand to the novice, for a person who wants the information and who understands the basic principles, they are easy to decipher and quick to understand.

Another example of a line graph is to show the number of people who attend sports matches over a period of time. Time would be illustrated on the x-axis, which means that the number of people who attend is dependent on the time. The graph would show how many people attended matches each year. It would show how the trend the amount of people who attended matches increased or decreased. It would also show specific slumps and rises in attendance numbers. These graphs can have future applications in determining the reasons for lack of attendance or for high attendance. This would require further research, such as looking at world events that may have caused particular slumps or high points.

Figure 5 depicts a line graph as a composite line graph. This means that it shows different types of data that have the same variables measured and are comparable. In this case, it is the growth of a normal mouse relative to that of mutant mice.

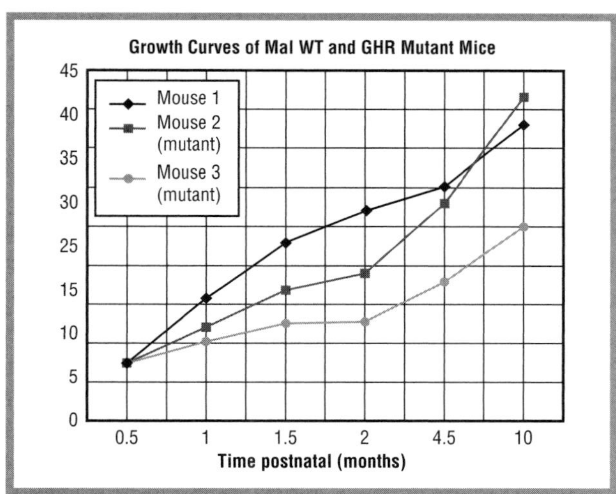

Figure 5: A line graph as a composite line graph.

A special type of line graph is a run chart. Run charts show the sequential measurements of a process over a specific period of time. One example is how many cookies are produced from a batch of dough that is presumably the same size each time. These types of graphs usually have limits, depending on what is being measured.

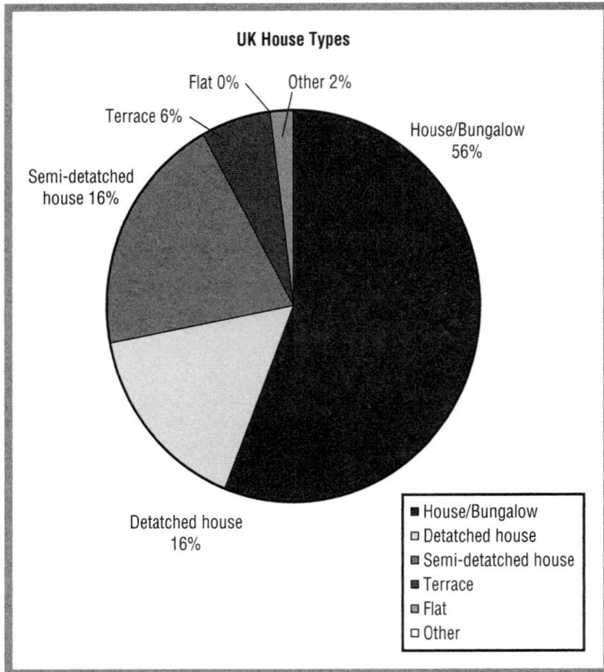

Figure 6: A pie graph can also be used to compare the different types of housing in which people live.

Another example could illustrate patient waiting time in a dentist's or doctor's office.

PIE GRAPH

Pie graphs can be used for such purposes as illustrating the different cultural backgrounds of people in a country. The categories in the pie chart are often linked, such as cultural backgrounds, as in the example of Figure 6 given. Pie graphs provide a more overall view of data and can be difficult to read if the data is not written on or near each pie section. However, they are useful in determining general trends or providing rough estimates. This is because pie charts are quite graphic and tend to be colorful, to make distinguishing between sections easy. They are useful in representing percentages of stock in a store of percentages of a population that have particular diseases. As shown in Figure 6, a pie graph can also be used to compare the different types of housing in which people live.

BAR GRAPHS

Bar graphs are used in a similar fashion to pie graphs in that they can express discrete data in specific categories with great efficiency. Bar graphs are more useful than pie graphs when exact numerical data is more important rather than a general overview. For example, the number of people who answered a particular multiple-choice question would be best represented in a bar graph. This would allow students to easily compare their own scores with other students' scores and also determine whether they were in a majority or not. A simple bar graph, as shown in Figure 7, quickly shows the frequency of specific heights of people which can then be used to compare people's heights.

Another example of how a bar graph can be used is depicted in Figure 8, a graph that compares technology access.

As shown in Figure 9, another method to depict data joins the successive midpoints of each of the bars to make another type of graph, a frequency histogram. Histograms provide information on the distribution of the data, a concept that relates more to statistics than to graphs and diagrams.

FISHBONE DIAGRAM

If a person in a business has a problem, such as low production, a fishbone diagram can be applied to determine the cause of the problem. The problem is written in the head of the fish. From the backbone of the fish are bones, each with a specific topic, such as psychological factors, company pressures, physical pressures, and home problems, for a set example. From each of these are the different contributing factors within those topics. This visual representation can help both the person and the company locate and identify the problem and thus start the process of solving the problem.

TRIANGULAR GRAPH

Each side of the triangle is labeled with a different soil type, for example. Along that line are marked the various percentages, from 0% to 100%. A point marked inside the triangle denotes a specific soil type. Its composition can be determined by drawing a line from it to each edge of the triangle. Where the line lands perpendicular to the side of the triangle is the percentage of that compound that can be found in that soil. This process can also be used in reverse by knowing the percentages of the composition and then determining the specific soil type from those percentages. Its most common use is the three-way split of sand, silt, and clay in soil and sediment samples. This type of diagram allows scientists to determine the type of soil that it is, and thus postulate where the region may have been developed (since land masses have moved due to continental drift). It can also help them determine what uses the soil may have. This means that this type of graph can help a person determine the

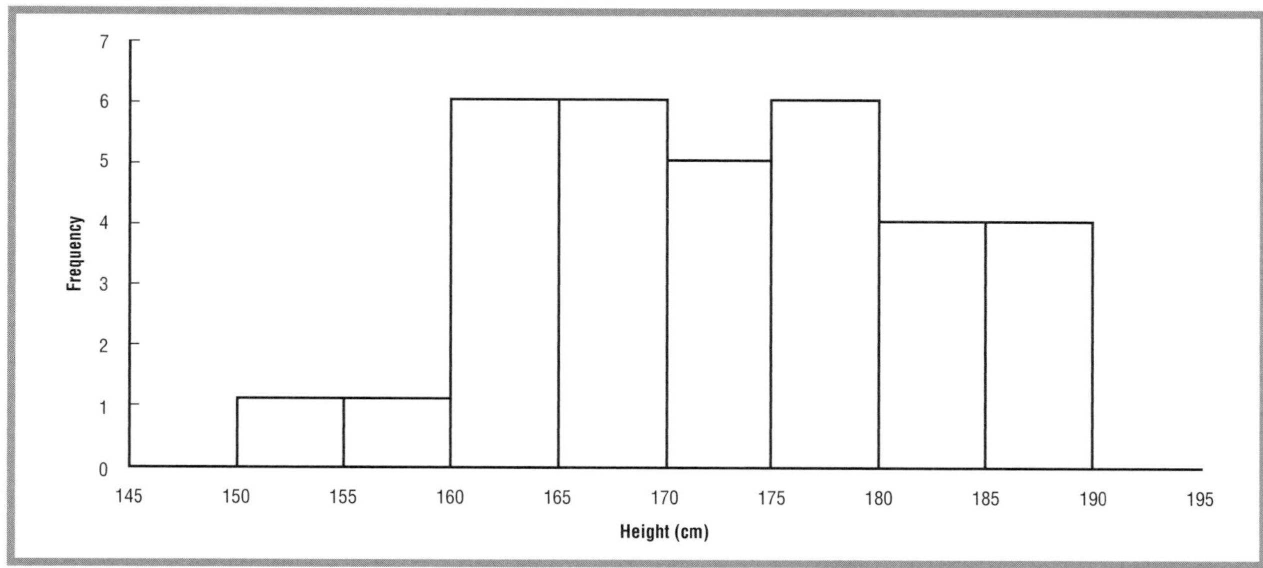

Figure 7: A bar graph.

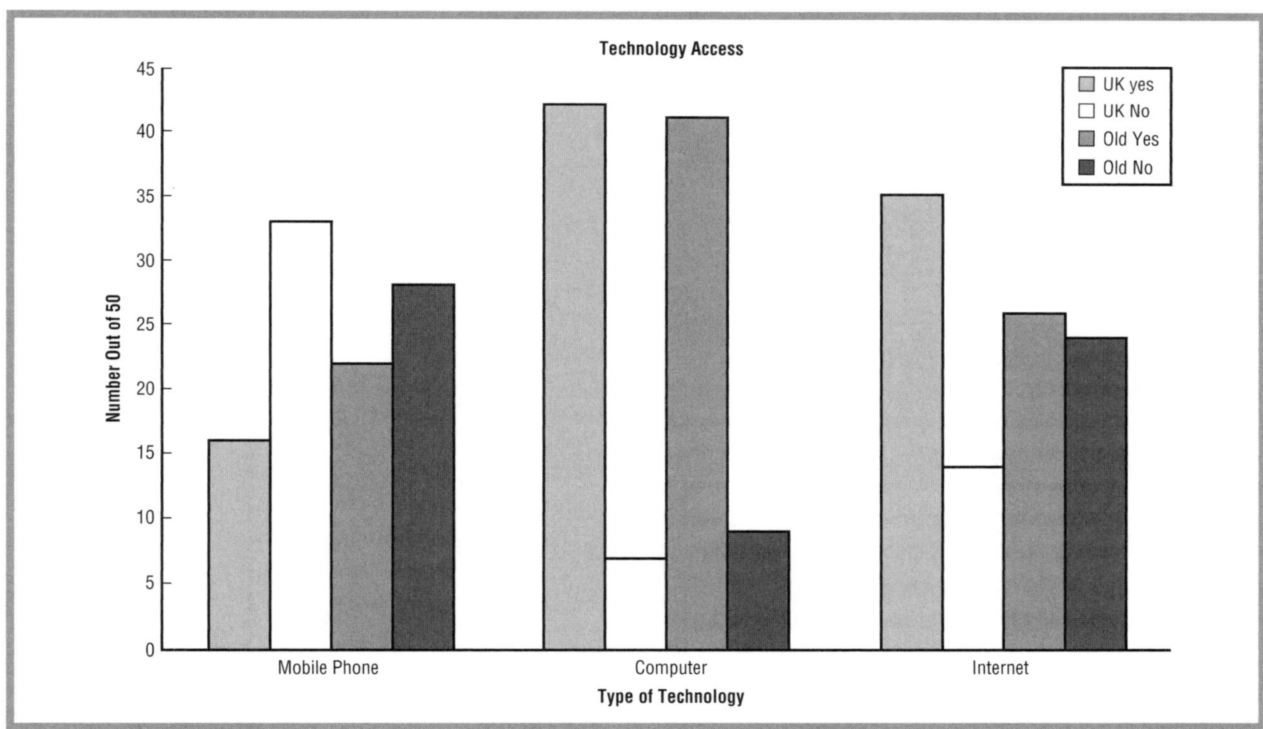

Figure 8: A bar graph comparing technology access.

type of soil they need for their garden to make sure their plants have the correct soil. However, this type of graph can also be used where three factors are needed to represent a whole and each is of a specific percentage.

FLOW CHART

An important type of diagram is a flow chart. Flow charts are useful because they show sequential steps in a process. They are good reference points for people learning

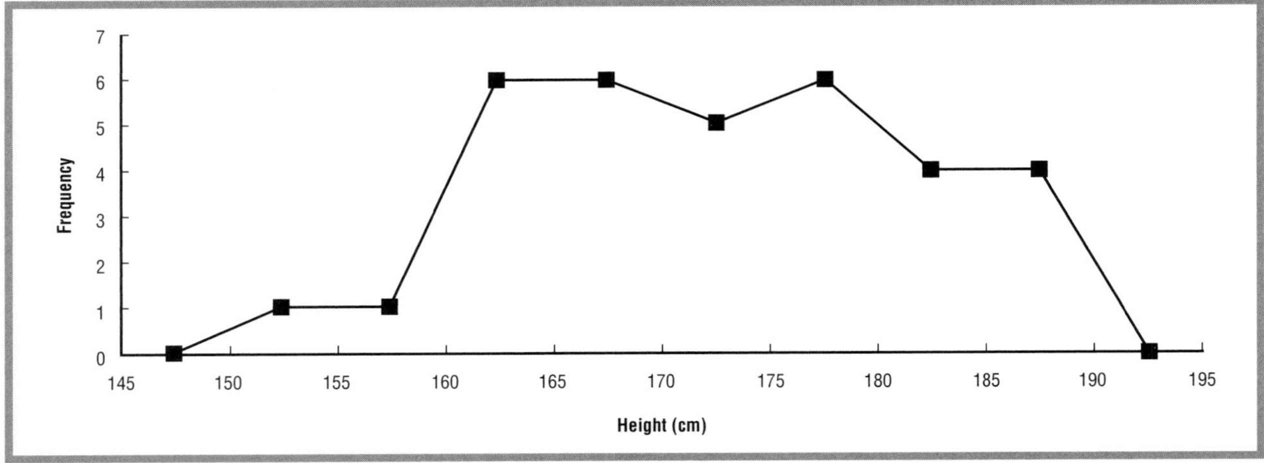

Figure 9: A frequency histogram.

a new skill or job. They are used to represent a group of connected components in a series where there may be more than one pathway through them. Occasionally, flow charts have a web-like structure. Quantitative aspects of the data flow can be distinguished by varying the line style or thickness between the components, and labeling each of the components thoroughly.

A flow chart has many applications. One of the most important applications is in computing where flow charts are used to represent the internal logical organization of computer programs. Examples of these charts can be seen on a personal home computer, or PC. When navigating through folders to find a file, one of a multitude of possible paths is followed. There are also methods for determining the number of different pathways, however, these are complex and are not discussed here. Flow charts are also used in business organization. One example is processing a sales transaction, from the flow of information and goods beginning at the receipt of the order, to the shipping and the invoicing of the customer.

New computer systems are larger, can process more data, and can accomplish this at a greater speed than older models. This means that more information can be stored on them and thus, more information needs to be accessed and processed at a comparative or faster speed. Alternative diagrammatic structures have been developed to aid system analysts in designing information processing systems that have quite complex internal information flows. The tools allow the analysts to model the relationships amongst the components of a system. These relationships can become quite complex, especially with new computer systems. These tools and structures are all encompassed under the general heading of relationship diagrams.

Representations of the World Wide Web, more commonly referred to as the Internet, are usually drawn as flow charts. This is due to the large amounts of interconnected data that exist on the Internet. The concept of virtual realities is based upon this interconnectivity. Virtual realities are becoming more and more a part of everyday life for people. Computer and arcade games use virtual reality concepts. Virtual realities have the ability to allow the user to experience things they may not be able to experience in real life, such as river rafting, snow skiing, or water skiing. Behind the graphical images of virtual realities are flow charts not unlike those that have been described. Each choice in a virtual reality game, for example, is a path that is followed though an interconnected web, in other words, a flow chart. One decision is simply just that, one decision, meaning that each time the game is played and one decision is made differently, then the entire game experience may be different. A flow chart is a diagram that can be used to map the progress of person's experience in a virtual reality game, or map Internet surfing or web structure as depicted in Figure 10.

This flow chart is an example of a Web page. Some flow charts illustrate a single flow. However, others can be drawn to represent the relationships between each component, in this case a web page, and thus, the various paths that can be followed through a web page via hyperlinks. Flow charts represent these complex data paths in simple, understandable ways.

TREE DIAGRAM

A tree diagram shows the relationships amongst main concepts and contributing concepts. Tree diagrams usually begin with a main concept or idea, such as an essay topic. From this main concept, several sub ideas branch off (hence the name tree diagram). From these, more information branches off until all the desired information is included. Tree diagrams can be vertical, which means they have the initial concept at the top, or horizontal, where the initial concept is on the right hand side.

A tree diagram can be used in statistics to determine the probability of throwing two heads and two tails when flipping a coin. A family tree is another example of a tree diagram. Some family trees are oriented to have the last person born at the bottom; this represents a reversed vertical tree diagram. However, the orientation does not detract from the purpose or meaning.

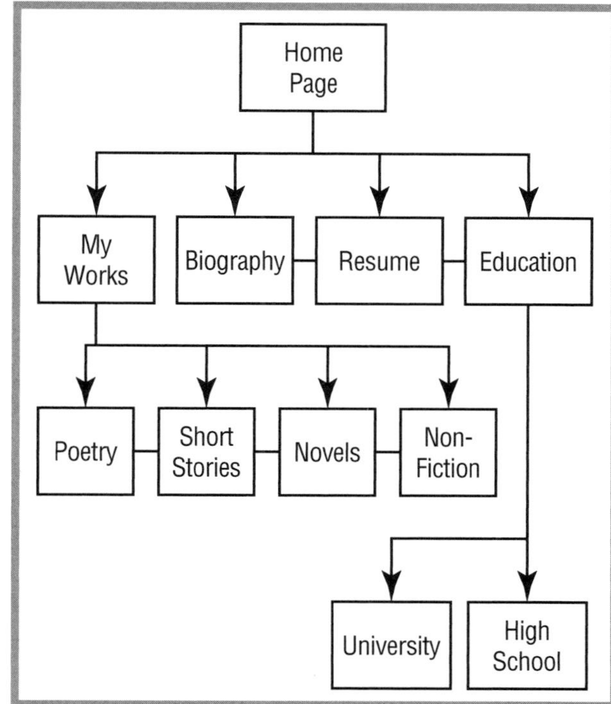

Figure 10: A flow chart.

ORGANIZATION CHARTS

Organization charts are slightly different than flow charts. Flow charts have a web-like structure, and organization charts tend to be less interconnected. Organizational charts usually have a hierarchal structure. This means that there is a leader at the top, and subordinates are illustrated below. Usually, as the chart progresses down the hierarchy, each level contains more entries.

The primary place where organization charts are used is in organizations or business firms. These present the hierarchal structure vividly. They are used best when there are more than three people in an organization. Organization charts can also be used to map the evolution of objects such as refrigerators, microwaves, and computers. They will show the earliest model at the top of the chart, and then year-by-year how new models evolved from the older models. Organizational charts are easy to understand and are much simpler to read than reading a paragraph or essay about the history of the appliance.

GANTT CHARTS

A Gantt chart is a time-management and product-management tool in a business that helps organize an individual or a group. It is a visual presentation of parts of a project and how they all relate to one another. Gantt charts show the progression of a project or a specific task to be completed. They use a timeline along the top and a list of tasks or people down the side. Different tasks may also be represented in different colors. It can help employees to determine who is relying on them and on whom they are relying to have work completed. Gantt charts also provide a visualization of project deadlines. They are most useful for tracking work, scheduling work, and planning work for a project, especially when there are several people involved.

MAPS

Maps are a type of diagram. They represent a certain area with a picture by showing the placement of one thing relative to another. In the case of road maps, they are usually labeled and contain important information about how to travel to a destination or where that destination is located. A map is a complex diagram consisting of various pieces of information portrayed in different colors and symbols that are explained in a legend.

A map is easier to read and understand than a complex set of instructions, especially for long distances. It provides an easy to view comparison of places relative to one another. Maps also provide specific information about amenities in a local area. They can show where parks are located, where libraries are, or where shopping centers are. World maps show the different continents and how they are arranged. Some maps show the world's ocean currents, others show the tectonic plates and the direction in which they are moving. Some maps are topographical, meaning that they have lines that represent different heights and they also show most, if not all, the specific details of the terrain.

There are many potential uses for maps. Every time a new estate or town is built, a new map has to be drawn. There is always going to be a need for maps. Now maps are being used in global positioning systems (GPS) to help determine specific locations. These are especially useful when a person cannot be found because they are buried under an avalanche or in a collapsed building. The maps incorporated into GPS help rescuers find these people.

WEATHER MAPS

Other commonly used maps are weather maps, which can illustrate areas worldwide, country-wide, and region-wide, along with local areas. Weather maps are a simple and easy to understand way to present data about the weather. Instead of having the wind speed and direction, the temperature and weather forecast for every city, and the isobars, the weather map presents this information in a single screen on the television or a single image in the newspaper. To look at a weather map is easier than to read all of the information.

Many types of maps are different from geographical maps. For example, a calendar is a map, and thus, by definition, a type of diagram. It shows the days in a month or year relative to one another and allows people to keep track of time. A diagram of the moon phases is also a diagram and can be thought of as a map. A map is an object that simply shows one thing relative to another. Therefore, the term map is quite versatile and includes many diagrams.

BODY DIAGRAM

Diagrams often seen in real life include those of the human body. For centuries, physicians attempted to treat patients without proper knowledge of human physiology. This led to physicians being feared by many people, and patients benefiting little from physicians. However, this was in the Dark Ages when medicine was a more superstitious than scientific practice. In the twenty-first century, physicians have many different options for treating patients, including x-ray images to determine where bones are broken. X rays are a type of diagram.

STREET SIGNS

Some of the most useful diagrams occur in day-to-day life. One of the most common diagrams seen by people every day is the street sign. Street signs denote instructions for traffic and pedestrians. Without street signs, traffic would be disorganized and safety would be compromised.

CIRCUIT DIAGRAM

Another vital diagram is an electrical circuit diagram. An electrician has a diagram of all of the circuits in a person's home. This allows the electrician to complete work. The diagram enables him to determine which circuit he must enter to fix an appliance or lighting fixture. It saves him from having to switch off all power to the house, and may also save his life. These diagrams need to be accurate and up to date; otherwise people's lives may be at risk.

Circuit diagrams are most often seen in computer-related work. Although most students have only experienced circuits that involve minimal components, such as a 9-volt battery and a small light bulb, circuits can become quite complex and difficult to decipher. Diagrams are infinitely useful in determining both the function and size of a circuit. Electrical circuits are everywhere, in computer, cars, refrigerators, washing machines, dishwashers, mobile phones, Discmans, and many more household items. The circuits found in everyday objects are formed from circuit diagrams. These diagrams allow manufacturers to manipulate the size and shape of circuits before they actually make them. This becomes more efficient with the use of circuit diagram drawing programs on computers, which save time and resources in the production stages. On a circuit diagram, different symbols denote different circuit components. Two vertical lines, one shorter than the other, denote a battery or cell, whereas two equal length lines denote a capacitor. A circle with a cross in it denotes a light bulb. There are thousands of components in electrical circuits. Each of them has a different symbol. A circuit diagram arranges these symbols in an order, so that they perform a particular function.

OTHER DIAGRAMS

Lighting designers use diagrams of lights to illuminate a Broadway spectacular. These diagrams show the type of lights and their placement. A lighting bar diagram allows a lighting designer to design the lighting for a show without the trouble of rigging and re-rigging the lights each time they want to make a change. This diagram helps the designer to work out where the lights should be initially placed. Through experience, the designer would be able to produce an effective design without having to rig any lights.

There also exist diagrams that are instructions on the use of a particular item. Instructional diagrams may be included in a manual for the operation of unfamiliar technology such as digital cameras. These manuals contain diagrams that provide important information on the functions of specific buttons. They also show what a button might look like and its placement on the camera in the diagrammatic instruction, so that the reader is easily able to discern the function and the button required for it. There are diagrams that explain the symbols used on a particular compact disc player. Diagrammed instructions also help when assembling consumer products such as furniture, toys, or bicycles. Putting together a bookcase, for example, would be difficult without a diagram showing where to attach the shelves.

Where to Learn More

Books

David, C., and S. Moore. *The Basic Practice of Statistics.* New York: W.H. Freeman and Company, 1995.

Weaver, Marcia. *Visual Literacy: How to Read and Use Information in Graphical Form.* New York, NY: Learning Express, LLC., 1999.

Web sites

Egger, Anne C. "Visualizing Scientific Data: An essential component of research." Visionlearning Vol. SCI-2 (1), 2004. <http://www.visionlearning.com/library/module_viewer.php?mid=10> (Mar. 30, 2005).

"Graphs." University of Rhode Island. <http://www.uri.edu/artsci/ecn/mead/306a/Overviews/overview.Graph.html> (Mar. 30, 2005).

Powers

Overview

The term "powers" is used to describe the result of repeatedly multiplying a number by itself. It is represented by: a × a × a × a . . . a, where 'a' is a variable, '×' shows the process of multiplication, and '. . .' shows that the process can be repeated some number of times. For example, nine (3 × 3), twenty-seven (3 × 3 × 3), and eighty-one (3 × 3 × 3 × 3) are all powers of 3. Two of the largest powers are called googol (10 multiplied by itself 100 times, or 10^{100}) and googolplex (10 multiplied by itself a googol number of times, or 10^{googol}).

Powers are used in virtually all areas of business, science, and education from helping schoolchildren with their studies of mathematic and defining the magnitude of earthquakes, to analyzing the amounts of acid rain that fall on manufacturing areas.

Fundamental Mathematical Concepts and Terms

A compact notation has been developed to represent powers. The notation is described by a^n, where the first variable a is the base that is successively multiplied by itself and the second variable n is the exponent which indicates the number of times the base is to be multiplied by itself.

The powers of 10 are: 10, 100, 1000, etc., represented as 10^1, 10^2, 10^3, etc. Many powers of ten have been given distinct names. For example, in the United States, thousand denotes 10^3, million represents 10^6, and billion stands for 10^9. As one example, 10^6 equals 10 × 10 × 10 × 10 × 10 × 10. When these six numbers are multiplied together, the result is 1,000,000 (or written out as one million). The definition of powers can be broadened to include zero, negative, and rational exponents. Powers are also used within scientific notation, logarithms, and series.

A Brief History of Discovery and Development

Using powers of numbers originated with the ancient mathematicians, most likely with the Egyptians (whose civilization lasted from about 3300 B.C. to 30 B.C. and the Babylonians (living from the eighteenth century B.C. to the sixth century B.C.). These ancient peoples encountered powers when developing formulas to describe geometric forms. The Egyptian Rhind pyrus, which dates from 1650 B.C., contains the concept of powers for numbers. The Pythagoreans (c. 450 B.C.) originated the use of x-squared for x^2 and x-cubed for x^3.

Diophantus of Alexandra (c. 200–284) used S for the square of an unknown, C for the cube, SS for the square square (fourth power), SC for the square cube (fifth power), and CC for the cube cube (sixth power). Ways to represent powers of unknowns began to spread throughout many countries in the fifteenth century. For instance, French physician Nicholas Chuquet (1445–1488) denoted successive powers of an unknown by placing numerical superscripts on the coefficients. He represented $4\times^5$ as 4^5.

The first mathematician to use letters for numbers as a way to perform mathematical calculations is generally said to be Francois Vite (1540–1603), an advisor to King Henri IV of France. Vite used vowels (A, E, I, O, U, and Y) for the unknowns and consonants (such as B, G, and D) for known quantities. The convention where letters near the beginning of the alphabet represent known quantities while letters near the end represent unknown quantities was introduced later by René Descartes (1596–1650). Descartes also introduced the notation of x, xx, xxx, etc.-where today mathematicians prefer x, x^2, x^3, etc.

Real-life Applications

AREAS OF POLYGONS AND VOLUMES OF SOLID FIGURES

The areas of polygons and the volumes of solid figures are expressed as powers of a particular length of the figure. For example, the area (A) of a square of side s is calculated as side s times side s, or $A = s^2$; that is, the second power of s. The volumes of solid geometrical figures are designated as the third power of a length. For example, the volume (V) of a cube with sides of length y is calculated as $V = y \times y \times y$, or $V = y^3$, where y^3 is the third power of y. In the case of a sphere with radius r, its equation for its volume is $V = (4/3)\pi r^3$, represented as 4^π times the third power of r.

EARTHQUAKES AND THE RICHTER SCALE

Everyday more than one thousand earthquakes occur around the world. Most of them are not noticed because they originate beneath the ocean, far underground, or are too small for humans to feel. The surface of the Earth consists of large pieces, or plates, which constantly grind against each other. When sufficient pressure builds up beneath two plates, it is released through cracks, or faults, between the plates. The result is an earthquake, or shaking of Earth's surface.

In order to calculate the magnitude of earthquakes, American seismologist Charles F. Richter (1900–1985) developed in 1935 a scale (now called the Richter scale) for measuring earthquake strength. The amplitude of the waves caused by the energy released in an earthquake increases by powers of 10 with respect to the magnitude numbers used by Richter. The released energy of an earthquake can be approximated by an equation that includes the energy magnitude of these waves and the distance from the measuring device, called a seismograph, to the earthquake's epicenter. Numbers for the Richter scale range from zero to infinity, although nine is generally the top limit ever reached. The Richter scale grows by powers of 10, where an increase of one point means that the strength of that earthquake is 10 times greater than the level before it.

For example, the famous San Francisco earthquake of 1906 (when later evaluated by the method developed by Richter) had a Richter reading of 7.8, which is 10 (10^1) times more intense than one with a reading of 6.8, 100 (10^2) times more intense than one with a reading of 5.8, and 1,000 (10^3) times more intense than one with a reading of 4.8.

COMPUTER SCIENCE AND BINARY LOGIC

In order to store digital information on modern computers, such as on the memory of hard-drives, computer hardware is made up of millions, or even billions, of tiny switches that can be either turned OFF or ON. The digits, 0 and 1, are used to stand for these two states of OFF and ON, respectively. Since these switches have exactly two different values, computer scientists work with a numbering system based on two digits. That numbering system is called the binary number system, which uses 2 as its base number. Each digit in a binary number represents a power of 2 ($2^0 = 1$, $2^1 = 2$, $2^2 = 4$, $2^3 = 8$, $2^4 = 16$, etc.).

Computers have been designed to use two voltage levels—usually 0 volts for logic-0 and either +3.3 volts or +5 volts for logic-1. With these two voltage levels, computer scientists can represent the two different values OFF and ON or, equivalently, values such as no and yes, false and true, low and high, and many other combinations. Since only two digits are used, any binary digit, or bit (the smallest unit of information inside a computer), can be transmitted and recorded electronically simply by the presence or absence of an electrical pulse or current. Even though it takes many more digits to represent binary numbers versus decimal numbers (for example, the decimal number 255 is represented in binary as 1111 1111), the greater speeds possible with the use of binary logic more than compensates for that fact.

ACIDS, BASES, AND pH LEVEL

Acidic and basic are two classes of chemical compounds that possess opposite characteristics. Acids are characterized as tasting tart, being able to change pink litmus paper to red, and often reacting with some metals to produce hydrogen gas, while bases taste bitter, turn litmus paper to blue, and feel slippery to the touch. Mixing acids and bases can cancel out their opposite characteristics, producing a substance that is neither acidic nor basic, but neutral.

In order to measure all the different chemicals found on Earth, the pH scale was developed to show how acidic or basic a substance is. The pH scale ranges from 0 to 14, with a substance having a pH of 7 considered neutral, one with a pH less than 7 being acidic, and one with a pH greater than 7 considered basic. The method of pH uses powers for comparing chemicals. Each whole pH value below 7 is ten times more acidic than the next higher value. For example, a substance with a pH of 4 is ten times (10^1) more acidic than a substance with a pH of 5, 100 times (10^2, or 10×10) more acidic than a substance with a pH of 6, and 1,000 times (10^3, or $10 \times 10 \times 10$) more acidic than a substance with a pH of 7. The same rationale is valid for pH values above 7, each of which is ten times more alkaline (basic) than the next lower whole value. For example, a substance with a pH of 10 is ten times (10^1) more alkaline than a substance with a pH of 9 and 100 times (10^2, 10 times 10) more alkaline than a substance with a pH of 8.

Knowing the value of pH is very important to many industries around the world. For example, the food industry relies on pH when dealing with all kinds of foods. The pH of carbonated colas (which contain phosphoric acid) is about 2.5, the pH of milk is about 6.5 (almost neutral), the pH of water is 7.0 (neutral), and the pH of bananas, garlic, and broccoli are all within the basic range.

The amount of pH in the atmosphere is important when acid rain falls on the Earth. Acid rain is a form of air pollution in which airborne acids, which are produced by electric power plants and other sources, fall to Earth in local and distant regions. Acid rain dissolves and washes away nutrients needed by plants, attacks trees, and damages bodies of water by making waters more acidic that then can harm fish and other aquatic animals. Because the corrosive nature of acid rain causes widespread damage to the environment, environmental scientists study acid rain in great detail. With an accurate measure of the pH of substances, based on the powers of numbers, scientists are better able to study and analyze the causes of acid rain and the ways to reduce or eliminate it.

ASTRONOMY AND BRIGHTNESS OF STARS

In astronomy, magnitude is a term used to designate the brightness of a star. The Greek astronomer Hipparchus (190 B.C.–120 B.C.) devised this system around 150 B.C. when he placed the brightest stars into the first magnitude class, the next brightest stars into second magnitude class, and so on until he reached the dimmest magnitude stars which were placed within the sixth magnitude class. By the nineteenth century, astronomers had developed the technology to objectively measure a star's brightness with the use of powers. Instead of abandoning the long-used magnitude system, astronomers modified it for their own use. They established that a difference of 5 magnitudes corresponds to a factor of exactly 100 times in intensity. For example, first magnitude stars are about $2.512^1 = 2.512$ times brighter than second magnitude stars, $2.512^2 = 2.512 \times 2.512$ times brighter than third magnitude stars, and $2.512^3 = 2.512 \times 2.512 \times 2.512$ times brighter than fourth magnitude stars, etc. Some very bright objects can have magnitudes of zero or even negative numbers and very faint objects have magnitudes greater than +6.

Potential Applications

THE POWERS OF NANOTECHNOLOGY

Powers is such a widely used term within mathematics that it will always be part of future applications. One promising new technology that will use powers in its development and application is nanotechnology, which is the research and development involved in manipulating materials on a very small scale so that microscopic machinery can be built. These nanotechnology materials and devices generally range from 1 to 100 nanometers, where one nanometer is equal to one-billionth of a meter (0.000000001, or 10^{-9} meter). Because scientists believe that nanotechnology will eventually give humans the ability to mold individual atoms and molecules into microscopic-sized biological, electrical, and mechanical machines, it may replace many current production processes.

Where to Learn More

Books

Berlinghoff, William P., and Fernando Q. Gouva. *Math Through the Ages: A Gentle History for Teachers and Others*, Expanded Edition. The Mathematical Association of

Key Terms

Decimal number system: A base-10 number system that requires ten different digits to represent numbers.

Logarithm: The power to which a base number, usually 10, has to be raised to in order to produce a specific number.

Scientific notation: A shorthand way to write very large or very small numbers.

America, Washington, D.C. and Farmington, ME: Oxton House Publishers, 2004.

Bluman, Allan G. *Mathematics in Our World.* Boston, MA: McGraw-Hill Higher Education, 2005.

Katz, Victor J. *History of Mathematics,* abridged edition. Reading, MA: Addison-Wesley, 2003.

Suzuki, Jeff. *A History of Mathematics.* Upper Saddle River, NJ: Prentice Hall, 2002.

Web sites

Bagenal, Fran, University of Colorado. "Algebra: Powers/Numbers, Variables, and Rules." Atlas Project. <http://dosxx.colorado.edu/~atlas/math/math61.html> (March 14, 2005).

Eames, Charles, and Ray Eames. "Powers of 10." <http://powersof10.com/> Eames Office. (March 16, 2005).

University of St. Andrews, Scotland "François Viéte." School of Mathematics and Statistics. January 2000. <http://www-groups.dcs.st-and.ac.uk/~history/Mathematicians/Viete.html>(March 14, 2005).

University of St. Andrews, Scotland "René Descartes." School of Mathematics and Statistics. December 1997. <http://www-groups.dcs.st-and.ac.uk/~history/Mathematicians/Descartes.html> (March 14, 2005).

University of St. Andrews, Scotland "Diophantus of Alexandria." School of Mathematics and Statistics. February 1999. <http://www-groups.dcs.st-and.ac.uk/~history/Mathematicians/Diophantus.html> (March 14, 2005).

University of St. Andrews, Scotland "Nicolas Chuquet." School of Mathematics and Statistics. December 1996. <http://www-groups.dcs.st-and.ac.uk/~history/Mathematicians/Chuquet.html> (March 14, 2005).

Prime Numbers

Overview

A prime number is a number that is larger than 1 and which can only be divided evenly by itself and by the number 1. Just a few examples of prime numbers are 2, 3, 5, 7, 11, 13, 17, 19, 23 and 29.

A Brief History of Discovery and Development

Prime numbers have fascinated people for centuries. When they were not battling Trojans and helping to devise philosophy and logic, the ancient Greeks were also tinkering with prime numbers. It was thought that these numbers held mystical power. The ancient Greeks were also interested in what came to be known as perfect numbers. These are numbers that can be divided evenly by other numbers (the divisors), with the divisors adding up to the original number. One example is the number 6. Six can be divided by 1 (to give 6), by 2 (to give 3), and by 3 (to give 2). Adding up the divisors $(1 + 2 + 3)$ equals 6.

Centuries before the modern era, mathematicians studied prime numbers. In 300 B.C., Euclid of Alexandria wrote an essay entitled 'The Elements' that collected the knowledge of mathematics up to that time. In 'The Elements', Euclid was able to demonstrate that prime numbers did not just stop at a predetermined value, but that they go on forever. In other words, prime numbers are infinite. Euclid also showed that if $2^n - 1$ is a prime number, then the number $2^{n-1} \times (2^n - 1)$ yields a perfect number.

Test Euclid's discovery by setting $n = 3$: $2^3 - 1 = (2 \times 2 \times 2) - 1 = 7$ (which is a prime number) so, $2^{3-1} \times (2^3 - 1) = 2^2 \times 7 = (2 \times 2) \times 7 = 28$. Twenty-eight can be divided into an even number by 1, 2, 4, 7 and 14; finally, $1 + 2 + 4 + 7 + 14 = 28$, so 28 is "perfect."

About 100 years later, another Greek mathematician, Eratosthenes, came up with a way of determining prime numbers. Among his other accomplishments, Eratosthenes was the first person to accurately estimate the diameter of Earth while serving as the chief librarian of the great ancient library in Alexandria. His prime-calculating invention was called the Sieve of Eratosthenes. This mathematical sieve drains away non-prime numbers from prime numbers.

To illustrate, Table 1 shows an arrangement of the numbers 1–100:

Perform the following steps:

- Cross out 1 (it's not a prime number)
- Circle 2 (the smallest prime number), then cross out every multiple of two (4, 6, 8, etc; in other words, every second number)

- Circle 3 (the next prime number) then cross out all the multiples of 3 (6, 9, 12, 15, etc.; some have already been crossed out)
- Circle the next number not circled or crossed out, which is 5, then cross out the multiples of 5 (10, 15, 20, 25, etc.; some have already been crossed out)
- Continue doing this until all the numbers have been circled or crossed out.

The circled numbers are the prime numbers.

Another prime number discovery made in the seventeenth century was made by Christian Goldbach, a historian and mathematician. He said that every even number could be expressed as the total of two prime numbers. As two examples, 6 can be expressed as 3 + 3, and 20 can be expressed as 17 + 3. His idea is known as the Goldbach conjecture. Even today, we are still not sure if his idea is true. But, scientists do know that the pattern is true for every even number between 2 and 400,000,000,000,000, and for some even numbers selected up to 10^{300} (10 followed by 300 zeros). In 2000–2002, a British firm offered a million dollars to anyone who could prove or disprove the Goldbach conjecture. No one did.

Since the 1700s, another a great challenge has been to determine the greatest prime number. Until the number-crunching power of big computers, this sort of activity did not get very far. However, with modern supercomputers the greatest prime number known now has over 4 million digits.

As recently as 2003, discoveries were announced regarding prime numbers. In that year, a team of physicists published a scientific paper in the prestigious journal *Nature* that provides evidence that the arrangement of prime numbers in amongst the other numbers is not just haphazard, but may have a pattern. Scientists and mathematicians are not sure what the significance of this might be. But, prime numbers may have a key role to play in the natural world.

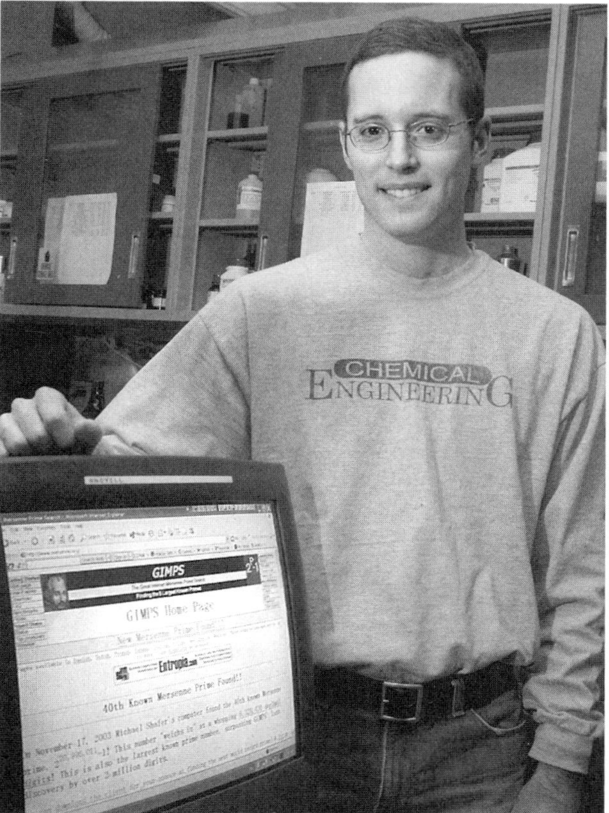

Michigan State University graduate student Michael Shafer stands next to the computer he used to discover the world's highest prime number. The number is 6,320,430 digits long and would need 1,400 to 1,500 pages to write out. AP/WIDE WORLD PHOTOS. REPRODUCED BY PERMISSION.

1	2	3	4	5	6	7	8	9	10
11	12	13	14	15	16	17	18	19	20
21	22	23	24	25	26	27	28	29	30
31	32	33	34	35	36	37	38	39	40
41	42	43	44	45	46	47	48	49	50
51	52	53	54	55	56	57	58	59	60
61	62	63	64	65	66	67	68	69	70
71	72	73	74	75	76	77	78	79	80
81	82	83	84	85	86	87	88	89	90
91	92	93	94	95	96	97	98	99	100

Table 1.

Real-life Applications

BIOLOGICAL APPLICATIONS OF PRIME NUMBERS

Plant-eating insects called cicadas spend a lot of their life underground in one form, before emerging as adults. In some types (species) of cicada, this appearance occurs at the same time for all the adults in the region, every 13 or 17 years.

Thirteen and 17 are prime numbers. Coincidence? Scientists who have studied the species of cicadas do not think so. Rather, they think, the use of a prime number for the life cycle has been a response to the pressure put on cicada population by other creatures who utilize them as food. In other words, the cicadas are the prey and the creatures lying in wait when they emerge to the surface are the predators.

Researchers have used mathematical ways to model the so-called predator-prey relationship. Modeling allows them to do experiments in their lab, on the computer, without having to actually go to nature and observe what is happening (which could be very hard to do).

In the mathematical model, the cicadas and their predators had life cycles that were randomly chosen to be different lengths. When both predator and prey were present in high numbers at the same time, it was bad news for the cicadas, as there were lots of hungry predators waiting for the cicadas as they came out of the ground. But, if the emergence of the cicadas occurred when there were not many predators, they had a much better chance of living long enough to mate.

In the computer studies, the researchers found that the best times for the cicadas to emerge from the ground was in life cycles that had prime numbers (e.g., 13 and 17 years). The researchers assert that a life cycle that is 13 or 17 years long increases the cicadas chances of avoiding population depletion. Consider what could happen if their life cycle was 12 years long. If cicada emerged every 12 years, any predator that had a life cycle of numbers that can divide into 12 (such as 2, 3, 4, or 6 years) could be around at the same time the cicadas emerged from the ground. There would be more chance of a hungry predator would be waiting. But, if a life cycle is 13 or 17 years long, a predator's life cycle also has to be 13 or 17 years long. The odds of that are much less.

Where to Learn More

Books

Cobb, C. *Cryptography for Dummies.* New York: For Dummies, January 2004.

Schneier, B. *Secrets and Lies: Digital Security in a Networked World.* New York: Wiley, August 2000.

Periodicals

Gole, E., O. Schulz, and M. Markus. "A Biological Generator of prime Numbers," *Nonlinear Phenomena in Complex Systems* (2000) 3: 208–213.

Web sites

Alfeld P. "Eratosthenes of Cyrene." <http://www.math.utah.edu/~alfeld/Eratosthenes.html> (September 7, 2004).

——— "Notes and Literature on Prime Numbers." <http://www.math.utah.edu/~alfeld/math/prime.html> (September 6, 2004).

Peterson I. "Prime-Time Cicadas." <http://www.sciencenews.org/articles/20030621/mathtrek.asp> (September 8, 2004).

Probability

Overview

Probability is the likelihood that a particular event will occur. Probability is used to estimate the chances of many different types of events happening. Insurance companies use probability to estimate how likely a particular driver is to cause an accident during the next year. Engineers use probability to predict how often critical pieces of equipment, such as jet engines on passenger planes, will fail. Gamblers in casinos routinely make wagers based on their understanding of the laws of probability, while investors make even riskier gambles on the rise and fall of the stock market or the price of a bushel of corn. Although probability is one of the most commonly used forms of mathematics in everyday life, many misconceptions exist about its formulation, meaning, and impact.

Fundamental Mathematical Concepts and Terms

Probability calculations are generally straightforward, though as the number of possible outcomes grows, the math required can become somewhat involved. Consider a simple example involving a single die (the singular form of dice), in which we wish to determine the probability of rolling a 4. The calculation for probability includes several elements. Outcomes are all the possible results we could achieve; since the die has 6 sides (1, 2, 3, 4, 5, and 6), and any of the six could land on top, the total number of possible outcomes in this experiment is 6.

Next, we must determine the total number of ways in which the event of interest could possibly occur; in this case, a roll of 4 can occur only one way. By dividing this value (the number of ways our desired outcome can possibly occur) by the total number of possible outcomes, we can determine the probability of the 4 being rolled, creating this equation: Probability = Desired outcome / Total Outcomes, or in numerical terms, $P = 1/6 = 1/6$. Thus we conclude that the probability of rolling a 4 on a single toss of the die is 1/6, or 1 in 6. We could perform the same calculation for each of the other values on the die, demonstrating that for each side of the die, the probability is also 1 in 6.

Interpreting this value is relatively straightforward: a probability of 1 in 6 tells us that if we roll the dice a large number of times, we will, on average, roll a single 4 for each six tosses of the die. If we wish to find out about how many 4s we will roll in 600 rolls of the die, we multiply the probability by the number of rolls, which are often called experiments or trials; in this case, we use the following equation: 1/6 $\times$ 600 = 100. This result tells us that over the course of 600 rolls, about 100 will be 4's.

This same procedure can be scaled up to evaluate events with thousands or millions of possible outcomes. If the names of each person living in the U.S. were written on slips of paper and one slip was randomly drawn, what chance would John Smith of Cloverleaf, Iowa, have of being drawn? In this example, only one John Smith exists in Cloverleaf, providing only one possible way to reach the desired outcome. The total number of people in the U.S. in 2005 was approximately 300,000,000, which is the total number of possible outcomes. John Smith's chance of having his name drawn is 1 in 300,000,000. If the drawing were held using the earth's entire population of 6,400,000,000, John's chance of being drawn would drop by a factor of 20.

A Brief History of Discovery and Development

While the very first game of chance cannot be specifically identified, historians are certain that these probability contests have been enjoyed for millennia. Ancient civilizations left behind small dice-shaped pieces of bone called astragalia, which apparently facilitated the earliest contests similar to modern dice games. Throughout the early history of man, gambling remained popular, with little apparent attention paid to the laws of nature and mathematics, which made the toss of the colored stones or polished bone fragments so maddeningly unpredictable.

During the sixteenth century, Gerolamo Cardano (1501–1576), a scholar of medicine, astrology, and philosophy made the first known attempt to explain the function of chance in gambling and other endeavors. Cardano was the first to deduce that an event's probability of occurring is determined by dividing the number of ways the event could occur by the total number of possible outcomes. Cardano explained that a roll of a single die has six possible outcomes, while a pair of dice can land thirty-six different ways; he also wrote about the statistical logic of a primitive ancestor of the modern game of poker. Unfortunately, Cardano's science was somewhat limited by his intense belief in astrology, which he used to predict future events of human lives. Perhaps his most successful prediction was naming the date of his own death far in advance; when the predicted date of his death arrived, Cardano insured his own correctness . . . by committing suicide.

A century later, mathematician Blaise Pascal (1623–1662) was asked why the odds of throwing a single six in four throws of one die do not equal the odds of throwing four sixes in twenty-four throws of two dice. Pascal accepted this challenge, then went on to devise the theory of probability as it is currently understood, in many cases applying these principles to the popular pastime of gambling. At age nineteen, Pascal also constructed the first mechanical adding machine.

While various other mathematicians added to the body of knowledge regarding chance and gambling in the decades that followed, the next major advance occurred in 1928, when John von Neumann put forward the basic concepts of game theory in a paper analyzing the probabilities associated with various poker hands. While game theory has found application in fields such as economics, its application to games of chance also continues, particularly given the advent of powerful, inexpensive computers.

Real-life Applications

SECURITY

The ability to conceal data from outsiders has been valued by military commanders for centuries; historians have uncovered evidence of military codes dating back more than 4,000 years. The Allied victory in World War II was hastened significantly when the Allies broke a presumably unbreakable code used by the Japanese, thus becoming privy to numerous confidential communications. Today, numerous applications for encoding and decoding data exist, most of them based on the fundamental principles of probability.

One critical use for this technology is data encryption, a technique for encoding data so that it is unreadable without a specific number, or key, which allows an authorized user to decrypt and read the message. Data encryption has become a critical technique as electronic transfers of sensitive financial data have become more routine. Many commercial websites now transfer buyers to a secure site, at which data such as credit card numbers is encrypted before it is transmitted from a user's computer.

Encryption works because of the laws of probability. An encrypted message can be read by any person with the proper numerical key, meaning that for a message to remain secure, the key must be virtually unguessable. Ever faster computers have made it possible for simple encryption schemes to be broken using a brute-force approach, in which the computer simply tries key after key until the proper one is located. Preventing this type of attack requires a large enough number of possible keys that the likelihood of guessing the proper key by chance becomes so small that it is not worth attempting. An encryption key's resistance to brute force attacks is measured as strength, with a more secure key being described as stronger encryption.

As of 2005, one of the most widely used encryption schemes is found in Microsoft Internet Explorer, where it encrypts data sent from computer users to the Internet. This encryption scheme uses a 128-bit encryption key, meaning that in order to read the encoded data, an interloper would have to correctly guess a 128-bit number. Since this length of key would theoretically take a modern supercomputer several hundred years to crack, 128-bit encryption is considered adequate for routine applications such as online shopping. In cases where additional security is desired, such as military applications, longer keys significantly increase the number of possible keys, producing a commensurate reduction in the odds of randomly guessing the key.

A related use for encryption techniques has recently appeared in the rapidly growing field of forensic computing. In the course of criminal investigations, law enforcement personnel frequently need to locate computer files related to a crime, a process much like finding the proverbial needle in a haystack. A typical computer hard drive contains hundreds of thousands of files, most of which arrive as part of the operating system or are installed with user applications; a basic installation of Microsoft Windows XP places between 10,000 and 30,000 separate files on a computer hard drive.

Unlike a computer user who knows where most of his important files are saved, a police investigator searching a computer for files with evidentiary value has no idea what the needed files are called, or in which directories they reside. Since it is impractical to manually open and read every file on the computer, encryption methods now allow investigators to automatically eliminate more than 90% of the files on a computer, permitting the investigator to focus on the remaining files.

This file-sorting system is based on the principle of encryption, in which any file can be processed to produce a unique identifying code. By creating these unique codes, or file signatures, of all the files installed by most operating systems and commercial applications, investigators have created a massive reference library for law enforcement purposes. Investigators can use this library to scan a suspect's hard drive, automatically eliminating any files which match the signature keys of known files while leaving the files which might have evidentiary value behind. The system works only because the number of potential file signatures is enormous; in the case of the MD5 algorithm, the total possible number of unique file signatures is 10^{38}, or a one with 38 zeros after it, making the odds of two files having the same file signature almost an impossibility. By reducing the number of files to be examined, this library enables investigators to more rapidly and more efficiently search hard drives, gathering evidence they might otherwise overlook.

GAMBLING AND PROBABILITY MYTHS

While the ancestors of today's dice games predate recorded history, the modern game of craps is far more recent, and is attributed to twelfth century Crusaders besieging a castle in Arabia. Most of today's other casino games also can be dated back to the Middle Ages, however one type of wagering can rightfully trace its lineage back more than twenty centuries. The longest-running wagering event practiced today is the ancient sport of horse racing.

Numerous archaeological finds support horse racing's claim as the most ancient form of gambling. A Hittite document dating to about 1500 B.C. describes in detail the process of breeding and training horses for the purpose of racing, while the Iliad provides a complete account of a chariot race. The Olympic games in 624 B.C. included specific rules for horse racing in contests of various distances, and the Romans soon added the concept of handicapping, or betting against the house. While the popularity of horse racing has risen and fallen over the centuries, today's racing, while faster and more refined, is virtually unchanged from the ancient contests held in Europe. While the advent of modern statistical analysis and computer equipment has provided the tools to analyze the mountains of statistical data available on past races, the ability to correctly predict the outcome of a horse remains an elusive goal.

While the interpretation of probability projections is fairly straightforward when applied to events which occur many times, the laws of probability become far less intuitive over short periods of time. One common probability myth, often cited by gamblers, is that numbers, horses, or players can become due, meaning that since they have not won in many plays of the game, they are now more likely to occur. This faulty line of reasoning is based on the understanding that over many thousands of plays, each number will appear a set number of times, hence the gambler assumes that the longer a value goes without appearing, the more likely it is to appear soon. Unfortunately, this belief is unfounded. In the case of completely unrelated events, such as the spin of a roulette wheel, the odds of the next spin are unchanged by the result of any previous spins. If the number 14 has not been spun on a particular wheel for six weeks, the odds of it appearing on the next spin are still exactly the same as they were before. The laws of probability do not provide for events to occur simply because they have not occurred previously.

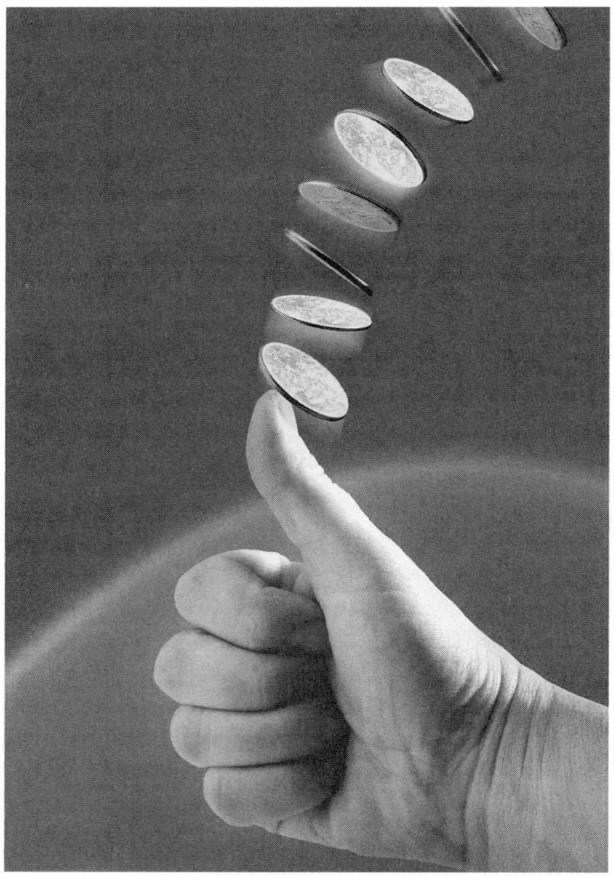

The probability of this coin landing heads or tails is easy to predict. ROYALTY-FREE/CORBIS.

A second probability myth, ironically, is the exact opposite perspective of the previous view. This perspective says that particular numbers can become "hot," or more likely to be spun. In adopting this philosophy, an observant gambler might notice that the number 27 had been spun on the wheel several times over the course of a short wagering session. The gambler, acting on the theory that numbers can become hot, now concludes that the number's frequent appearance in past spins makes it more likely to appear in a future spin, and he will wager heavily on this particular number. Once again, the laws of probability and chance dictate that, assuming the roulette wheel is functioning correctly, the chance of a future spin cannot be predicted by how often a particular number has appeared in recent spins. Regardless of how hot a number appears, it is no more likely to appear on the next spin of the wheel than any other value. Ironically, the theory of hot numbers, which says that the same number will come up many times together, is the exact opposite of the theory of coming due, which says that a number will appear when it has not been spun for some time. While gamblers subscribe to both philosophies (and back up their philosophies with their wallets), both theories cannot simultaneously be right; in truth, probability theory says that neither theory is correct, and that past events do not impact future spins of the wheel.

PROBABILITY IN SPORTS AND ENTERTAINMENT

Many sports rely on probability to predict future events. Baseball is among the most statistically-oriented sports, with numbers available for almost every aspect of the game. A player's batting average is a measure of the percentage of times he hits safely, expressed as a 3-place decimal value such as .333. While this value allows an assessment of a player's past performance, it is also useful in predicting his future effectiveness. For instance, a player batting .200, which can also be expressed 2:10, 1:5, or 20% can be predicted to hit safely 20% of his times at bat, or 1 time in 5 attempts. For this batter, the odds against him hitting safely on any given trip to the plate will be 4:1. Baseball batting averages are calculated using an involved set of rules, meaning that a player batting .200 will generally make it to first base safely more than 20% of the time; for this reason, some managers prefer to use a player's on-base average, which includes walks and errors in the player's success ratio.

Bowling is a popular sport in which players actually receive two chances to succeed, in the form of two shots (if necessary) to knock down all ten pins. Statisticians have used mountains of data from previous bowling competitions to calculate the odds of a professional bowler making a variety of shots. For example, when a professional bowler steps up to roll his first ball in a frame, the objective is to knock down all ten pins, scoring a strike. For the second shot, the odds of clearing the lane depend on which pins remain standing; three pins standing close together have much higher odds of falling than two widely separated pins. For most bowlers, a split is one of the hardest shots in the game, requiring the player to slice the ball to the outside of one pin, knocking it across the lane to hit the other one. Even for a professional, splits are long-shots. According to the Professional Bowlers Association, the 7-10 split, in which two pins remain at opposite sides of the lane, has been attempted 400 times in televised matches. In all these attempts, the professionals have managed to convert only three, putting the odds of a professional making this shot at 3:400, or about 1 time in 133 attempts.

How often do miracles occur? The term miracle has several different meanings; in theological language, it refers to an act of God that defies the laws of nature, though its most common use today refers to any seemingly impossible event that actually occurs. When the Boston Red Sox finally broke the decades-long curse of the Bambino and won the World Series in 2004, fans proclaimed the victory a miracle. When a jet airliner crashes and one or two passengers walk away without injury, many label their survival a miracle. And in a handful of cases where a single individual has won a state lottery, not once but twice, writers routinely throw out the term to describe this odds-defying run of luck. Ironically, the term is applied almost exclusively to positive events like those described, ignoring equally improbable turns of probability which lead to unexpected death or injury.

While no statistical definition of miracle exists, an estimate can be made based on common language. To most people, the expression "one in a million" describes something quite rare, though still achievable. People frequently use this expression to describe a job they truly love or a dear family member or friend, suggesting that this level of probability does not rise to the level of miracle status. For this discussion, we must conclude that a miracle is much rarer than 1 in a million; for simplicity's sake, we will assume that a probability of 1 in one billion qualifies an event as a miracle. In other words, a miraculous event is one which occurs only once in every billion opportunities. To get some sense of this level of probability, one billion seconds would take more than 30 years to elapse.

In determining how often these miraculous events occur, it becomes important to recognize that while odds of 1 in one billion are almost unimaginably low, these odds apply not just to a single person, but to many millions of individuals. For example, assume that any single person in the United States has a miraculous, or one in a billion, chance of being struck by lightning in a given day. With odds like these, any single person can safely go on with his life without worrying about storm clouds. But when these odds are applied to the entire 300 million people in the U.S., the equation changes dramatically since each of the 300 million provides another opportunity for the miraculous event to occur. Now, across the entire population, the odds of a lightning strike in a day become 300 million in one billion, or roughly 1 in 3.33. At this probability level, some individual in the U.S. would be struck by lightning every three days, making the miraculous seem almost routine, since many of these strikes would undoubtedly be covered on national news. Fortunately, lightning strikes appear to be infrequent enough to reach even the so-called miraculous level proposed here. But given the large number of citizens in the U.S., it seems statistically likely that one in a million events actually occur on the North American continent several times each day.

PROBABILITY IN BUSINESS AND INDUSTRY

Some business endeavors require a calculation of probabilities, even though little data on which to base the calculation is available. Complex pieces of machinery like the NASA space shuttle are notoriously hard to estimate reliability projections for, due largely to the massive number of components involved. Some components are simple; for example, a tire on the space shuttle is one of the more dependable components. Other components contain thousands of parts; the shuttle's main engines are among the most complex propulsion systems ever designed. In order to calculate the odds of an accident occurring in a single shuttle flight, the chances of failure for each individual component must be calculated, then combined with those of the other components to produce a composite estimate of the ship's chances of returning safely.

As the number of components rises, the process becomes increasingly difficult; because of the shuttle's complexity this process becomes virtually impossible to carry out accurately for such a machine, sometimes forcing engineers to make an educated guess. Unfortunately, these guesses are sometimes given more credibility than they deserve. Prior to the shuttle *Challenger's* loss on the twenty-fifth shuttle mission, engineers had assessed the shuttle's chance of a catastrophic failure at 1 in 100,000, meaning the ship could have flown every day for 300 years while suffering only one major failure during that time. Unfortunately, these overly optimistic assessments appeared to ignore previous experience with unmanned solid rockets, which suggested an accident rate closer to 1 in 25 or 1 in 50 for the boosters alone. To date, actual experience with the shuttle system has led to 2 shuttle accidents in 113 missions, suggesting that the probability of loss is far closer to the 1 in 50 value than the 1 in 100,000 estimate.

Most consumer products sold today include a warranty period, during which the manufacturer agrees to either repair or replace the product if problems occur. For most products, users expect the item to last far beyond the warranty period; new automobiles typically include a three to five year warranty, even though most buyers expect to drive a new car for twice that long. In some cases, manufacturers attempt to estimate the likely service life of a product by providing a measure called mean time before failure, or MTBF. For example, a computer monitor might be sold with an advertised MTBF of 50,000 hours, which equates to 10 hours of use, 5 days per week, for more than nineteen years. For most customers, nineteen years is longer than they typically keep a monitor, so they will feel comfortable with this purchase. However,

MTBF is not the same as a warranty or a minimum lifetime; rather, MTBF provides the mean, or average lifetime of this product model before failure. In other words, half of the products will last longer than MTBF (50,000 hours in this case), but the other half will fall below the average, failing at some point less than the advertised lifetime.

If MTBF does not give a minimum lifetime, how should it be interpreted when trying to assess a product's potential service life? First, if the MTBF has been correctly calculated, the buyer can expect that the item will provide the rated service life or more half the time, so if he buys twenty of the monitors, he can expect at least ten to last 50,000 hours or longer, in some cases perhaps much longer.

The other monitors in the group can be expected to last for varying periods of time, with most of them lasting close to the average lifespan of 50,000 hours and a few failing as the time-span grows further from the mean. In a few cases, monitors might actually quit working within the original warranty period, meaning they would be replaced by the manufacturer. Unfortunately, MTBF calculations for complex electronic equipment can be impractical or impossible to calculate mathematically, meaning that in some cases the MTBF is based largely on engineer intuition and experience with similar parts, rather than actual experimentation.

One of the most exciting moments in a teenager's life is when she finally receives her driver's license. But soon after this triumph may come a rude surprise: car insurance for young drivers is often several times as expensive as for older adults. Why do insurers charge teen drivers more?

Insurance companies are among the largest users of statistical and probability data. Specialists called actuaries spend their days determining exactly how likely events are to occur, allowing the insurer to charge correctly for its policies. Actuarial tables provide summaries of this data; for example, an actuary could use one of these tables to determine that a 45-year-old man in good health is likely to live to be 82 years old, and that his odds of dying next year are 1 in 14,400. Using these probabilities, the insurer can then determine how much to charge the man for a life insurance policy which pays $100,000 to his family in the event of his death.

These probability tables allow insurers to provide discounts to specific customers, such as those who don't smoke, since they have a higher probability of living longer. Automobile insurers also use actuarial data to predict which drivers are more likely to be involved in an accident, in which case the insurer will be obligated to pay for repairs. Using this information, insurers then give lower rates to drivers who have lower odds of having an accident and higher rates to those with higher odds. Based on past experience with millions of drivers, insurance companies know that the odds of a teenage driver, especially a male, having an accident are much higher than for a 30- or 40-year-old. Since the company is more likely to pay a claim for these young drivers, it is forced to charge higher premiums in order to cover the expected losses. As long as young drivers continue to have more accidents in general, even safe teenage drivers will continue to pay higher premiums for auto insurance. In a few cases, actuarial data has shown that certain groups, such as Honor Roll students, are less likely to have accidents, and some insurers now offer discounts to students with strong academic performance.

OTHER USES OF PROBABILITY

In 2001 Russian engineers fired braking rockets to bring the aging Mir space station back to earth. The reentry was carefully orchestrated to insure that most of the station would burn up in the earth's atmosphere, and any surviving pieces would land harmlessly in the Indian Ocean. Recognizing the incredibly long odds of losing, restaurant chain Taco Bell made an astonishing offer. The company floated a 40-foot square target featuring the words "Free Taco Here" in the Indian Ocean off the coast of Australia. The company then widely advertised that if the remains of the Mir station hit the target, Taco Bell would give one free taco to every person living in the United States. Mir eventually landed thousands of miles from the target, and the company avoided having to serve 300 million free tacos. However, executives at Taco Bell apparently recognized that even the unlikeliest of events occasionally occurs; the company took out an insurance policy in advance just in case the falling station defied the exceptionally long odds and hit the target.

Sometimes the seemingly impossible can be accomplished due to an audience's lack of statistical savvy. Consider this simple magic trick. A magician, claiming to have psychic powers, stands before a crowd and announces that he has noticed an odd coincidence: although there are 365 possible birthdays in a year, he has psychically observed that two of the individuals in this particular audience happen to share the exact same birthday. He then asks a series of questions to help locate the unlikely pair, and after confirming this fact, moves on with his act. Was it psychic power, or simple probability?

To most casual observers, the large number of possible birthdays seems to make the prediction a long shot at best. But considered in terms of probability theory, it begins to look far less magical. Assume that the crowd consists of 12 people. The magician has a 0.5073 chance of

Key Terms

Actuary: A mathematical expert who evaluates the statistical likelihood of various insurable events for underwriting purposes.

Encryption: Using a mathematical algorithm to code a message or make it unintelligible.

Gambling: A popular form of entertainment in which players select one of several possible outcomes and wager money on that outcome.

Probability: The likelihood that a particular event will occur within a specified period of time. A branch of mathematics used to predict future events.

being correct, one better than in two. With a bit of showmanship, most psychic performers are able to easily dismiss the predictions they miss using a variety of explanations. But in close to half of this performer's appearances, he will shock the crowd by appearing to do the impossible, when in fact he has simply made a smart bet based on the simple laws of probability.

In a few instances, probabilities are used to attract attention or create fear. Newspaper and magazine headlines during the mid-1990s warned air travelers to avoid planes with fewer than thirty seats, based on statistics which seemed to indicate that these smaller planes were several times more likely to crash than larger jets. But this probability was based on a classification system which grouped small commercial planes in the same category as helicopters and some other types of planes, unrealistically inflating the numbers for the category and making the commuter planes seem less safe. Eliminating the other types of equipment from the equations produced probability figures demonstrating that the smaller commercial planes are approximately as likely to crash as their larger cousins.

Potential Applications

As computational power continues to double every two years, the ability to apply probability theory in new ways will lead to further applications for this powerful tool. In some cases, these applications may involve major improvements in current applications, such as forecasting weather patterns or predicting when and explaining why freeways suddenly become congested. The ability of faster computers to crack increasingly complex codes will lead to an escalating battle between code-writers and code-breakers.

In other cases, advances in probability theory may well result in unforeseen applications. Based on mathematical advances made by eighteenth century mathematician Thomas Bayes, scientists are just beginning to develop software which is comfortable dealing with concepts such as "probably" and "more likely" rather than the simple yes or no typically required in computer programming. Google and other search engines already use rudimentary forms of Bayesian reasoning to answer search queries. Potential future applications include cameras which would visually examine a patient and warn a physician of symptoms making the person more likely to suffer a stroke.

Where to Learn More

Books

Epstein, Richard A. *The Theory of Gambling and Statistical Logic.* New York: Academic Press, 1977.

Glassner, Barry. *The Culture of Fear; Why Americans are Afraid of the Wrong Things.* New York: Basic Books, 1999.

Orkin, Mike. *What Are the Odds? Chance in Everyday Life.* New York: W.H. Freeman and Co., 2000.

Web sites

Feynman, Richard P. "Personal Observations on the Reliability of the Shuttle." <http://www.virtualschool.edu/mon/SocialConstruction/FeynmanChallengerRpt.html> (March 27, 2005).

Singh, Simon. "Crypto Q & A." <http://www.simonsingh.com/Crypto_Q&A.html> (March 28, 2005).

Proportion

Overview

Proportion is an equation used to compare the magnitudes of quantities. It can be defined as an equation that presents equality between two ratios. In other words, if the ratio between two characteristics of an object is equal to the ratio between the same two characteristics of another object, the two objects are considered to be in proportion. These characteristics could be anything that can be measured (such as size, quantity, dimension, etc.). For example, consider two rectangles, the first having length and width equal to 8 in. (20 cm) and 4 in. (10 cm) respectively, and the other having a length of 6 in. (15 cm) and width of 3 in. (7.5 cm). These two rectangles are in proportion as the ratio of the length and width of each rectangle is equal.

Although proportion is a concept mainly used in design, it is widely applied to other aspects of daily life as well. One of the most common examples is grocery shopping, where proportion is frequently used to compare prices of items with different sizes. In addition, proportion finds uses in numerous other fields, including architecture, art, maps, astronomy, business, imaging, technology, and even cooking.

Fundamental Mathematical Concepts and Terms

As stated earlier, proportion is indicated by the equality between two ratios. Mathematically, it can be expressed in two ways—a/b = c/d or a:b = c:d. The outer terms of the equation are known as extremes, while the inner (or middle) terms are known as means. For example, in the above equation "a" and "d" are extremes, whereas "b" and "c" are means.

SOLVING RATIOS WITH CROSS PRODUCTS

One way to test equality is by simply calculating the values of the ratios. However, a more commonly used method involves the use of cross products. Cross products can be calculated by multiplying the outer terms (or extremes) and then the inner terms (means). If both values are equal, the ratios are in proportion.

Consider the ratios 2/5 and 3/7.5. In this case, the cross product of the extremes is $2 \times 7.5 = 15$, while the cross product of means is $5 \times 3 = 15$. Hence, the ratios are in proportion. Note that simple division here would have been far more complex and time consuming, as compared to calculating cross products.

This is one of the reasons for the popularity of the cross product method.

The cross product method also has another significant benefit. Real life applications use the concept of proportion mainly to compare two things or objects. In many cases, there may be a missing term in the proportion. For example, a grocery store owner charges $1.50 for 1 lb. (0.4 kg) of beef roast. He wants to set the price of a 3 lb. (1.2 kg) roast, such that it is in proportion with the price of the 1 lb. (0.4 kg) roast. This can be easily done by writing the proportion equation, and then using cross product to determine the price.

The equation can be written as—$1.50/1 = x/3$, where "x" is the price of the 3 lb. roast. By calculating the cross products of the means and extremes, the value of "x" comes out to be 4.50. In other words, the 3 lb. (1.2 kg) roast should be priced at $4.50 for it to be in proportion. Simply put, you can calculate a missing term from a ratio if this ratio is in proportion to another known ratio. This underlying concept of proportion is extremely useful in real-life applications.

DIRECT PROPORTION

If change in one component causes a change of equal magnitude (size, percentage) in another component, the two components are said to be in direct proportion. Another way of expressing this is by stating that the first component is directly proportional to the second component. In a nutshell, direct proportion is a concept that pertains to the change in the values of two (or more) components that are already in proportion.

For example, imagine the price of a candy bar is $0.50. The number of candy bars is always proportional to the total price of the bars—the ratio of the number of candy bars to the total price always remains same. One bar costs $0.50, two bars cost $1.00, four cost $2.00, eight bars cost $4.00, and so on. Put simply, a change in the number of bars causes a change in the total price. Moreover, the magnitude of the change is also the same. In other words, the change in the number of bars as well as the price can be represented by a common factor. The number of bars keeps doubling (or $1 \times 2 = 2$, $2 \times 2 = 4$, $4 \times 2 = 8$). Similarly, the price also doubles ($0.50 × 2 = $1.00, $1.00 × 2 = $2.00, $2.00 × 2 = $4.00). Hence, the number of candy bars is directly proportional to the total price of the bars. Also the change is represented by a common factor (two in this case).

Mathematically, direct proportion is indicated as $a \propto b$ (a is directly proportional to b). The main advantage of direct proportions is that they can be expressed in the form of an equation. For example, the relationship between the total number of bars and the total price, in the above case, can be shown as:

Total number of candy bars = k × Total Price, where k is the common factor.

The common factor is known as the proportionality constant. This equation may be used to easily calculate the total price if the number of candy bars is known, and vice versa. All direct proportion relationships can be expressed by such equations. Consequently, they are used extensively in various real-life activities and applications.

INVERSE PROPORTION

Like direct proportion, inverse proportion also pertains to the change in two (or more) components. However, in the case of inverse proportion, an incremental change in one component causes a decrement in the other component. In other words, if the magnitude of one component increases, the value of the other component decreases, and vice versa.

Consider, for example, a car traveling from one place to another. If the car has a constant speed (and assuming it does not stop anywhere), the more it travels, the less the remaining distance to the target destination. Hence, in this case, as the total travel time increases, the distance to the destination decreases—travel time is inversely proportional to distance remaining.

Similar to direct proportion, the change can be represented by a factor. However, the factors that represent change for both components are multiplicative inverses of each other. In simple terms, if the value of one component changes by a factor of three, the change in the value of the other component will be 1/3. Consequently, inverse proportion is also known as reciprocal proportion, and is mathematically indicated as $a \propto 1/b$ (or travel time ∝ 1/distance remaining, for the above example).

An inverse proportion relationship can also be expressed in the form of an equation. For instance, the two components (travel time and distance remaining) in the above example can be shown as:

Travel time × k/distance remaining, where k is the proportionality constant.

A Brief History of Discovery and Development

Throughout history, proportion has been used extensively in numerous areas. The Greek mathematician Pythagoras (580 B.C.–500 B.C.) who is most well known for the Pythagorean theorem, developed the Theory of

Michelangelo's marble statue of *David* in Florence, Italy. Measurements of the statue debunked long-held notions that the 13.5-ft (4.1-m) high statue was out of proportion to the human form. AP/WIDE WORLD PHOTOS. REPRODUCED BY PERMISSION.

Proportion to relate music with mathematics. He established musical scales that were based on the concept of proportion.

Subsequently, evidence of proportion can be seen in many works of art and architecture, especially in ancient Greece and Rome. Some of the most popular paintings by renowned artists such as Michelangelo (1475–1564), Raphael (1483–1520), and Leonardo da Vinci (1452–1519) were based on proportion. The concept of proportion is vital to art and architecture as it describes the size, location, or amount of one element to another within the entire work (e.g., *Vitruvian Man* by Leonardo da Vinci). The proportion of various parts of the body in this painting is very similar to the proportion seen in an average human body.

Similarly, much like modern architecture, ancient structures and buildings also incorporated proportion. The ancient Egyptians used it in the construction of the pyramids. The Parthenon in Athens, Greece, is another structure where proportion, along with ratio and scale is used extensively to create a "harmony" among various elements.

Interestingly, Isaac Newton's (1643–1727) second law of motion states that the acceleration of an object in motion is directly proportional to the force applied on it—a classic equation indicating direct proportion between two properties, acceleration and force.

Historians and mathematicians also believe that the great musicians Mozart (1756–1791) and Beethoven (1770–1827) used proportion to compose music. Proportional scaling allows the composition of harmonic, pleasant-sounding, music—a concept initially put forward by Pythagoras.

Subsequently, by the nineteenth century proportion was applied to numerous applications including those in business and sciences.

Real-life Applications

ARCHITECTURE

Architecture uses mathematical concepts such as proportions and ratio extensively. Since ancient times, architects and designers have been building various parts of a structure in proportion to attain visual appeal, unity, stability, and order. These principles hold true even today. Proportion is employed in a number of ways in architecture. Most popular buildings and structures—ancient as well as modern, are based on what is commonly known as the divine proportion or golden proportion.

The divine proportion consists of two or more ratios that are equal to phi (or 1.618). In other words, if the ratio (also known as divine ratios) of various parts of a building (or a structure) is equal to the number 1.618, then the proportion of these various parts is known as the divine proportion. Throughout the world, monuments, famous buildings, and other structures have been created using the divine proportion. This includes the pyramids of Giza, the Parthenon in Greece, the Colosseum in Rome, numerous cathedrals including St. Peter's Cathedral in the Vatican, the Taj Mahal in India, the Pentagon in the United States of America, and many more.

As stated earlier, proportions are used on various elements (or parts) of the entire structure. For example, the front elevation of the Parthenon is built to the divine proportions:its width is 1.618 times its height. Besides divine proportion, basic principles of proportion are also used. For example, the Pentagon is made up of five internal (or concentric) pentagons. Each of these internal pentagons is in proportion to the outer pentagon.

The concept of proportion is used widely in modern architecture as well. Apartment buildings, or houses within the same community may have different sizes of apartments (or houses). However, they are typically in proportion to each other. Sports stadiums also incorporate proportion: the distance between the bases in a baseball field is always proportional to the length (or width) of the field. Similarly, the width of a goal post in a soccer field is proportional to the width of the entire field.

In addition, architects design miniature models before building the actual structure. These models, known as scale models, serve as detailed representations of the final structure. These scale models are much smaller in size, but are in proportion to the final structure. For example, if the scale model of a house is a hundred times smaller than the actual house, every room (or part) of the model would also be a hundred times smaller than the corresponding room (or part) in the actual house—all parts of the model are in proportion with the actual house. Similarly, different parts within the model are also in proportion. If the actual house should be built such that there are two rooms—one room twice the size of the other, the model would also depict two rooms, where the size of one room is twice that of the other. Simply put, the ratio of the sizes of the two rooms is equal in both cases.

The main advantage of a model is that it allows the architect to visualize a structure before it is built. Also, once the model is created, using proportion, various measurements of the final structure can be easily determined and constructed accordingly.

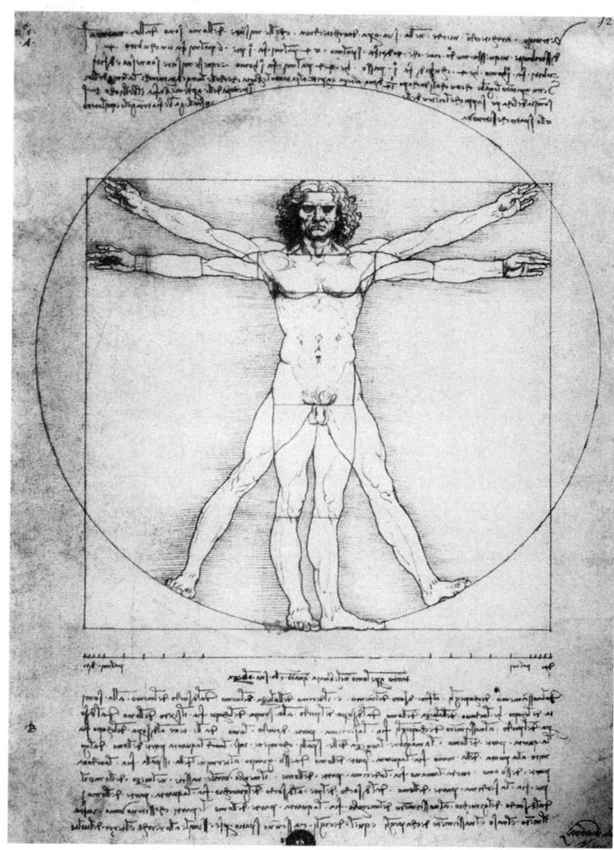

The proportions of man are carefully delineated in the drawing *Vitruvian Man* by Leonardo da Vinci. CORBIS CORPORATION. REPRODUCED BY PERMISSION.

ART, SCULPTURE, AND DESIGN

Like architecture, painting and sculpting also relies on the concept of proportion. Some of the great painters and sculptors, for centuries, have used mathematical models of proportion to attain visual appeal and symmetry (balance) in their work. Portraits and paintings depicting natural scenery are, more often than not, in proportion with the real thing. For a portrait of a person, a good painter would ensure that the measurements of body parts in the painting are in proportion to the actual measurements of the person. This can be seen in most of the ancient as well as modern day portraits.

In addition, different elements within the same painting are also in proportion. In a painting of natural scenery depicting a house, trees, fences, and mountains, the size of each of these is not similar. A house in the painting would be bigger than the size of the fence (unless they are supposed to be at different locations far away). In other words, depending on their location, the sizes are always in proportion—similar to what we see in the real world.

The same holds true for sculptures as well. Like a painting, the sculpture of a person may be bigger (or smaller) in size than the person. However, in most cases, the measurements are in proportion. The advantage of proportion for creating sculptures is evident when the difference in size of the actual object and that of the sculpture is large. Mount Rushmore, in South Dakota, is a classic example. The design and development of the famous memorial to the four presidents—George Washington, Thomas Jefferson, Abraham Lincoln, and Theodore Roosevelt, is based on a number of mathematical concepts, such as ratio, proportion, and scale.

Prior to sculpting the faces of the presidents on the mountain itself, the designer of the memorial, Gutzon

Borglum, developed a smaller model. The size and measurements of the memorial on the mountain are in proportion to the model. Carving the faces on the mountain directly would have been an extremely difficult task for the designer and his team. However, a smaller but proportional model greatly simplified the process. Many technical aspects such as distance between the faces, size of each face, measurements within a particular face, could be easily calculated in the model. Once all measurements were recorded, the designer used the proportion equation to calculate the actual measurements of the memorial (in order for it to look exactly like the model itself).

The principles of interior design also rely on proportion. Furniture, for example, is designed so that its parts are proportionate to each other. This is critical in achieving stability and balance. Furniture that is out-of-proportion is not considered visually appealing. The parts of a chair—the arms, legs, seat, and back—are in proportion to each other, and the chair as a whole.

MEDICINE

Medicines are very essential to treat many illnesses and diseases. Medicines are also used during surgeries and medical diagnosis. They often contain more than two ingredients or compositions that are essential to have desired effect. The proportion of this composition becomes very important. In other words, every medicine contains a specific proportion of its ingredients.

Prescriptions, as well as over the counter drugs, require the mixture of various chemicals, and other additional constituents, to be in certain proportions. For example, over the counter medicines for pain relief often contain aspirin, a required drug to relieve pain, along with other drugs. The proportion of each constituent present in medicine is important as they are meant to treat a certain type of disease, illness, or ailment.

Changing the proportion of the constituents can have different effects. Several common ingredients are used to treat different types of illnesses. The reason for this is that medicines, when prepared using different proportions of the same drugs (or ingredients), act differently, and hence are meant for different diseases.

Proportion is also used frequently by doctors and nurses, while preparing dosages for patients. Patients may require dosages of drugs that vary in quantity and strength. For example, some times a patient may need a dosage that contains 200 mg of a drug that comes as 100 mg diluted in 1 ml of fluid. The technical specifications associated with dosage measurement are beyond the scope of this article. However, for our purpose, the above dosage can be thought of containing a drug in specific quantity (200 mg), having specific strength (100 mg diluted in 1 ml of fluid). The quantity of a drug is proportional to its strength. Using this relation, health care professionals can calculate the quantity of the drug to be administered for a particular strength.

MAPS

Maps may represent a large geographical area and can be of various types depending on the features they emphasize. The area represented by a map can vary from a small room to the entire universe.

There exists a relationship between a specific distance on the map and its actual distance. This relationship is defined by the mathematical concept of scale (or map scale). However, it is important to note that the map scale is based on proportion. In simple terms, the size of the map and the size of the area it shows are always in proportion.

Consider a map that depicts an area that is a hundred times larger than the size of the map. In this case, the relationship between the map and the actual area can be shown as the map scale (a ratio in this case) 1:100—one unit of measurement (cm, inch, feet, etc.) on the map is equal to hundred units in the actual area. The ratio between any part of the map to its actual size remains the constant (1:100). Therefore, every part of the map is in proportion to its actual size. For example, if the actual distance between two points is 100 inches, then the distance between the same two points on the map would be 1 inch. Similarly, if the actual distance between two points is 500 inches, the distance between these two points on the map is 5 inches—the distance between any two points on the map is proportional to the actual distance between them.

Maps can be categorized into two types—the large scale map, and the small scale map. The large scale map shows a smaller area but in greater detail, whereas a small scale map shows a larger area in less detail. The map scale for these maps would differ; however, the maps are always in proportion to the actual size. A city map would be an example of a large scale map as compared to a world map (small scale).

ERGONOMICS

Ergonomics is a science that studies technology and how well it suits the human body. Ergonomics involves understanding basic body parts, their functions and abilities to operate equipments, machinery, products, and other technological devices. Ergonomics is commonly used while designing cars, among other things. Ergonomic car designs are based on the principles of proportion.

Consider, for example, a car seat for drivers. Its height from the surface, inclination, and movements patterns are all designed in proportion to the human body. The size of the seat has to be in proportion with the size of an average human driver. In addition, you do not expect a person to have a giant steering wheel in front of him/her—the size of the wheel (the diameter of the wheel) has to be in proportion to the size of the hand grip, shoulder width, and distance between the wheel and person driving the car.

Ergonomics is used extensively in many areas as well. This includes design of kitchen and appliances, design of home and office furniture, bathroom appliances, electronics, computer systems, airplane and train interiors, and much more. Every ergonomically designed object is proportional to the size of the human body (or a part of it).

For example, a bed is usually designed in proportion to the human body. The length of a bed is proportional to the average height of a person. Many beds in Europe are around seven feet (2 m), whereas those in Asia are around six feet (2 m) long. This also influences other design standards such as height of the bed from the floor, and width of the bed.

ENGINEERING DESIGN

Engineers apply the principle of proportion in many ways including when designing automobiles, airplanes, and trains. Representative two-dimensional models (similar to scale models discussed earlier) are designed before finalizing and manufacturing a car, plane, or train. These are detailed models depicting each and every characteristic. The automobile is then built such that its size and other measurements are directly proportional to the model. In other words, a relation based on proportion is established between the model and the actual object.

The main benefit of creating models for automobiles (as well as airplanes and trains) is to easily study design issues. For example, after calculating the measurements of a seat in the car model, using proportion, the actual size of the seat can be calculated. This will enable the designer to analyze whether the size of the seat is appropriate for a person.

As the dimensions and size of the car are proportional to the model, any change in the model would affect the car. Besides, parts of the model (or car) are also proportional to the model (or car) as a whole. Put simply, if for example, the size of the leg room is changed, the change in the total size of the car can be calculated. if leg room needs to be increased, and at the same time the size of the car must remain constant, the designer would have to reduce the size of some other part of the car.

Once a model with ideal measurements is created, manufacturing the final object becomes a lot easier.

MUSICAL INSTRUMENTS

Since ancient times, mathematicians have always established relationships between principles of mathematics and music. Pythagoras was the first people known to study and apply concepts of proportion and scale to music. These principles are also valid for most musical instruments.

It is widely believed that instruments designed using specific proportions produce superior music. This can be seen in both ancient as well as modern day instruments. For example, to achieve better quality of music, the distance between strings on a guitar (or a violin) is proportional to its entire width. In fact, proportion is used for designing every part of the instrument. Similarly, for a piano to function properly, all its parts have to be in proportion to one another.

CHEMISTRY

Chemicals are often a mixture of a variety of substances. These substances are present in certain ratios. For example, the chemical composition of ammonia is NH_3. Here, the amount of nitrogen (N) is directly proportional to the amount of hydrogen (H)—the ratio of nitrogen atoms to hydrogen atoms is 1:3. In other words, if the number of nitrogen atoms increases by one, the number of hydrogen atoms have to be increased by three. Similarly, if two nitrogen atoms are added, six hydrogen atoms must also be added to continue for the substance to be ammonia. The ratio between nitrogen and hydrogen is always maintained.

Setting up equations as proportions is one of the most effective ways of solving a number of problems in chemistry. For example, to prepare chemical solutions, the chemicals are usually dissolved in water or alcohol. The quantity of chemical present in the solution is known as the strength of the solution. In simple terms, a 70% solution would contain 70% of chemical and 30% of alcohol (or water). While preparing the solution of a specific concentration, the amount of chemical is always proportional to the amount of alcohol (or water). This relationship is especially useful while preparing solutions in different quantities but the same concentration.

A 50 mL (four tablespoons) of chemical solution contains 20 mL (a little more than one tablespoon) of alcohol. If the amount of chemical solution has to be increased to 80 mL (a little more than five tablespoons), what would be the amount of alcohol present in this solution? This

Key Terms

Ratio: The ratio of a to b is a way to convey the idea of relative magnitude of two amounts. Thus if the number a is always twice the number b, we can say that the ratio of a to b is "2 to 1." This ratio is sometimes written 2:1. Today, however, it is more common to write a ratio as a fraction, in this case 2/1.

Proportion: Two quantities with equal ratios.

Scale: The ratio of the size of an object to the size of its representation.

Symmetry, or balance: A design is symmetrical if its two opposite sides divided by a line in the center are identical, or nearly identical.

can be calculated by setting up a proportionality equation as shown below:

20 mL alcohol / 50 mL solution = x mL alcohol/ 80 mL solution, where x is unknown amount of alcohol. The quantity of alcohol should be 32 mL (two tablespoons) for an 80 mL solution.

Such equations are used widely by doctors, scientists, and students.

DIETS

Dieticians and fitness experts often apply mathematical approaches to developing "balanced" diets. They indicate that every meal should have proteins, carbohydrates, and fats in a certain proportion to each other (and the entire meal). This relationship helps greatly in calculating the amount of proteins, carbs, and fats for different meal portions.

For example, a particular meal amounts to 400 calories—160 calories from proteins, 160 calories from carbohydrates, and 80 calories from fat. If another meal is equivalent to 600 calories, the amount of proteins, carbs, and fats would increase to 240 calories, 240 calories, and 120 calories respectively. Note that the amount of proteins, carbs, and fats is in proportion.

Most food items list the amount (in grams) of protein, carbohydrate, and fat content. For instance, 100 grams (3.5 oz) of ice-cream may contain 20 grams (0.7 oz) of fat. The amount of fat in 50 grams (1.7 oz) of the same ice-cream would be 10 grams (0.3 oz), and so on—fat content is proportional to the total quantity. Food items are always available in specific quantities. Put simply, by applying proportion equations, the content of proteins, carbohydrates, and fats can easily estimated for different quantities.

The same concept is also applied to cooking. While preparing a food item, the ingredients are in proportion to each other (and to the total quantity of the food item).

STOCK MARKET

Mathematical concepts such as proportion and ratio have a lot of business applications. One such example is in the stock market. There are factors that contribute to the share value of a company. However, more often than not, a company's share value fluctuates based on profit it makes. Besides, the value also depends on the number of buyers of the company shares. Simply put, the value of a share is proportional to a combination of factors, including the profit and number of buyers.

Most companies divide a percentage of profits amongst all its shareholders (people who own the company's shares). The amount given per share is known as dividend. Higher the number of shares a person owns, higher the dividend. Another way to look at this is that the total dividend is proportional to the number of shares owned.

PROPORTION IN NATURE

The number Phi is an unusual number with astounding mathematical properties. As explained earlier, the golden section, a principle on which ancient Greek architecture was based, is derived from a ratio that further results in the number phi. Phi appears in proportions of the human body as well as the proportions of various other animals. The renaissance artists referred to the golden section as the divine proportion and used it for achieving balance in arts. The divine proportion principle is found in abundance in nature. The spirals of a sea shell, the galaxy, the body of a dolphin, the structure of a butterfly, a peacock feather, the patterns of flowers and plants, the rings of Saturn, all follow the divine proportion principle.

The average human face is also an example of divine proportion. The head forms the golden rectangle with eyes exactly at the center. The mouth and nose are each placed at golden sections of the distance between the eyes and the bottom of the chin. Assume that the eyes are represented by A, nose by B, mouth by C and chin by D. The ratio of line AC to line AD is the same as ratio of line BC to line AC. This means that the ratio of distance between eyes and mouth to the distance between eyes and chin is in proportion with the ratio of distance between nose and mouth and eyes and mouth. Some scientists who study psychological reactions to faces assert that concepts of beauty may be related to facial symmetry and proportion.

Interestingly, the average human face, when viewed from side also reflects the divine proportion principle. Even the dimensions of human teeth are based on this principle. Some dentists are even considering the knowledge of this principle to enhance their aesthetic dentistry skills. The human hand is also an example of the divine proportion.

Where to Learn More

Books

Elam, Kimberly. *Geometry of Design: Studies in Proportion and Composition.* New York: Princeton Architectural Press, 2001.

Padovan, Richard. *Proportion: Science, Philosophy, Architecture.* London: E & FN Spon, 1999.

Quadratic, Cubic, and Quartic Equations

Overview

An equation often describes a function, a rule that relates numbers in one set to numbers in another. Rather than listing all the numbers related by a function, letters, also termed variables, are often used to stand in for the numbers.

Fundamental Mathematical Concepts and Terms

The function $y = 2x$ says that for every number x in some set there is some other number, y, in some other set that is twice as large as x. Some functions consist of a sum of powers of x, like $y = x^3 + 3x^2 + 2x + 1$.

Here the number just above each x tells us how many times to multiply x times itself: that is, $x^3 = x \times x \times x$, and so forth. Functions of this form are named by the highest power of x they contain, which is the rank or order of the equation. For example, the highest power of x in $y = 2x$ is 1 (because $x = x^1$), so this is a first-order equation. The highest power of x in $y = x^3 + 3x^2 + 2x + 1$ is 3, so this is a third-order equation.

The first four orders have special names, namely linear, quadratic, cubic, and quartic. Quadratic and higher-order equations appear constantly in science, engineering, and business mathematics. They are used literally millions of times a day in these fields, designing electronics, analyzing data, implementing codes, predicting profits, and performing other tasks.

Examples of equations of the first four orders are given in Table 1. In the examples, the letters A through E are used to stand for any constants (fixed numbers), with the exception that A cannot equal 0. These constants are called the coefficients of the equation.

A "solution" to an equation is an x, y pair for which the equation holds true. For example, a solution to the linear equation $y = 2x$ is $x = 5$, $y = 10$, because $10 = 2 \times 5$. In this equation—in fact, in all linear equations—there is one x for each y. Finding solutions to equations is one of the most common tasks in the mathematics of science, engineering, and business. Often we know what y is, or what we want it to be—the cost of an item to be manufactured, say—and we want to know what x (or x's) will produce that y. The variable x often stands for something that we can chose or control, such as the length of an assembly line or the amount of a chemical added to a reaction.

For equations where y is equal to a sum of powers of x, including linear, quadratic, cubic, and quartic equations, the x's for which the equation is true are called its roots. Often the y value is subtracted from both sides of the equation to produce a nice, neat 0 on the left-hand side of the equation, but this is a minor detail. What is important is that the number of roots is equal to the order of the equation. A linear (first-order) equation has one root, a quadratic (second-order) equation has two roots, and so on.

We can find the roots of any linear, quadratic, cubic, or quartic equation by writing down certain equations containing the coefficients of the original equation. This cannot be done for equations of order higher than 4, as mathematicians have known since the 1820s. The first four orders are therefore special. The equation that gives the roots of a quadratic equation, $y = Ax^2 + Bx + C$, is one of the most commonly used formulas in all math and science, and has been known since mathematicians of Babylon discovered it some 4,000 years ago:

$$X = \frac{-B \pm \sqrt{B^2 - 4AC}}{2A}$$

This formula is known as "the quadratic equation." In the equation $0 = 2x^2 + 3x - 1$, we have $A = 2$, $B = 3$, and $C = -1$ and the quadratic equation gives us the two roots:

$$x_1 = \frac{-3 + \sqrt{3^2 - 4 \times 2(-1)}}{2 \times 2} = .28 \qquad x_2 = \frac{-3 - \sqrt{3^2 - 4 \times 2}}{2 \times 2}$$

These roots are the two values of x for which $0 = 2x^2 + 3x - 1$ is true. If you plug either of them in for x and do the arithmetic on a calculator, you'll see that 0 really is the answer. (The small numbers hanging off x_1 and x_2 are just labels to set them apart.)

Type of equation	General form	Example
Linear	y = Ax + B	y = x + 10
Quadratic	y = Ax²	y = 2x² + 3–1
Cubic	y = Ax³ + Bx² + Cx + D	y = 12 x³ + x + 5
Quartic	y = Ax⁴ + Bx³+ Cx² + Dx + E	y = x⁴ – 12x³ + x² + 100

Table 1.

Real-life Applications

AREA AND VOLUME

The most basic uses of quadratic and cubic equations are for determining area and volume. In fact, it was the need to calculate land areas that motivated the Babylonians to discover the quadratic equation to begin with. You already know that the area of a square with edges x units long is $x \times x$ or x^2. If we call the area of a square S, then we have the quadratic equation $S = x^2$ (which can also be written $0 = x^2 - S$). The formulas for the area of a circle, a triangle, or even of the surface areas of solids like spheres and cubes, all contain x^2; all are quadratic equations. Surface area is important in real estate, medicine, physics, and engineering. It affects how fast an object cools off (greater area equals quicker cooling), which is why machines that need to get rid of extra heat sometimes have little metal fins stuck on them to increase their surface area. It affects how quickly a droplet evaporates (greater area equals quicker evaporation). It affects how quickly a chemical reaction proceeds (greater area equals quicker reaction).

Cubic equations come up just as naturally. Recall that the volume of a cube with an edges x units long is x^3. If we call the volume of the cube V, then we have the quadratic equation $0 = x^3 - V$. And, just as with surface area, this cubic relationship pops up not only in the formula for the volume of a cube, but in the formula for the volume of a sphere or cylinder or any other three-dimensional object.

The fact that area is described by a quadratic equation and volume by a cubic equation affects many things in nature. Any object's surface area is proportional to x^2—where x stands for how wide the object is—but its volume is proportional to x^3. And as you make the object bigger, that is, increase x, x^3 will always grow faster than x^2. This is why insects can't (lucky for us) grow to the size of dogs or whales: they breathe using surface area (x^2) but their need for oxygen goes by body volume (x^3). This is why elephants have fat legs: the strength of a leg-bone goes by cross-sectional area (x^2), but the weight the bone has to bear goes by the volume of the elephant (x^3).

ACCELERATION

Quadratic equations are needed to predict the paths of accelerating objects. Acceleration is any change in speed. When the driver of a car steps on the gas or hits the brakes, the car accelerates (goes faster or slower). When you drop a ball or throw it up in the air it accelerates. And almost any time a machine with moving parts is designed, from a CD player to a car engine to a jet plane, the people designing the product must deal with accelerations.

CAR TIRES

Car tires are made of rubber-like plastics derived from petroleum and interwoven with metal wires, and must work well despite thousands of miles of use, violent blows from bumps, fast turns, and other stresses. Your life depends on them every day, and their design is a complex art. Computer calculations are used to predict how a new tire design will behave, as this is much cheaper than casting actual tires in a trial-and-error way. One of the most important factors in modeling a tire using calculations is describing the mechanical properties of the "rubber" used in the tire: how it responds to stretching, squeezing, and twisting. In a class of new synthetic tire materials called "carbon black filled rubber compounds," it has been found that a cubic equation best describes the stress-strain relationship—that is, how much the material gives in response to a certain amount of force. This cubic equation is used in writing a computer program that will accurately predict how a tire made with these compounds will behave.

JUST IN TIME MANUFACTURING

Traditional economics treated supply and demand as the two factors deciding profitability in manufacturing. However, in the 1990s some Japanese manufacturers introduced a philosophy called "just in time" (JIT) manufacturing. In this approach, a manufacturer—say of cars, computers, or cameras—tries to produce as many items as possible just in time to deliver them to a buyer. Manufacturing a product and then having it sit in a warehouse, waiting to be sold, reduces profit. But a manufacturer must balance certain variables: they must announce a price and stick to it, they must guess at how much delay or "lead time" they will need to deliver a product, and they must guess at how much demand for the product there will be. The goal, as always, is to earn maximum profit. It turns out that the solution of a cubic equation is central to solving the equation for maximizing profit.

HOSPITAL SIZE

Since the 1980s, hospitals have found it increasingly difficult to make a profit—or even to stay out of debt. Mathematical cost-profit analysis has therefore been brought into play to help hospitals make more profit. One basic decision that a hospital must make is how many beds to have. Having too few or too many beds makes it harder for a hospital to be profitable. Traditionally, profitability has been described as a quadratic function of bed size (the number of beds in the hospital, not how big each bed is); more recent work has shown that a cubic equation works even better. (Other factors are involved, such as where the hospital is located and how affluent the surrounding population is. But if these assumptions are held steady, profitability is a cubic function of bed size.) Using a cubic equation, researchers have found that there isn't just one bed size that is most profitable, but two; or, rather, a point this is typical of a cubic equation, which can have two maximum points rather than one (as a quadratic equation does). From 0 to 238 beds, profit increases. After 238 beds it decreases until 560, after which it goes up indefinitely (but other factors prevent us from building infinitely large hospitals). A hospital is therefore most profitable, in the United States under current conditions, if it is either medium-sized (about 238 beds) or as big as it can be (560 beds or larger).

GUIDING WEAPONS

In steering weapons such as missiles and planes, it is necessary to tell the computer that guides the weapon where it is. Each position is coded as a set of numbers, the "coordinates" of the weapon or vehicle. These can be given in traditional terms as latitude and longitude (numbers derived from a network of imaginary lines laid down on the Earth's surface by map-makers) plus altitude (height above the surface), or in terms of an "Earth-centered coordinate system." Since one type of coordinates is better for some purposes and the other is better for other purposes, it is sometimes necessary to translate between them—to take position information given in one form and turn it into the other form. Going from latitude-longitude coordinates to Earth-centered coordinates is mathematically easy, but going the other way requires the solution of a quartic equation.

Where to Learn More

Web sites

Budd, Chris, and Chris Sangwin. "101 Uses of a Quadratic Equation." *Plus Magazine.* March 29, 2004 and May 30, 2004. Part I: <http://plus.maths.org/issue29/features/quadratic/index-gifd.html>. Part II: <http://plus.maths.org/issue30/features/quadratic/index-gifd.html> (Oct. 22, 2004).

Ratio

Overview

A ratio defines the numerical relationship between two comparable quantities. Examining the ratios between two or more values often provides valuable insight into the patterns and behaviors of numbers.

Ratios exist naturally throughout the universe. The ratio of the size of one planet to another nearby planet can affect the orbits of both planets. The ratio of owls to mice plays a big role in the survival of both species. The ratio of height to trunk width limits the growth of trees. Humans have used ratios in almost all of our creations throughout history. The physical stability of a building depends on several ratios—involving height, width, angles, and the strength of materials that must be carefully analyzed to ensure the safety of the people inside. The accurate mixing of chemicals that allows us to create stronger materials is also reliant on ratios that define how much of each substance is needed with respect to the other materials. People around the world use ratios on a daily basis to organize time and finances.

Fundamental Mathematical Concepts and Terms

A ratio between two numbers X and Y is usually expressed in one of three ways:

- X/Y (much like a fraction)
- X:Y
- "X to Y"

Each of these expressions represents the ratio of X to Y.

For example, if there are 12 cars for every three trucks, then the ratio of cars to trucks can be written as 12/3, as 12:3, or as "12 to 3." Given this information about cars and trucks, it is also true that the ratio of trucks to cars is 3/12, 3:12, or "3 to 12".

All of these expressions for the ratio of cars to trucks (or trucks to cars) state exactly the same thing: for every 12 cars, there are three trucks. Suppose that people in a certain neighborhood always keep their cars in their garages, but leave their trucks out in the driveway. If three trucks are visible in the neighborhood, then there are 12 cars in the neighborhood, even though they are hidden in garages.

The foundation of the idea of a ratio is that whatever happens to one of the numbers also happens to the other. Suppose that six trucks can be seen in driveways around the neighborhood. This means that there are 24 cars hidden in garages. The number of trucks was doubled (multiplied by 2) so the number of cars must have doubled as well. Division of ratios works in the same way. If there was only one truck in the entire neighborhood, then there would be only four cars. Here, the number of trucks and cars are both divided by two to arrive at the ratio 1:4. In fact, this is the simplest form of the ratio of trucks to cars. In a case such as this, the ratio can be simplified so that one of the values is one, which is a good illustration of how ratios work: no matter how many trucks are in the neighborhood, the number of cars is four times as large. Not all ratios can be simplified this neatly—2:3 for example. In cases like this, a decimal can be used as 2:3 simplifies to 1:1.5. In any case, it is easiest to understand the relationship between the two values when the ratio is simplified.

Ratios can be multiplied together to discover new ratios. For instance, if there are two cars for every truck, and three trucks at every house, then there are six cars at every house. That is, 2:1 multiplied by 3:1 is equal to 6:1. Perhaps money provides a better illustration of this concept. There are four quarters to every dollar and five nickels to every quarter; so there are 20 nickels to every dollar. This can be verified by multiplying the five pennies in each nickel by 20 (the number of nickels in a dollar) to get 100 pennies to every dollar.

Although often expressed as a quotient (one number divided by another, such as 2/3), ratios are not the same thing as fractions. For example, if Otis has two dogs and four cats, then the ratio of dogs to cats in his house is two to four, which simplifies to 1:2 or 1/2. This indicates how many dogs there are compared to cats (there are half as many dogs as cats). However, the fraction of animals in Otis' house that are dogs is two out of the total number of animals or 2/6, which simplifies to 1/3. This means that one third of all of his animals are dogs. Be careful to understand how fractions are related to ratios when using the quotient style of notation. To avoid confusion, this text most often uses the X:Y style of notation for ratios.

A Brief History of Discovery and Development

The term ratio stems from an early sixteenth century Latin word meaning reason or computation. However, the mathematical concept of ratios helped people understand the universe around them long before that.

For example, the relationship between a circle's diameter (the length of any line connecting one side of the circle to the other through the center of the circle) and circumference (the length of the boundary of the circle) was approximated for thousands of years before the Greek mathematician Archimedes discovered a way to define the relationship exactly. This ratio can be used to determine the circumference of a circle if its diameter is known, and vice versa. The circumference of any circle is equal to the diameter multiplied by this ratio, commonly represented by the Greek letter pi, and approximately equal to 3.14159265.

Ancient Egyptians approximated pi (though they did not call it pi) as 3.1605. The Old Testament of the Judeo-Christian Bible contains a reference to an approximation of 3:1 for the ratio of a circle's radius to the circumference of a circle. Although ancient Babylonians generally agreed with this approximation throughout most of their history, a stone tablet believed to have been created by Babylonians sometime between 1900 and 1680 B.C. referred to a slightly more accurate approximation of 3.125 for pi.

Early approximations of pi were dependent on approximations of the circumference of circles. It is believed that most approximations of circumference were found using methods similar to those used by Archimedes. First a circle was placed inside of the smallest hexagon (a polygon with six sides) that it could fit into. The length of the perimeter of the hexagon was calculated by measuring one side and multiplying this value by six. Next, the perimeter of largest hexagon that could fit inside the circle was calculated. Because the smaller hexagon just barely fits into the circle, and the circle just barely fits into the larger hexagon, the circumference of the circle is somewhere between the lengths of the perimeters of the two hexagons. To arrive at a better approximation, the number of sides of the two surrounding polygons was increased. As more sides were added, the two polygons fit the circle more snugly and the perimeters became closer and closer to the circumference of the circle. Archimedes used these approximations as clues that eventually led him to find a way to define the ratio of diameter to circumference exactly.

Another important ratio studied throughout history is the Golden Ratio, also known as the Golden Mean, the Divine Section, the Golden Section, the Golden Cut, the Divine Proportion, and many other names. The main reason that this ratio has so many names is that it has been discovered at different times by civilizations that use different languages and, most importantly, different numbering systems. The Golden Ratio is approximately 1.6180339887498948482 to 1 (how the Golden Ratio is calculated is beyond the scope of this text). The Golden Ratio is usually denoted by the Greek letter *phi* (ϕ).

The Golden Ratio can be found throughout nature—from the patterns found in leaves, pinecones, and seashells, to the reproductive patterns of certain animal species. It is also argued that the Golden Ratio provided a basis for the architecture of the ancient Egyptians (including the designs of pyramids and tombs), Greeks (the Parthenon), and Romans. Some ancient Egyptian hieroglyphics show signs of the Golden Ratio as well. Leonardo da Vinci, Mozart, and Beethoven purposely incorporated this ratio into their works. The seemingly endless applications of the Golden Ratio provide brilliant illustrations of the fascinating relationships between numbers.

Real-life Applications

LENGTH OF A TRIP

Ratios can be used to estimate length. For an example let us assume that Tom needs to drive from New York to Miami for a business convention on Saturday evening. He has never driven that far and wants to figure out about how long it will take, so he buys a map of the United States. He notices two bars labeled Scale in the corner of the map. The longer of the two bars represents 100 miles, and the shorter bar represents 100 kilometers. He uses his ruler and finds that the 100-mile bar is one inch long; so the ratio of inches to miles on Tom's map is one to 100. Using the other side of his ruler, he finds that the 100 kilometer bar is one centimeter long; so the ratio of centimeters to kilometers is also one to 100.

Tom is more comfortable thinking in terms of miles, so he chooses to approximate the length his trip based on the inch to mile ratio of 1:100. All he needs to do is find out how many inches separate New York and Miami on the map. Tom lays his ruler on the map, with the beginning of the ruler (representing zero in inches) at New York. The shortest driving route is not a straight line, so he must approximate how long, in inches, his route is on the map. Starting from New York, he measures one inch in the direction of the route that he will take, and marks the spot on the map with a pencil. Then he moves the beginning of the ruler to the mark he just made and measures another inch, following his intended route as accurately as possible. Continuing in this way, he makes 13 marks. The last mark is a little past Miami on the map, so he figures that the route is a little less than 13 inches long. He can't be late to his convention, so he decides to use 13 inches as the base of his calculations. As he found before, the ratio of inches to miles represented on the map is 1:100.

Tom then wants to figure out how many miles are represented by 13 inches, so he must multiply the ratio through by 13 to get a ratio of 13:1,300. This ratio indicates that 13 inches on the map represents 1,300 miles in the real world. So Tom's trip will be about 1,300 miles in distance (length).

Tom now needs to utilize another ratio to help him decide when to leave New York. Without exceeding the speed limit, he can drive about 500 miles in a day. So his mile to day ratio is 500:1. This means that he can drive 500 miles in a single day, 1,000 miles in two days, 1,500 miles in three days, and so on. He needs to go a total of 1,300 miles, so he cannot make it in two days. He can make it easily in three days. He may be a little early but he will not be rushed. He decides that if he leaves on Thursday morning, he will get to the convention with time to spare.

COST OF GAS

In the previous example, Tom calculated 1,300 miles as a slight overestimate for the length of his trip from New York to Miami. He now wants to calculate how much money he will need for gas so that he can plan the budget for his trip. His car gets an average of 25 miles per gallon, which is a mile to gallon ratio of 25:1. Tom uses this ratio to calculate how many gallons of gas his car will need to go 1,300 miles. 1,300 miles is 52 times as long as 25 miles, which means that Tom must multiply both sides of the ratio by 52. In this way, he calculates the mile to gallon ratio 1,300:52. To go 1,300 miles, Tom's car will need 52 gallons of gas.

Next, Tom looks on the Internet and discovers that the average cost of gas along his route is two dollars per gallon. So the ratio of dollars to gallons of gas is 2:1. To find out how much 52 gallons of gas will cost, Tom multiplies both sides of the ratio by 52 to get a dollars to gallon of gas ratio of 104:52, meaning that Tom needs $104 to buy 52 gallons of gas for his car. After working this figure into his budget, he finds that he has plenty of money for his trip to Miami.

GENETIC TRAITS

In 1866, Austrian monk and geneticist Gregor Johann Mendel (1822–1937), published his results from an extensive series of experiments that investigated how characteristics are passed to offspring. One such experiment involved the cross-pollination (transferring the pollen of one plant to another) of two different varieties of pea plants, a green wrinkly pea plant and a yellow rounded pea plant. In this experiment, Mendel discovered that the ratio of yellow rounded offspring to green wrinkly offspring was 3:1, meaning that the cross-pollination process produced three yellow rounded pea

China's one-child family planning program, in combination with a preference for male children, has created an unbalanced boy-girl ratio according to U.S. State Department officials. AP/WIDE WORLD PHOTOS. REPRODUCED BY PERMISSION.

plants for every green wrinkly pea plant. This suggested that the yellow characteristic is three times as likely to appear in the offspring as the green characteristic, and the round characteristic is three times as likely to appear as the wrinkly characteristic. This dominance of yellow and rounded characteristics led Mendel to believe that there are two different types of genetic traits: dominant traits and recessive traits. In these two types of pea plants, the yellow color and round shape dominate the green color and wrinkly shape.

Mendel used this idea of dominance to explain why the ratio of dominant traits to recessive traits is always 3:1. For instance, if the dominant yellow color trait is represented by Y, and the recessive green color trait is represented by g, then the four possible combinations of these traits are YY, Yg, gY, and gg (where each parent plant provides either a Y or a g). A dominant trait only needs to appear once in the combination in order for the dominant characteristic to appear in the offspring. Y appears in three of the combinations, and the only combination that results in a green pea plant is gg. Therefore, the ratio of offspring that show a dominant characteristic to offspring that show a recessive characteristic is three to one, or 3:1. This is true for the shape trait as well. Keep in mind that Mendel's actual experiments and results were more complicated than described here.

Mendel's experiments and conclusions explained a phenomenon that had confused people for thousands of years, that traits can appear after skipping generations. For example, cross-pollination of two yellow pea plants can result in a green offspring as long as at least one parent had a Yg or gY gene.

The importance of Mendel's results was not truly recognized until the beginning of the twentieth century when multiple researchers independently rediscovered Mendel's conclusions in their own experiments. Since then, Mendel's findings have been the foundation of many genetic studies and practices, including the creation of new flowers and new species of pet fish, and enhancements in farm production that improve the quality of produce found in most grocery stores.

STUDENT-TEACHER RATIO

The student-teacher ratio compares the number of students to the number of teachers at a given school. For example, if a school has 44 teachers and 968 students, the student-teacher ratio at this school is 968:44, which simplifies to 22:1. This can be interpreted in a few different ways. A mother might see it as an indication of how much attention her child will receive in school (e.g., her child will share each teacher with 22 other students). This ratio enables a teacher to predict how many students he or she will teach and how many papers she will be grading (about 22 students per class multiplied by the number of classes he or she teaches). A school official may see it as an indication of how many teachers need to be hired in the following year. A prospective college student will usually take the student-teacher ratio into consideration when choosing a school for higher studies.

MUSIC

Ratios can be found in every facet of music. Rhythm, the speed and pattern of beats, has been the foundation of music dating back to ancient drum circles. Rhythm determines how many beats there are in a measure (the standard unit in the arrangement of song). For example, if there are four beats in a measure, then the ratio of beats to measures is 4:1. This means that the musician considers four beats—possibly counted by tapping a foot—to be a single standard unit in the arrangement of the song. Different ratios of beats to measures affect different types of music. For example, a typical waltz has a ratio of 3:1.

The relationship between the pitch of two musical notes is also a ratio. Whether created by your vocal chords, a guitar, or a finger moving around the rim of a wine glass, the sound of a note is determined by the frequency (speed) of the vibration causing it. As you move from left to right on the keys of a piano, the difference from one note to the next is determined by the ratio between the frequencies of the two notes: the ratio between the frequencies of two subsequent notes is always the same. Harmony (whether or not two or more notes sound good when played together) is determined by the ratios of their frequencies as well.

Ratios in music allow songwriters and musicians to communicate the intended shape and feel of a song. The many ratios in a composition define how the various sounds relate to each other in time and, whether consciously noticed or not, give the music both structure and beauty.

AUTOMOBILE PERFORMANCE

The safe and efficient operation of any automobile is dependent on many ratios. Oils and fluids must be present in certain ratios to keep the engine and brakes operating properly. The relationships between the size, weight, and position of various parts ensure that a car or truck can make turns while traveling at reasonable speeds and can stop quickly when necessary. Two ratios found in automobiles are compression ratios and gear ratios.

The compression ratio is used to predict how efficiently an engine will perform. In general, a higher compression ratio indicates better engine performance. High compression ratios are often associated with requirements of more expensive fuel and frequent engine maintenance. The determination of an engine's compression ratio involves the relationship between the sizes of the parts of the engine that cause combustion (the small explosions that provide an engine with power).

The speed and power of an automobile depend partially on the ratio between the sizes of gears that cause the wheels to turn. A larger gear turns slower because it has more teeth and takes longer to complete a full revolution. If a gear that is powered by the engine is attached to a smaller gear, the smaller gear will turn more quickly than the large gear. This increases the speed of revolution without increasing the need for power. Given certain gear ratios for an automobile, a specialist can determine how many revolutions per minute (RPM) are required to go a certain speed, or how many tons can be pulled without overexerting the engine. A typical car has multiple sets of gears intended to perform different actions. The first gear has a high gear ratio in order to provide the car with enough power to get the car started. In higher gears, the gear ratio is increased in order to enable faster speeds. Also, a car would eventually get stuck without an additional gear set that caused the car to move in reverse.

SPORTS

Ratios are often used to assess the performance of an athlete or athletic team. The relationship between two or more statistics often proves a better indication of performance than a single statistic alone.

As an example, a point guard's contribution to a basketball team is partly measured by his assist-to-turnover ratio. This ratio is determined by comparing the number of assists (passes that lead to an immediate basket) to the number of turnovers (anything that causes the ball to be lost to the other team). Suppose Gary has had 53 assists this year, and has turned the ball over to the other team 44 times. Gary's assist-to-turnover ratio is 53:44.

Gary's talents could be judge based on turnovers alone. If Gary had more turnovers than anyone else this season, sports analysts might think that he is the worst point guard because he gives the ball up more often than anyone else. But what if he also happened to have the most assists? Would the analysts still think so poorly of him? The converse is true as well: if Gary's talents were judged based only on the number of assists that he has without taking into account the fact that he turns the ball over quite often, the analysts would not have a very accurate picture of how Gary actually performs on the court.

AGE OF EARTH

In 1905, New Zealand/English physicist Ernest Rutherford (1871–1937) announced a discovery that would forever change the approximations of the age of Earth. He suggested that the age of rocks could be computed by analyzing one of two ratios: the ratio of uranium to lead or the ratio of uranium to helium. These ratios can be used to determine how long radioactive materials have been decaying, and in turn, to determine how long ago rocks were formed. Prior to this discovery, the process of radioactive decay was poorly understood, and guesses at the age of Earth were just that: guesses. Since Rutherford's discoveries, new tools and methods have been derived to improve estimations of Earth's age. For example, calculations in the dating process, including values for decay rates, have been repeatedly improved upon. As of 2005, the best estimation for the age of Earth is in the neighborhood of 4.5 billion years.

HEALTHY LIVING

A person's height-to-weight ratio is the relationship between how tall that person is and how much that person weighs. If a person is six feet tall and weighs 180 pounds, then his height-to-weight ratio is six feet to 180 pounds, or 1 foot per 30 pounds. This ratio can be seen as an indication of how healthy a person is. There are, of course, many other important considerations—including body type, bone thickness, and muscle density—that help determine an individual's optimal weight. All of these factors can be put into terms of ratios.

COOKING

Whenever a chef follows a recipe, he uses ratios to determine how much of each ingredient to stir in. Suppose a chef is cooking his favorite soup for a large dinner party. He has a recipe that tells him how much of everything is required for making enough of the soup to serve 20 people, but there will be 140 people at the dinner party. The ratio of people served by his recipe to the actual number of people that he needs to serve is 20:140, which simplifies to 1:7 (by dividing both sides by 20). This tells the chef that he needs to buy seven times the amount of ingredients suggested by the recipe in order to make enough soup for the dinner party.

The chef can also use ratios to determine how much of one ingredient will be needed based on the required amount of another ingredient. For instance, the chef knows that the ratio of sugar to butter in this recipe is 1:3. This means that the amount of sugar needed to make any amount of this recipe is a third of the amount of butter needed. The chef has already calculated that he needs six cups of butter to make the soup for 140 people. With no further calculations, he knows that he needs two cups of sugar to make this amount of soup.

CLEANING WATER

Chlorine is the main chemical that is used to clean both drinking water and the water in swimming pools. The biggest difference between the processes for cleaning drinking water and swimming water is the concentration of the chemicals, the ratio of the amount of chemicals to the amount of water. This ratio is much lower in drinking water than in swimming water. That is, the water you drink has a smaller amount of chemicals in it than the water in most swimming pools. The concentration of chemicals in drinking water must be precisely monitored in order to ensure that enough chemicals are present to kill bacteria, but not enough to be harmful when swallowed by humans. Water in a swimming pool must contain a higher concentration of chemicals because the water is constantly in contact with contaminants from swimmers and the air above. The fact that a swimming pool is open to the air also allows the chemicals to evaporate, so new chemicals must be added periodically. These ratios between water and chemicals are essential for the different uses of water. Water from a swimming pool is not safe to drink in large quantities; and swimming in water with the concentration of chemicals found in drinking water would quickly result in the growth of algae and bacteria in the pool.

Potential Applications

STEM CELL RESEARCH

Stem cells are special cells in the human body that have the ability to become any type of human cell. This single type of cell can create skin and muscle tissue, bones

Key Terms

Concentration: The ratio of one substance mixed with another substance.

Percent: From Latin for *per centum*, meaning per hundred, a special type of ratio in which the second value is 100; used to represent the amount present with respect to the whole. Expressed as a percentage, the ratio times 100 (e.g., 78/100 = .78 and so .78 × 100 = 78%).

Rate: A comparison of the change in one quantity, such as distance, temperature, weight, or time, to the change in a second quantity of this type. The comparison is often shown as a formula, a ratio, or a fraction, dividing the change in the first quantity by the change in the second quantity. When the changes being compared occur over a measurable period of time, their ratio determines an average rate of change.

Ratio: The ratio of a to b is a way to convey the idea of relative magnitude of two amounts. Thus if the number a is always twice the number b, we can say that the ratio of a to b is "2 to 1." This ratio is sometimes written 2:1. Today, however, it is more common to write a ratio as a fraction, in this case 2/1.

and bone marrow, and organs such as the liver and lungs. This characteristic has made stem cells the main focus of regenerative medicine, a field of research involving the recreation of cells in the human body. The regeneration of cells may be the solution to many problems that have been unsolvable in the past. Potential uses of cell regeneration include regaining skin and muscle tissue lost in physical accidents; allowing someone bound to a wheelchair to walk; and curing diseases such as Parkinson's, cancer, Alzheimer's, and diabetes. Unfortunately, it may be many years before stem cells are regularly used in routine medical procedures.

Scientists have much to learn about manipulating stems cells to create a desired part of the body. Ratios play a big role in many stem cell research projects. For example, the ratio of blood cells in a donor to blood cells in the recipient may be an important factor in the success of stem cell transplants.

OPTIMIZING LIVESTOCK PRODUCTION

In nature, the sex ratio (ratio of males to females) of many species remains close to 1:1 (often referred to as 50:50 or fifty-fifty), meaning that about half of the population is male and about half is female. In species with males that can mate with multiple females, this may not seem a very efficient ratio. Nevertheless, the sex ratio remains approximately 1:1.

Farmers have for millenia artificially kept the ratio of female cattle (cows) to male cattle (bulls) high, because a single male cow can fertilize multiple female cows. Suppose a single male cow can regularly fertilize up to 20 cows. If a dairy farm with 100 cows had 50 males and 50 females, then only half of their cows would be producing milk, and the male cows would not be performing to their capacity. But if there were 5 males and 95 females, then the farm would have more cows producing milk, and the males would be able to do their job at a rate closer to their limit.

DETERMINING THE ORIGIN OF THE MOON

In 2003, German scientists compared the ratios of two elements present in rocks from the Earth, Moon, Mars, and various meteorites to arrive at a better approximation of how and when the moon was formed. The two elements compared were niobium (a metal commonly found in alloy steels) to tantalum (an acid-resistant metal commonly found in dental and surgical instruments).

Most astronomers have long subscribed to the theory that the moon was formed when a celestial body (roughly half the size of Earth) crashed into Earth causing a large mixture of rocky debris from both bodies to fly into space, some of which lumped together to form the moon, while the rest dropped back to Earth. The amount of the Moon that is made up of material from the body that struck Earth has long been passionately debated; as has the amount made up of material from the Earth itself. The percentage of the Moon that is made up of material from the impacting body, for example, was approximated at as low as 1% by some scientists, and as high as 90% by others. By comparing the ratios of niobium and tantalum, the German team of scientists was able to determine that the amount of the moon that is composed of material from the body that struck Earth is somewhere between 35% and 65%. The rest of the Moon is composed of material from Earth.

The approximate age of the moon, another value that scientists have had a hard time agreeing about, was also refined during these studies. Calculations based on the ratios of niobium and tantalum suggest that the Moon must have been created at about the same time as Earth: about 4.5 billion years ago. As scientists continue to study the moon, the approximations of its composition and age will become increasingly accurate.

Where to Learn More

Books

Livio, Mario. *The Golden Ratio: The Story of Phi, the World's Most Astonishing Number.* Broadway Books, 2003.

Periodicals

Jacobsen, Stein B. "Geochemistry: How Old Is Planet Earth?" *Science* 300, No. 5625 (2003): 1513–1514.

Web sites

National Institutes of Health. "Stem Cell Basics." The official National Institutes of Health resource for stem cell research. June 10, 2004. <http://stemcells.nih.gov/info/basics/basics6.asp> (March 12, 2005).

Rounding

Overview

Rounding is a way of simplifying a number to an approximate value. The intent is to create a number that is easier to envision conceptually and is more practical to use. When a result close to the actual value is sufficient, rounding can be a useful operation.

Fundamental Mathematical Concepts and Terms

WHOLE NUMBERS

Numbers can be rounded to the nearest unit or a larger scale across multiple units, as is appropriate.

The rules of rounding are rudimentary. Irrespective of, for example, the power of ten under consideration, rounding is based on the equal number of incremental increases between one power of ten and the subsequent power of ten.

For example, the single digit incremental pattern between 0 and 10 (0, 1, 2, 3, 4, 5, 6, 7, 8, 9, 10) is mirrored by the pattern between 10 and 20, 100 and 200, 10,000 and 20,000, 100,000 and 200,000 and so on.

In addition, all the incremental series display a common pattern of internal symmetry. As a representative example, between 1 and 10 there are as many numbers below 5 as there are above. This aspect is crucial to rounding. To round a number to the nearest ten, the last digit of the number is the determinant. If that digit is 1 through 4, then the number is rounded to the next lower number that ends in 0 (the next lower even ten). For example, the number 14 is rounded to 10. If the last digit is 5 or more, then the number is rounded to the next higher even ten. For example, 17 would be rounded to 20.

The same pattern is followed when rounding numbers to the nearest hundred. Numbers that end in 1 through 49 are rounded to the next lower number ending in 00. Thus, 624 is rounded to 600, as is 648. Numbers that end in 50 or higher are rounded to the next even hundred. Thus, 650, 675 and 688 are all rounded to 700.

Similarly, when rounding to the nearest thousand, numbers whose last three digits are 001 through 499 would be rounded to the next lower number ending in 000, and numbers whose last three digits are 500 and higher would be rounded to the next even thousand. As examples, 4,390 and 4,450 are rounded to 4,000, while 4,600 and 4,835 are rounded to 5,000.

The same pattern carries through the increasing powers of ten.

DECIMALS

Rounding decimals is no more complicated than rounding whole numbers. If the thousandths place of a decimal (i.e., 0.00x) is four or less, that place is dropped and the value of the other digits remains unchanged. As an example, rounding 0.574 to the nearest hundredth produces 0.57.

If, however, the thousandths place is five through nine, then the hundredths place is increased by one digit. For example, 0.577 is rounded to 0.58.

Decimals can also be rounded to the tenth. In this operation, if the hundredths and thousandths places are less than forty-nine, they are deleted and the value of the tenths place remains unchanged. As an example, 0.638 is rounded to 0.6.

But, if the hundredths and thousandths places exceeded fifty, then these would be dropped and the value of the tenths place is increased by one. For example, 0.679 is rounded to 0.7.

Decimals that extend to more decimal places can also be rounded. Then, they utilize the preceding two decimal places to make the rounding decision. Thus, 0.647756 is rounded to 0.6478 and 0.32434612 is rounded to 0.324346.

PI

One prominent case is the value of the ratio of the circumference of a circle to the diameter of the circle. This ratio, which is called "pi," has been computed to approximately 100 billion digits, with no repeat or end of the decimal sequence found. The first few digits of pi are 3.14159265 . . . Since there is an infinite number of digits in pi, it has to be rounded to be used at all.

Using the aforementioned rules, pi could be rounded off to 3.142 (since the next digit is 5), 3.1416 (since the next digit is 9), or 3.141593 (since the next digit is 6). Obtaining all the rounded versions of pi would literally be a never-ending task. The decision of where to round off pi depends on the how precise the number needs to be, with more decimal places producing a more precisely accurate approximation of pi.

A Brief History of Discovery and Development

References to rounded-off values of pi can be found in the Bible and other sacred scripture. Many books contain apparently incorrect derivations resulting from the use of a rounded estimate of pi, indicate that rounding has been practiced since antiquity.

Real-life Applications

LENGTH AND WEIGHT

Rounding is a way of simplifying numbers to make them easier to comprehend and use. Precise measurements (at least to the nearest 0.1 inch) are necessary in carpentry to provide a correct fit between components. However, in many other applications an estimation of dimensions is sufficient.

Length measurements provide numerous examples of rounding. For example, a brochure advertising a house for sale might want to note that the house is far back from a busy highway. Instead of noting that the driveway is 221.5 feet long, rounding off the distance to 200 feet will still convey the desired impression to the prospective buyer.

Rounding off lengths can also be a more descriptive way of comparing two objects. As an example, it is accurate to describe a 39-inch-long board as being 1.77 times the length of a 22-inch-long board. However, this ratio is hard to conceptualize. Rounding off the lengths of the longer board to 40 inches and the shorter board to 20 inches produces a ratio of 2. This is close to the actual ratio, but is a much easier difference to understand.

This ease of comparison also applies to weights. An object that weighs 262 pounds is 4.3 times as heavy as an object that weighs 61 pounds. While descriptive, and certainly accurate, this weight distinction is more difficult to gauge than if the weights are rounded off to 250 and 50 pounds, producing a ratio of 4.

BULK PURCHASES

When contemplating the bulk purchase of an item, rounding off permits a quick estimation of the total purchase price. This can be an important factor in making the purchase decision.

An item may be advertised with a price (i.e., $10.65) based on a single purchase. In considering a bulk purchase (i.e., 21 items), rounding can be used in several ways. The item cost can be rounded (i.e., $11.00) and multiplied by the number of items to give the total purchase cost.

The number of items can also be rounded (i.e., 20). The purchase price can then be determined by multiplying the rounded single purchase price by the rounded number of items.

Either of the estimates, which typically are close to the actual (non-rounded) value, provides the information necessary for the purchase decision.

POPULATION

In a census, the population is determined as accurately as possible. However, such an exact tally is not always necessary or convenient. Indeed, in the case of a city of town, maintaining an exact tally can be difficult with the population shifting daily.

Rounding the population can be a more convenient way to present the information. This is especially true when the population figure can be rounded up, as would be the case for a civic population of 47,724. The rounded population of 48,000 would be a hedge for the natural increase in population number over time.

Many communities that post welcoming signage will use a rounded population number. At the very least, this saves constant modification or replacement of the sign to keep the population number current.

The rounding of population is also extends to the state level. For example, according to 2003 figures from the United States Census Bureau, the population of California was 38,484,453. Because this number will be constantly fluctuating with births, deaths, immigration and emigration, such an exact number may not be useful in some instances. Instead, the population can be rounded to the nearest million (38,000,000) or nearest hundred thousand (38,500,000).

Similarly, rounding can be done for selected categories of census data. Using the California example, the 2003 Census Bureau figures established that women made up 50.2% of the state's populations. Rounding the number to 50% makes discussion of the female segment of the population easier, without comprising the possible significance of the figure.

LUNAR CYCLES

"Thirty days has September, April, June and November," begins a well-known nursery rhyme. Calculations of lunar cycles are based on 30 days a month. This is despite the fact that there are only four such months in the year.

The derivation of the monthly period in our Julian calendar (which consists of three 365-day years followed by a 366-day leap year) is based on the monthly transit of the Moon around Earth. However, the lunar cycle actually covers some 29.54 days. The 30 day period that is enshrined in our calendar is a rounded estimate of the actual lunar cycle time.

Similarly, the 365-day length of the normal year—based on the transit of Earth about the Sun—is itself a rounded number. Because every forth year is a leap year to maintain the synchronicity of the calendar, each year actually comprises 365.33 days. Because the decimals are less than 0.50, the rounded number becomes 365.

ENERGY CONSUMPTION

Figures on the consumption of oil and gasoline that are released by agencies such as the United States Department of Energy (DOE) are rounded. For example, in 2004 the DOE reported that the United States used an average of 20 million barrels of oil per day. This number has been rounded up or down from a more precise estimate. The result is still a potent number, which serves as a reminder of just how much of a non-renewable resource is claimed.

WEIGHT DETERMINATION

Trucks that haul produce, freight and other loads on the nation's highways are designed to hold a maximum weight. Exceeding that weight can make a load dangerously unstable, which can lead to an accident. As well, trucks that exceed the weight limit for bridges and roadways can damage these routes.

To try to ensure compliance with weight limits, transport trucks are periodically required to pull into roadside inspection stations where the vehicles are weighed.

Part of the process used to determine weight compliance involves the establishment of several weight categories; for example, under 30,000 pounds, 30,000–80,000 pounds and more than 80,000 pounds. When a particular truck is weighed, the result can be rounded off to the nearest 10,000 pounds. So, a truck weighing 57,650 pounds can be recorded as 60,000 pounds. If the truck is meant to be in the intermediate weight category, then it is in compliance.

ACCOUNTING

When people successfully sell an item on the eBay Internet auction site, a portion of the sum they receive represents the company's fee. As more transactions are conducted, a running balance is kept of the amount owing.

eBay uses six decimal places to charge on an account. For example, a calculated balance might be $6.333560. However, the balance tally that appears on a customer's account is rounded to two decimal places ($6.33). As with other examples of rounding, there can be a slight difference between the actual and rounded values. But, rounding produces a balance that conforms to accepted billing practices.

Cash registers are programmed to round off a sum to the nearest hundredth. For example, if the sales tax charged on purchased items is 0.015%, then the sales tax added to an item sold for $17.50 is 17.50×0.015, or 0.2625. The final purchase price for the item would be $\$17.50 + \$0.2625 = \$17.7625$.

Key Terms

Decimal: Based on the number ten; proceeding by tens.

Pi: The ratio of the circumference of a circle to the diameter: $\pi = C/d$ where C is the circumference and d is the diameter. This fact was known to the ancient Egyptians who used for π the number 22/7 (3.14159) which is accurate enough for most applications.

Whole number: Any positive number, including zero, with no fraction or decimal.

This four decimal place tally would never appear on the cash register receipt. Instead, the internal programming of the register rounds $17.7625 to $17.76.

TIME

Clocks and watches allow time to be measured to the second. With digital display capability, the estimation of time is needless.

Yet often time is rounded and expressed in more general terms. Instead of expressing a time as 6:43, a common practice is to note the time as "quarter to seven." Because an hour can be conveniently divided into four 15-minute segments, time can be rounded within a 15-minute bracket of time. Thus, a time of 6:32 could fairly accurately be rounded and expressed as "half past six."

The slight loss in the precision of the expressed time has not come at the loss of meaning.

MILEAGE

Mileage is typically expressed as the average of the distance traveled and the time taken to travel that distance. The result can be a multi-decimal number that is unwieldy and conveys too much information than is necessary.

As an example, if someone drove 335 miles in 4.75 hours, their average mileage would be 335 / 4.75, or 70.526315 miles/hour. For practical purposes, such as to calculate gas mileage, a simpler answer is best. Considering the above mileage to the first decimal place (70.5) allows the mileage to be rounded up to 71 miles/hour. The rounded answer carries the suitable depth of meaning for the problem.

Rounding is also practiced by drivers when calculating the distance from one local to another using a road map. Often maps will display distance to one decimal place (i.e., 25.5 miles). But, when adding a number of distances mentally, it is easier to round each of the distances to whole numbers and add the series of whole numbers together. This can usually be accomplished easily, quickly, and will provide the driver with the answer that has the appropriate level of meaning.

PRECISION

Precision is indicated by the number of significant figures in a number. For example, using a meter ruler, it is possible to measure a length to the millimeter (one thousandth of a meter, or 0.001). However, expressing some measurements to the thousandth of a meter can be imprecise.

For long distances, it would not be reliable or even honest to report a distance to this level of precision. Rather, by rounding the number to the tenth decimal place, a value is reported that represents a more reliable estimate of the distance.

Scientific notation is valuable in improving the precision of rounded numbers. For example, 363.6 meters can be expressed as 3.636×10^2. The number can also be rounded off and expressed in scientific notation as 4×10^2 or, more precisely, 3.6×10^2.

Where to Learn More

Books

Niederman, Derrick, and David Boyum. *What the Numbers Say: A Field Guide to Mastering Our Numerical World.* New York: Broadway, 2003.

Pickover, C.A. *The Mathematics of Oz: Mental Gymnastics from Beyond the Edge.* Cambridge: Cambridge University Press, 2002.

Pickover, C.A. *Wonders of Numbers: Adventures in Mathematics, Mind, and Meaning.* Oxford: Oxford University Press, 2002.

Web Sites

Arnold, Douglas N. "The Patriot Missile Failure" <http://www.ima.umn.edu/~arnold/disasters/patriot.html> (February 12, 2005).

British Broadcasting Corporation. "Skillswise factsheet: what is rounding?" <http://www.bbc.co.uk/skillswise/numbers/wholenumbers/whatarenumbers/rounding/> (February 12, 2005).

Purplemath. "Rounding and Significant Digits I." <http://www.purplemath.com/modules/rounding.htm> (February 12, 2005).

Rubric

Overview

The word rubric sounds like something from a poem. Indeed, the term comes from the Latin word *rubrica*, meaning 'red earth'. Historically, rubric refers to the prompts that were written in red ink in some documents during the Middle Ages. Later, red ink was used to highlight noteworthy sections (often rules) within legal documents.

But, contrary to the wordplay of poetry, a rubric is grounded in logic and math. A rubric is defined as a set of rules that allow tasks or activities to be scored. By doing so, both students and teachers are more aware of what achievement benchmarks there are in a task, and what a score really means in terms of what was achieved and what was not. This sort of knowledge can help students improve, since they are more aware of what specific tasks need to be done to improve.

Not surprisingly, the scoring involved in a rubric is tied to mathematics. The means used to assess the performance of the particular task and the detailed descriptions of the various levels of performance can involve math. Rubric math is certainly real life math, and is in action every day in most every classroom.

Real-life Applications

SCORING RUBRICS

A good rubric can detail out the various grades of quality of each of the criteria that have been picked as being important in the performance of a task. Often, this detail takes the form of a point scale. For example, in an oral presentation, a rating of 1 on a 1-5 point scale could be understood to mean 'a poorly organized presentation, with a poor use of voice and props.' A grade of 5 would be given for 'a presentation that is excellently organized and presented with a riveting use of voice and props.' The rubric has clearly detailed the expectations of the students and how their efforts will be scored. There is not any confusion over what a certain score means.

Another example will help to show the clarity that rubrics can provide. A poor rubric would be 'Students will show an understanding of how to use a ruler.' Instead, a proper rubric for this situation could read 'Students will demonstrate that they can use a ruler to accurately measure the target length and width of large and small objects.'

In the non-rubric classroom, scoring an assignment by a percentage value can lead to confusion. For example, two students hand in an essay on the same topic. One of

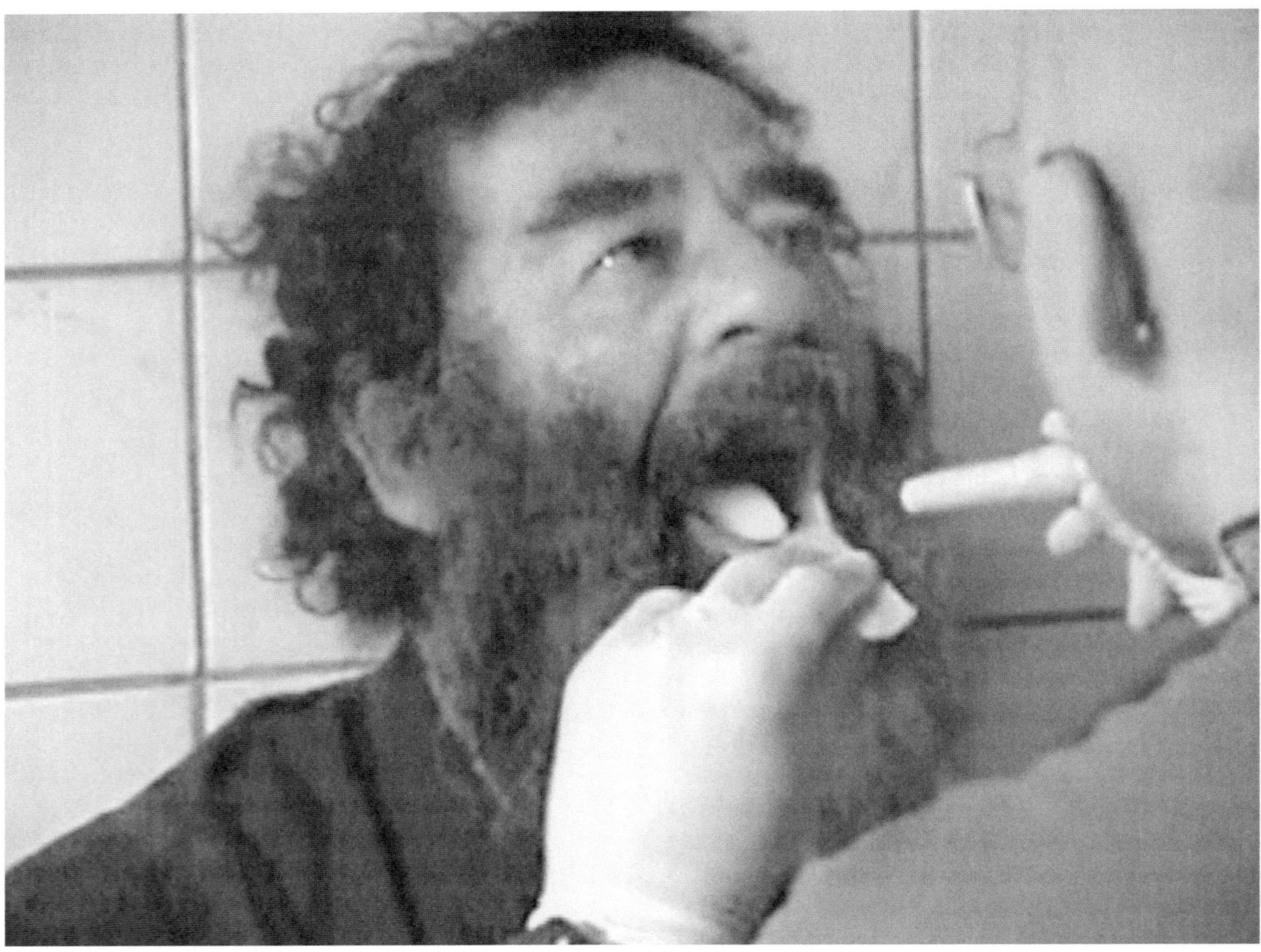

Captured former Iraqi leader Saddam Hussein undergoes medical examinations in Baghdad. Aspects of medical exams and other biometric tests can be converted to numerical data through the use of rubrics that allow examiners to match subjective observations such as gum color to a numerical scoring system. AP/WIDE WORLD PHOTOS. REPRODUCED BY PERMISSION.

the students receives a grade of 72 percent and the other student's grade is 76 percent. It can be difficult for a student to understand why one essay is four percent "better," and how another essay can be improved upon in the future, since percentage evaluations seldom have criteria associated with them. It may even be difficult for the teacher to explain the basis of the marking to the student!

Rubrics can also consist of questions. 'Does the student understand how to measure centimeters and millimeters?' and 'Can the student produce measurements that are two- and three-times as long as an example dimension?'

Scoring rubrics can also be used to gauge a student's performance. These can involve a checklist or using a number-based rating scale of performance. They specifically itemize the performance and provide a number that indicates how well or poorly a task was achieved. In other words, such rubrics provide quantitative results. Rubrics can also provide qualitative results. Examples include ratings like 'excellent,' 'good,' and 'poor.'

When different people look at an assignment or a test document, they can have different rankings of what are the important things to note. Scoring rubrics helps avoid ambiguity and differences in assessing an assignment or document, since the criteria for scoring will already be set out. Deciding on which criteria to include and in what order can be the toughest stage. But, once the rubric is established, scoring becomes routine.

There are several different types of scoring rubrics. Which one is used depends on the purpose of the evaluation that is to be done.

South Korean high school students bow in Seoul, as they pray for their seniors' success in college entrance exams. School exams that require essays, such as the revised College Scholastic Ability Test (SAT exams), are often scored by rubrics that define criteria that allow scorers to place numbers in otherwise subjective evaluations. AP/WIDE WORLD PHOTOS. REPRODUCED BY PERMISSION.

ANALYTIC RUBRICS AND HOLISTIC RUBRICS

These terms sound complicated, but they are not really. An analytic rubric is one that details how several different criteria are to be used in the evaluation of something. For example, an analytic rubric in an essay presentation could detail what aspects of grammar are going to be assessed and how the structure of the essay contributes to its impact. Each aspect could have a checklist of points that will be evaluated, and different scoring lists for the different criteria.

Sometimes it is not possible or beneficial to evaluate a project on separate criteria. Then, an overall view is best. That is where the holistic rubric comes in handy. In a holistic rubric a single scoring scale is used.

Generally, an analytic rubric is suitable for quantitative scoring (that is, where numbers are actually assigned as scores), whereas a holistic rubric is best suited for qualitative assessments.

The two rubrics are not mutually exclusive. It can happen that one of the criteria in an analytic rubric is more general. So, the assessment of that one criterion can involve a holistic rubric.

GENERAL RUBRICS AND TASK-SPECIFIC RUBRICS

In setting up the scoring of a task or activity, the first step is in creating what is called the general rubric. This is an overview or an outline that helps create the more detailed rules of the scoring. For example, if a course in school is designed to improve a student's skill in performing microbiology experiments, a general scoring rubric could be developed to evaluate each experiment that a student does. The feedback that a student got following the completion of one experiment could help them to carry out better experiments in the future.

A task-specific rubric would be concerned with the evaluation of an individual experiment. The criteria would be different from the general rubric, and would focus on that particular experiment.

A task-specific rubric lays out the details of how a single task is to be approached. As well, the rubric provides a basis for scoring how well (or not so well) the particular task was done. Put another way, the task-specific rubric details what really counts about the particular task being done.

As another example, a rubric for an oral presentation could tell students that their presentation will be judged on the originality of the topic, the organization of the information, and the presentation itself; how informative and entertaining the delivery is; the use of voice; and the constructive use of props.

Developing a Scoring Rubric The very first step in developing a scoring rubric is figuring out what aspects of the project, report, lesson, or other item being evaluated are important to the evaluation. It is fruitless to focus on something that will not provide any feedback that can be used for future improvement.

The qualities that are identified as being important form the framework on which the rubric is made. For example, in assessing a report such qualities could be spelling, grammar, organization, presentation style and use of language. The details of the rubric would be compiled using these as the starting points.

There should be enough qualities to make for a meaningful assessment, but not too many qualities. If there are many qualities, it can become difficult to score any one of them. It is better to have fewer qualities with several scoring criteria in each one than a lot of qualities with only one criterion in each.

Ideally, there should be three criteria per criterion, since typically there will be indicators of poor, average, and standout performance for each quality. The criteria should not depend on each other. Each should be able to be evaluated on its own.

When developing the criteria, it is better to have a definite indication of how each criterion will be determined. For example, it is better to say 'Student's writing will be free of spelling errors,' than to say 'Student's writing will be good.' 'Free of spelling errors' is something that can be quantified. 'Good' is hard to quantify.

If the evaluation involves assigning a score (1,2,3, . . . or A,B,C, . . .) then the same score should mean the same thing for different categories. It would be confusing to have a score of 2 pertain to merely satisfactory in one category and outstanding in another category.

Finally, the rubric needs to be tested in action. Typically, the first run-through of a rubric will show that some revision is necessary. This is to be expected. The math involved in a rubric is not the more straightforward math of an equation. Rather, the mathematics of scoring is part of a more subjective evaluation. So, some tinkering may be needed to make the rubric as good an instrument of assessing performance as it can be. But, the effort will be worth it.

The math that is part of a rubric can help create a tool that assesses the performance of a task in a way that is clear to the teacher and, most importantly, to the student. The student will be able to use the information to improve. Different teachers will be able to use the same rubric effectively. The real life math of a rubric is thus an important part of a great classroom.

Where to Learn More

Books

Arter, J.A., and J. McTighe. *Scoring Rubrics in the Classroom: Using Performance Criteria for Assessing and Improving Student Performance.* New York: Corwin Press, 2000.

Moen, C.B. *25 Fun and Fabulous Literature Response Activities and Rubrics: Quick, Engaging Activities and Reproducible Rubrics the Help Kids Understand Literature.* New York: Scholastic Professional Books, 2002.

Web sites

Middleweb.com. "Just what is a rubric?" <http://www.middleweb.com/CSLB2rubric.html> (November 9, 2004).

Moskal, B.M. "Scoring Rubrics: What, When and How?" <http://pareonlne.net/getvn.asp?v=7&n;=3> (November 9, 2004).

Smith, J. "Base Arithmetic." <http://www.jegsworks.com/Lessons/reference/basearith.htm> (October 30, 2004).

Sampling

Overview

Sampling is the statistical process of analyzing a group of items selected from a bigger set. It is not always feasible to perform a study on all members of a population. For example, it is impossible to interview all AIDS patients throughout the world to study the stress and problems they endure in their daily life. A better approach is to interview a group of such patients and generalize the common results to all HIV positive people, the world over.

In sampling, a smaller manageable group of items, elements, members, or individuals representing the entire population is studied. Observations made from the analysis are generalized for the larger set to which the sample belongs. Sampling helps identify and understand the group's dynamics, ongoing trends, and their implications.

Sampling is a widely implemented concept used to perform demographic studies, environmental research, marketing analysis, and soil testing, to name a few applications. It will not be incorrect to say that sampling is utilized in all aspects of life ranging from medicine, social behavior, business, music, sports, and technology to ecology, and the balance in nature.

Fundamental Mathematical Concepts and Terms

PROBABILITY SAMPLING

Probability and non-probability sampling are the two commonly used forms of sampling implemented in various sciences. In probability sampling, every member (or object) of the sample group gets an equal opportunity (in other words, they are all given equal weight). Probability sampling begins by listing the traits and features to be studied. Identifying these traits helps in defining the populations to be researched. For example, to study the effects of smoking on women of reproductive age, female smokers in the age group of twelve to fifty years are most likely to show traits identifying the sample to be studied. If however, this study is to be conducted for women from a particular ethnicity or geographic region, then the subject's background and location will also form a part of the features defining the population to be analyzed.

Once the group to be studied is identified, all individuals belonging to that group have the same opportunity to participate in the research effort, thus reducing bias and error. However, at times, scientists randomly choose their subjects from the selected group. This constitutes unrestricted or simple random sampling.

Another type of random sampling is restricted or stratified sampling, in which the population is categorized into homogeneous segments with the idea that maximum possible variations can be accounted for, thereby minimizing the chances of arriving at biased results. Samples representing each unit are then identified and studied.

In contrast to random sampling is the more frequently used systematic sampling, wherein the first element is selected randomly and the remaining elements are identified on the basis of a calculated sampling interval.

For example, a student might want to interview store owners of all the malls in a particular location. If the identified area has several malls, then it would be extremely time-consuming to talk to all the owners of all the stores of all the malls. To make the task easy and still get a good representation of the owners, the student can determine the total number of malls and stores. Assume there are a total of ten malls and 250 stores. The student decides to interview 20% of the population, which means fifty vendors. The sampling interval can be easily computed by dividing population size (250 in this example) by the sample size (50 in this example). Accordingly, the sampling interval in the illustrated example is five.

Another form of probability sampling is cluster sampling, in which the investigator selects subjects in a "phased manner"; first identifying clusters to be studied, and then randomly or systematically identifying individuals to participate in the study. Using the example discussed earlier, a researcher may, for example, first identify cities from where to select women smokers to perform the study. The examiner then randomly or systematically selects individual participants from the identified locations.

In the real world, none of the sampling techniques mentioned above are employed in isolation. Instead, researchers use a suitable combination to perform studies. This strategy of utilizing one or more techniques to investigate an issue is known as multi-stage sampling. It is helpful in carrying out elaborate research involving huge populations spread over large geographical areas.

NON-PROBABILITY SAMPLING

In non-probability sampling, the researchers typically select subjects depending on their availability. The basic assumption of non-probability sampling is that any sample available would be sufficient to accurately represent the entire population, thus leading to correct results. With non-probability sampling, not all members of the group receive an equal opportunity to participate in the study.

Some forms of non-probability studies are conducted with individuals easily available. For example, visitors to malls or other public places might find a television crew interviewing passers by, thus offering an inexpensive way of understanding public opinion on a particular subject. This form of non-probability sampling, ideal for quick, economical, investigative, and narrative researches, is therefore referred to as convenience, accidental, or haphazard sampling.

The biggest disadvantage of convenience sampling is the high degree of bias because it is completely at the analyst's judgment to select members for the study. However, errors occurring due to bias can be minimized if the study is conducted on a uniform population that shows consistent features through its expanse. This way, all members experience and perceive almost the same things and represent an accurate picture. For example, soil samples from a smaller field are likely to yield similar results, but the probability of samples collected from a large field showing greater variation is high.

Another form of non-probability sampling is volunteer sampling that subjects a volunteer to participate in the study. A good example of volunteer sampling is the call-in surveys that various television and radio channels conduct to assess public sentiment. The downside of volunteer sampling is that only those interested enough in the issue participate in the study, thereby introducing strong biases. It does not take into account the opinions of those who might be concerned about the issue but avoid active participation.

Yet another method of non-probability sampling is judgment sampling, wherein the investigator uses his/her judgment to select the members for a study. The biggest disadvantage of this sampling method is that the selector's judgment might be heavily biased and inaccurate. In such a situation, results of the investigation carried out will be erroneous, no matter how elaborately the study is performed. These biases can however be reduced if judgment sampling is utilized in controlled environments, such as a life sciences laboratory where variables are few and limited by the issue to be studied.

Apart from these types of non-probability sampling, the most commonly used is quota sampling, in which members are selected from various sub-groups of a population until they satisfy a pre-calculated quota that is typically in proportion to the total population size. Several market researchers use the principle of quota sampling. For example, a mattress manufacturing organization might want to know the opinion of senior citizens who constitute 5% of the population. If the sample to be studied is of 1,000 people, then of those 1,000 candidates, 50 must be senior citizens. To meet this quota, the interviewer will then approach any 50 people fulfilling the age criteria.

Snowball sampling, another form of non-probability sampling, involves employing few subjects for a study. These members, in turn, enlist their acquaintances (people they know), who on their part sign up their friends and colleagues for the study. The basic idea here is that the individuals who have signed up initially would be the best to know about more of their kind. For example, a quilt supplies store aiming at updating their offerings in accordance with the new products available in the market, and the precise needs of their patrons, will be better off by interviewing quilters. Here, the best way to interview the maximum number of people would be to request regular customers to suggest names of fellow quilters they might know.

As is obvious, snowball sampling is useful in studies targeting small or inaccessible populations. The situation may further become difficult if the members of such populations are scattered everywhere.

A Brief History of Discovery and Development

Sampling is almost always a part of scientific testing. Rarely are all objects or situations examined, so by testing selective samples, scientists are able to make broad conclusions.

It was only through careful sampling and segregation of pea plants that Gregory Mendel (1822 A.D.–1884 A.D.), known today as one of the founders of modern genetics, discovered the laws of heredity. His methods since then have been followed to produce improved varieties of not only crops and plants but also award-winning breeds of dogs, horses, cattle, and other animals.

Similarly, Englist naturalist Charles Darwin (1809 A.D.–1882 A.D.) studied samples of animals and birds living on the Galapagos Islands. Diligent analysis of those limited population samples resulted in the theory of evolution, the unifying principle of biological science.

Starting from 1920, the *Literary Digest*, in circulation from 1890 to the late 1930s, correctly predicted the winner of the presidential campaigns for four elections in a row. Their method was simple. They collected the name of the voters' favorite candidate in six states. They however subsequently failed the fifth time because they selected samples from telephone directories and auto registration records, thus approaching only the wealthy strata of population. Beginning 1936, Gallup presidential polls used quota sampling to successfully predict presidential elections. Today, politicians and pollsters use sophisticated mathematical sampling to predict elections and to shape policy.

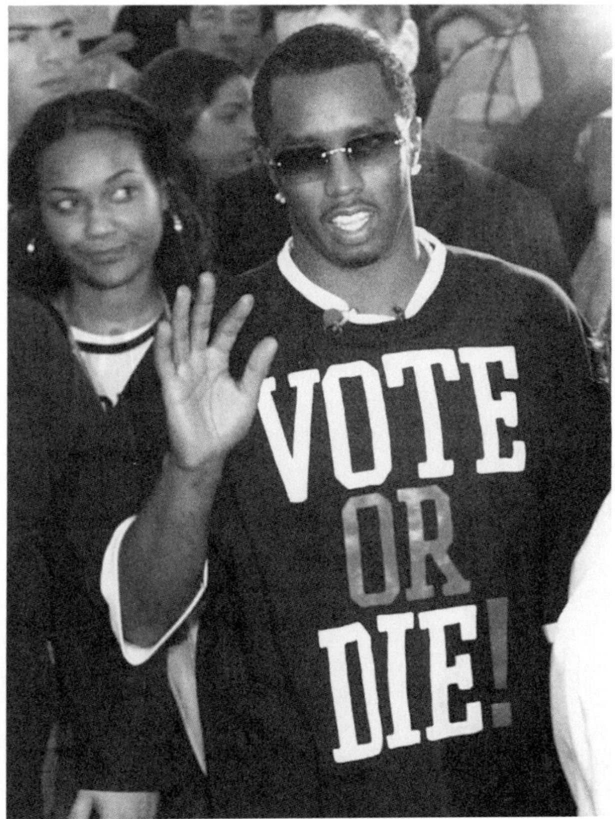

Sean "Diddy" Combs and others used sampling to determine where best to target efforts to encourage young people to vote during the 2004 presidential election. AP/WIDE WORLD PHOTOS. REPRODUCED BY PERMISSION.

Real-life Applications

AGRICULTURE

Sampling has been implemented to improve agricultural practices since ancient times. It is a well-known fact that not all kinds of plants grow in similar kinds of soils and climates. Meticulous sampling and testing over the centuries has now made it clear the type of soil a crop requires for thriving.

Today, samples are collected and put through extensive tests to assess a variety of factors affecting plant growth, soil quality, pH balance, nutrient levels, and concentration of micro-organisms. As mentioned earlier, most real-life applications use a combination of sampling methods. The same is the case here. However, the most prominent sampling method used in agriculture is the cluster sampling, wherein groups are identified and then they are either systematically or randomly used as samples.

PLANT ANALYSIS

Sampling analysis of plants is useful in identifying any nutrient deficiencies or excessive accumulation of a particular nutrient that proves harmful to plant growth. By collecting samples of plant tissue and studying them carefully, scientists can also assess the effects of insecticides, pesticides, new chemicals thought to have beneficial effects, and the proximity of an industrial area or waste dumping grounds on the general health of the plants of a specific region.

Plant analysis can give crucial insight into the kind of plants a given piece of land will support in the future. For example, a random plant analysis of a green area near a waste chemicals removal yard will clearly show that the existence of the greenery is threatened. From this, one can easily deduce that the land might become completely barren in a few years and may actually become a wasteland. Accordingly, the authorities can take measures to save the fertile land.

Similarly, random sampling of plants showing a sudden drop in development can give important clues about the cause of the weakening. The reasons can be numerous, ranging from introduction of a new insect or bug into the plant community and increased human interference to a sudden rise in the local temperature. All this investigation however depends heavily on a best possible collection of enough samples representing plants of the identified region.

Sampling enough plants gives a good indication of the general health of other similar plants in a field. The entire sampling process followed here is based on the mathematical concept of cluster sampling. Clusters are first identified. For example, plants near a waste chemicals removal yard form one cluster; plants showing a sudden drop in development become another cluster, and so on. Random samples from each of these clusters are then studied.

SOIL SAMPLING

A regular soil analysis helps farmers and people engaged in commercial agriculture in assessing the quality of the soil. Depending on the findings of the soil analysis, they can enrich their fields before sowing the next crop. Alternatively, if the analysis shows increased occurrence of a harmful element, they can take due measures to prevent its growth.

Plants continuously absorb minerals and other nutrients from the soil. An annual soil analysis before the next sowing can tremendously help farmers in selecting the most appropriate crop to be grown and the right fertilizer to be used, thereby bringing down the costs, reducing avoidable harmful effects of unsuitable chemicals, and increasing crop yield to the maximum. All this leads to enhanced profits in terms of resources as well as money.

Typically, a soil analysis requires systematic sampling requiring the collector to visualize the field to be divided into grids. Based on a sampling interval is determined. For example, samples are gathered at a rate of 15–20 soil samples from a 40 acre field. Collecting fewer samples runs the risk of producing inaccurate results. To investigate nutrient concentration, especially that of nitrogen, samples from varying depths are taken. In other words, the sampling interval is critical.

SCIENTIFIC RESEARCH

Modern research depends largely on intelligent ways of implementing sampling procedures; so much so that no exploration can be carried out without collecting and analyzing samples. Numerous ongoing studies are a case in point.

By carefully studying samples of available material, scientists have discovered microorganisms thriving in the most extremes of locations, such as hot sulphur springs that have very low levels of oxygen. These studies involve a combination of cluster sampling with systematic and stratified sampling. Interestingly, the evidence supporting this discovery is so strong that the theory about all life needing oxygen for its sustenance is now being questioned.

Clinical trials testing effects of new drugs on human patients are possible only because of large numbers of patients volunteering for the tests. A combination of volunteer and judgment sampling are used to conduct these tests. Volunteers are required for the tests, and selection of the volunteers depends on the judgment of the person doing the sampling.

Geologists also depend on regular random sampling to continue with their explorations of Earth.

Scientists from all disciplines of studies put in collective efforts to discover any evidence throwing light on living beings that walked the surface of Earth before humans. Fossil samples are studied to find missing links in the evolutionary process. This process is based on concepts of stratified sampling. Dinosaur fossil studies can prove to be the key in determining the cause of their sudden extinction.

DRUG MANUFACTURING

Sampling is an extensive and essential part of the process of designing and manufacturing drugs. Right from the beginning, research assistants accompany surgeons in operation theaters to collect samples of diseased tissues and conduct experiments to determine the cause of the ailment. As a part of these experiments, they compare the symptoms of a diseased animal with those of a healthy animal. To arrive at an authentic conclusion, they use judgment sampling to maintain batches of ailing and healthy tissue so findings can be generalized for the human population. However, this application requires both cluster sampling to identify the samples, and then analyzing the selected samples based on judgment.

Supporting these studies are the patient interviews or surveys regularly conducted by people involved in health care management and social sciences. Such investigations give the emotional, societal, and mental perspective to the effects of the disease. Depending on these studies, communities may shun or accept patients of a particular disease. For example, the stigma with AIDS is so strong that even the family members of an HIV+ person begin to avoid him or her, though everyone knows that AIDS does not spread through merely touching, hugging, or talking with the patient. This misconception however is gradually losing ground after countless interviews, medical reports, and proper promotion of facts about the basic respect that HIV positive people crave.

After understanding the causes and symptoms of a disease, scientists propose solutions and conduct experiments to study the practicality of the proposed remedies. This time round, carefully selected samples of diseased animals are treated with chemicals or therapy thought to be beneficial and the results are studied. Different treatments may affect different aspects of the disease. For example, a prospective drug to combat hair loss may influence hair growth on the whole body while another may have localized effects on certain body parts.

In addition, it is also important to study the side effects of these treatments. For instance, the so-called hair growth promoting medicine with localized effects may sound good, but might cause extreme nausea and dizziness, in which case it is not user-friendly and loses its marketability. On the other hand, researchers might accidentally stumble upon some positive side effects. A probable flu medicine may help reduce obesity, in the event of which drug manufacturers can either start a new investigation exploring the weight-reducing effects of the drug or market the flu tablet as it is while highlighting its desirable side effects too. Each of these can be thought of as clusters that undergo specific types of mathematical sampling.

Once the drug is ready for human trials, patients volunteering through non-probability sampling are invited to try it out, and studies are performed vigilantly to assess its impact. Volunteer sampling is the key here because members chosen through other sampling methods would literally force them to try the medication and thus prove to be unethical.

Quite often, new medication may not show immediate side effects, but its long-term use may cause unwanted results. Therefore, extensive data spanning several years of drug consumption are maintained and further examined. If possible, patients are grouped by different criteria, including age, sex, race, and region, to identify if a particular population responds uniquely to the treatment.

In a nutshell, independent of the stage of drug manufacturing, sampling is an important step in the introduction of new cures. Different forms of sampling may be employed at different times but it would be impossible to pioneer new drugs without the painstaking task of assembling samples of tissues and subjects that match the required factors—a concept based on the statistical principle of cluster sampling.

WEATHER FORECASTS

Weather predicting organizations receive loads of related data, which is then used for weather forecast. In spite of all the predictions carefully arrived at, the actual weather conditions are invariably somewhat different. Weather information analysts now compare samples of this difference along with current information to statistically arrive at the best weather forecast model for the next day, week, or month.

In the event of an approaching snowstorm or a thunderstorm, its estimated force can be compared with a past event of similar nature. Information from such a study is used to caution the public about an oncoming natural catastrophe, and estimate the extent of loss. This can be thought of as a type of convenience sampling, where information is presented based on availability (in this case, past data in a similar situation).

A different form of sampling referred to as matched sampling is used to calculate the risk ratio of accidents occurring on a bad weather day, particularly those related with some form of precipitation. In this type of sampling, a given period of unfavorable weather conditions is matched with the same duration of otherwise desirable weather. For instance, a heavy snowfall period starting at 9:00 a.m. on a Friday morning lasting two hours is compared with an identical two-hour period of another Friday morning with clear weather conditions. The control duration is ideally selected from a couple of weeks before or after the time period to be studied. Number and types of accidents occurring during the experimental duration are compared with those occurring during the control period to arrive at risk ratios. Comparing a specific weather related duration with a control is essential in matched sampling.

ENVIRONMENTAL STUDIES

Sampling finds widespread use in environmental studies, especially those related with measuring pollution. To study air and water pollution levels at different places, researchers collect samples and put them through numerous scientific procedures to draw conclusions.

Levels of pollutants in air and water can be studied from various perspectives. For example, assessment of harmful airborne microbes is particularly useful for food processing units, organizations handling any kind of living organisms, pharmaceutical companies, and hospitals. Larger air samples are required from places considered having relatively cleaner air, because they have fewer pollutants.

Studies indicate that collecting air samples and testing them is a far better and more accurate approach than traditional methods. Additionally, sampling is quicker and can be done in a shorter duration.

Similarly, studying water samples helps identify water contaminants following which corrective measures can be taken to save the water body. Trained personnel and investigators take samples of water from the source to be examined. Random water samples, collected in special containers, are carefully scrutinized to determine healthy and harmful elements contributing to the general health of the water resource.

It is interesting to note that results of water sampling may differ depending on the time when samples were collected. For results to be an accurate representation, due care must therefore be taken to collect all samples at the same time.

Environmentalists and ecologists regularly assess the delicate balance maintained within a system by performing various tests on the objects and living beings specific to that natural environment. For example, researchers evaluate soil quality of an exceptionally fertile area by studying samples of its soil, water, flora, and fauna. Similarly, productivity of fishing grounds can be judged by doing water analysis and a study of the fish inhabiting the place.

All of the above are processes based on sampling. Some involve random sampling, while some are applications of stratified sampling. The foundation of each of these is, however, identifying different groups of samples—similar to what is done in cluster sampling. Thus sampling in environmental studies can be thought of as a combination of cluster sampling with other kinds of probability sampling.

DEMOGRAPHIC SURVEYS

Demographic surveys involve counting the number of people who match the criteria to be studied. The most well known form of demographic survey is the census, wherein the government launches a mass scale activity, typically every ten years, of counting the number of people living in the country.

Though sampling is not an inherent part of census, it was used in 2000 to calculate the number of people belonging to minority groups, and the homeless since they lack an address. This involved a mix of judgment sampling, quota sampling, and convenience sampling.

Once the total population is calculated, different types of sampling, be it probability or non-probability sampling, are used for various purposes. For example, comparison of samples from the latest and past census results can be used to assess ongoing trends within the country. A study in the early 2000s showed that an increasing number of women drivers are getting involved in road accidents. Though this study may throw some light on the increasing stress levels among women drivers, but one must also note that over the years, more and more women are acquiring driving licenses. In fact, the number of women drivers is increasing faster than ever before. This, to a large extent, explains more women being involved in accidents.

Census results also give key information about future requirements and society make up. For instance, if a given region shows increased levels of education, it can be safely assumed that people from that region have a higher probability of performing well in the future. If on the other hand, census results from a flourishing area show an increase in the number of elderly people and a relatively lower number of young adults and children, then that area may suffer setbacks or show diminished development after a few years.

Demographic surveys can thus be used to deduce population dynamics. The objectives of such analyses range from identifying problems of minority groups, advancements in society, general progression of a group of people, and population health to predicting growth rate in the coming years and expected development responsiveness.

ASTRONOMY

Astronauts have brought back samples of soil and rocks from the moon. Because of the bulky spacesuits the astronauts wear, their mobility is quite restricted. To enable them to successfully collect samples, NASA has used a variety of tools such as improvised rakes, tongs, scoops, hammers, and electric drills on their trips to the moon.

The procedures that are to be followed to collect these samples depend on several factors. For example, while collecting samples from near a crater, astronauts follow what is known as radial sampling (a mathematical concept). The basis of radial sampling is that materials thrown out of greater depths are deposited closer to the rim of the crater, while substances coming out of shallower depths accumulate far away from the rim. The astronauts therefore collect samples from varying distances from a crater, thereby ensuring collection of matter from different depths from the surface.

Similar to the astronauts, space vehicles and rovers have been involved in collecting rock samples form the surface of Mars. While the space machines collect and send the samples back to Earth, scientists in laboratories study them to derive conclusions about Mars.

Mars samples include photographs assisting investigators in identifying objects of interest. High-quality electronic imaging equipments are required for this purpose. Later, after identifying the rocks to be examined further, sophisticated drills and other instruments are used to retrieve the appropriate samples. Somewhat varying from astronauts and rovers, but essentially achieving the same objective, are space shuttles that continue to send pictures of planets and other objects in the universe. A thorough examination of these pictures and their comparison with collected data unfolds new information literally everyday.

Aside from this, astronomers use standard sampling procedures to devise theories about the universe. Sampling plays a key role in the development of astronomy as a science because it is impossible to perform laboratory experiments the way it is possible for other science streams. Astronomers must use sampling to draw inferences about changes going on in the universe.

Sampling further becomes feasible in astronomy because any event in the universe takes millions of years. However, different objects in the universe can be observed in various stages of the event, thus making it possible for astronomers to predict the complete cycle.

ARCHEOLOGY

Archeologists use cluster sampling to identify and decide the sites to be dug for archeological findings. If an area of interest shows ancient artifacts on the surface, such as the arid and semi-arid regions, it is easy to explore and ascertain the particular spots to be excavated.

A problem, however, arises if objects of archeological value are embedded deep under Earth's surface and are covered with soil, grasslands, ice, human neighborhoods, and other objects on the ground. In such a situation, archeologists typically utilize the principle of probability sampling to pinpoint the specific areas to be dug up.

Systematic sampling is used to divide the region of interest in a grid-like fashion into excavating units that lie adjacent to each other without overlapping. Of these units, those identified to be explored further are dug with the help of machines or sometimes manually to detect any findings of an archeological nature. Measures are taken to employ non-destructive, though more time-consuming, digging methods.

MARKET ASSESSMENT

Marketing is the heart of any profit-making business and all businesses aim at making huge profits year after year. In spite of this, some companies perform exceptionally well while others fail to even capture consumer attention.

One of the key reasons for varying business performance is an understanding of the market. Organizations generating huge revenues typically have a sound understanding of their customers' needs. In other words, they have a grip over the market, possible only through market assessment.

There are different ways of studying market trends and all of them employ various sampling techniques. You might have come across company employees, typically students and other part-timers, working from a stall in a mall or a department store promoting a new product or collecting feedback from the visitors. People interested in the goods being endorsed or the parent company always make it a point to spend a few minutes at the stall either to know more about the product or give their input. This is an effective sampling strategy as it not only gives the customers the option of interacting with the organization, but the manufacturer also comes in direct contact with the target clientele. Gathering information otherwise can be an expensive, time-consuming activity that might drain an organization's resources.

The success of sampling depends largely on careful planning and interpretation of the collected data. Before embarking on such an activity, analysts should first identify objectives of the study, design a robust sampling process that includes defining characteristics of the population to be studied, the best suited method of selecting subjects, the ideal number of members constituting the sample, and the most effective way of conducting marketing research and assembling information. Additionally, researchers should devise a competent way of analyzing all the accumulated data. Without a serious approach to conducting market research, even the most extensive sampling strategy fails.

MARKETING

Several companies promote their new products by flooding the market with samples of the product.

Key Terms

Convenience sampling: Sampling done based on the easy of availability of the elements.

Simple random sampling: A sampling method that provides every element equal chance of being selected.

Stratified sampling: In this type of random sampling, elements are grouped together before sampling.

Systematic sampling: In this type of sampling, there are intervals between each selection for sampling.

This gives the consumers an excellent opportunity to try out the product in little quantities at nominal rates. If the users like the product, it is easier for them to switch to the regular packing. Sampling thus gives the producers a chance to gauge public response while users test the new product.

Even for products that have a strong market share, companies often retain their samples so retailers can offer them to prospective clients. Cosmetics manufacturers frequently use this marketing technique. Their outlets promptly offer sample sachets or trial packs to interested customers. Otherwise too, travelers go for smaller packing because of their convenient size.

For goods targeting a specific section of the market, producers send samples through mail. Organizations involved in producing baby products, for instance, send sample formula, diapers, diaper rash ointments, shampoo, and parenting magazines to new parents and parents-to-be. Similarly, crafters receive samples of new crafting products specific to their craft.

Typically, receivers of these samples constitute a special group of consumers for their needs form a niche market. They usually first join the manufacturers' mailing list or at least show an interest in trying out the products by filling out a form either on the Internet, or in a magazine, newspaper, or some other source. This way the organizations making specialized products get in touch with the clients who are actually interested in their products, without spending money in extensive advertising targeting everyone in general.

Offering samples is a win-win situation for both the consumers as well as the producers. While the clients can test the product either for free or at reduced prices, manufacturers reach out to consumers who are truly interested in their goods and are instrumental in giving them feedback.

Where to Learn More

Books

Thompson, Steven K. *Sampling.* Wiley-Interscience, 2002.

Thompson, William. *Sampling Rare or Elusive Species: Concepts, Designs, and Techniques for Estimating Population Parameters.* Island Press, 2004.

Web sites

NC State University. "Sampling" <http://www2.chass.ncsu.edu/garson/pa765/sampling.htm> (May 09, 2005).

Scale

Overview

Scale defines the relationship between the actual size of an object and its representation in the form of a prototype. It is used extensively in a variety of real world scenarios and is useful in modeling extremely large objects (or even tiny objects) into an easy to comprehend size. Scaling is done with respect to certain properties of the object, such as length, temperature, or mass. This concept is employed widely in architecture, astronomy, and imaging. For example, a scale model of a part of the solar system can provide a clearer understanding of the relative size and distances of planets and other objects in it.

Scale is also used in numerous other aspects of daily life including music, art, sports, fitness, business, technology, aviation, and a whole range of sciences, such as physics, chemistry, and engineering. The most common application of scales is found in maps.

Fundamental Mathematical Concepts and Terms

LINEAR SCALE

Scales can be associated with various properties of an object. Accordingly, there are several types of scale. The most basic form of scale is the linear scale. The linear scale follows a linear pattern and is used to quantify distance. The foot-ruler and the measuring tape are most well known examples of linear scale.

A key characteristic of the linear scale is that the length represented by two equidistant marks is always the same. Take, for example, a scale marked as 100, 300, 500, 700, and 900. As shown below, this would be a linear scale, as the length between any two equidistant marks, say 100 and 300, or 700 and 900, is always 200. (See Figure 1.)

Maps are the most prominent applications of linear scale.

LOGARITHMIC SCALE

One of the biggest limitations of linear scale is that it becomes difficult to manage if the quantity (or length) represented by the scale has a large range. This is where the

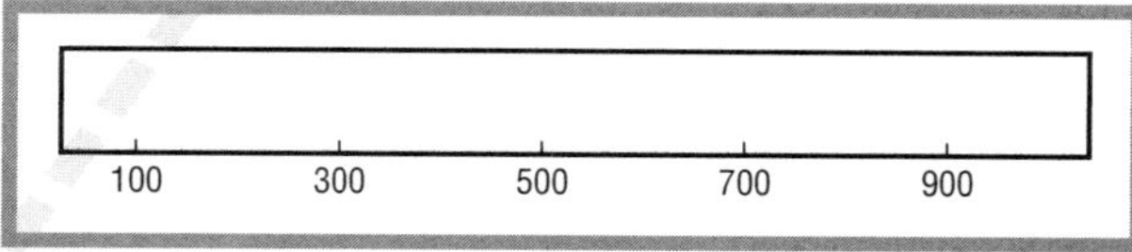

Figure 1: A linear scale.

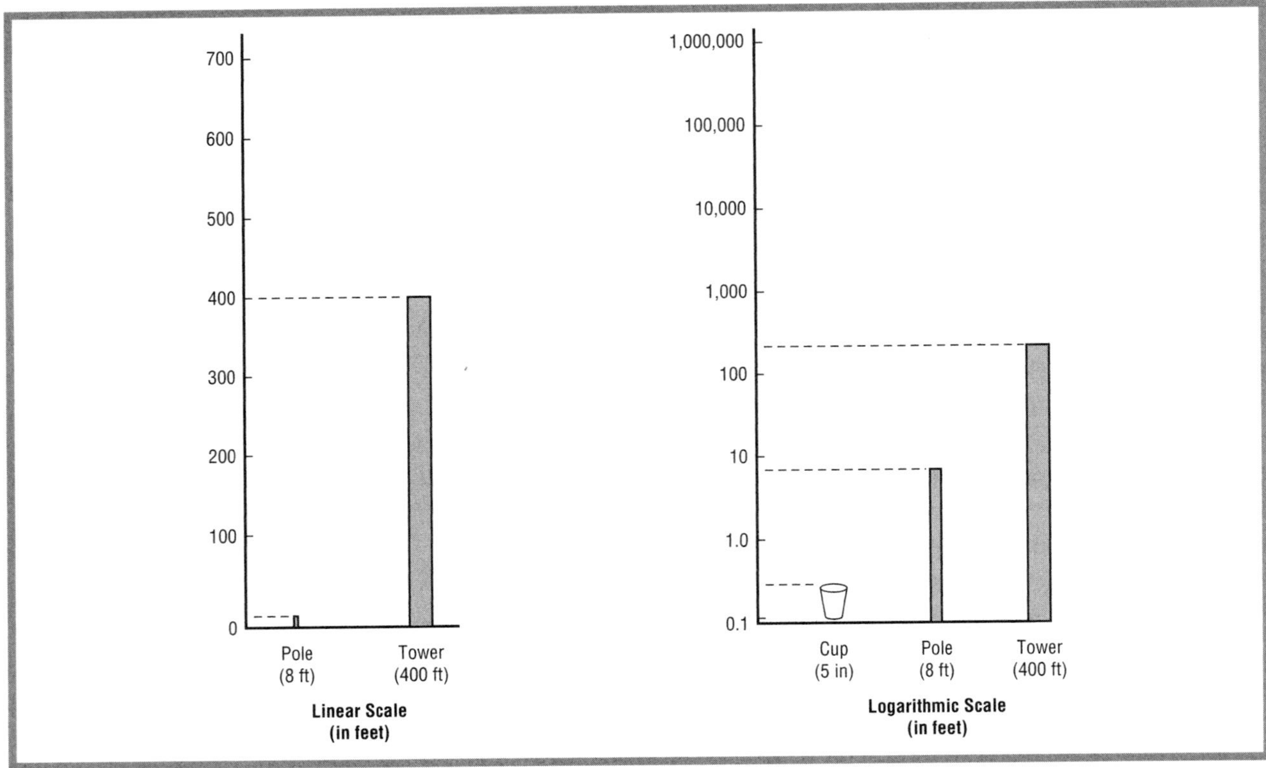

Figure 2.

logarithmic scale comes in. Simply put, the logarithmic scale represents the logarithm of the quantity rather than the quantity itself. In other words, the natural steps on a logarithmic scale increase in a multiplicative fashion rather than an additive or linear fashion. For example, a scale marked as 50, 500, 5000, 50000 would be a logarithmic scale as any succeeding mark is ten times the preceding mark ($50 \times 10 = 500$, $500 \times 10 = 5000$, and so on).

Differences between linear and logarithmic scales is shown in Figure 2.

On the linear scale in Figure 2 it is not possible to indicate the cup. The 8-foot pole is also not accurately shown. However, all three objects—the tower, the pole, and the cup—are clearly indicated on the logarithmic scale. Logarithmic scales thus become extremely useful in such scenarios.

INTERVAL SCALE

In an interval scale, the adjacent points are at equal intervals and also represent the magnitude of the underlying quantity. The most common example of interval scale is the thermometer (for measuring temperature in Celsius). The difference between 15°C (Celsius) and 20°C is the same as that between 30°C and 35°C—5°C. In other words, we can say that 20°C is warmer than 15°C by 5°C, and similarly for the other two points.

However, the interval scale does not have an absolute zero point. Consequently, we cannot say that 30°C is twice as warm as 15°C. The reason for this is that 0°C is not "absolute" zero (there is some heat at this point as well). The Celsius thermometer, thus, qualifies as an interval scale application.

In comparison, a scale indicating the percentile values of a few students may not qualify as an interval scale. The percentile value specifies the percent of total distribution that is equal to or less than that value. For example, in our case, saying that a student scores in the 75th percentile would indicate that 75% of the total students are either ranked equal to or below this student. Consequently, the difference, in terms of number of students between the 80th percentile and 75th percentile, may not be the same as between the 50th percentile and 45th percentile.

RATIO SCALE

The ratio scale includes all features of the interval scale. In addition, it also has an "absolute" zero or "true" zero point. Such scales allow for measurement of magnitude between equal intervals (as in the case of interval scale) and permit calculation of ratio between two points as well. As a result, we may say that a certain value on the ratio scale is twice as much as another value.

Take, for example, the Kelvin temperature scale. Due to the presence of an absolute zero—0 K (Kelvin), 100 K is certainly twice as warm as 50 K. Moreover, the difference between 15 K and 20 K is also same as the difference between 40 K and 45 K.

The Kelvin temperature scale can also be used to show how our earlier statement (30°C is not twice as warm as 15°C) is true.

For example 0°C = 273 K, yet 30°C = 273 K + 30 = 303 K and 15°C = 273 K + 15 = 288 K. Accordingly, 303 K is not twice of 288 K and 30°C is not twice as warm as 15°C.

NOMINAL SCALE

The nominal scale is perhaps the most primitive model of measurement. It is a classification tool more than anything else. For decades now, mathematicians have been questioning the authenticity of this scale. Nevertheless, we will still discuss this concept in brief.

The numbers presented on a nominal scale do not indicate values or quantity. They simply indicate a category or a type. For example, consider a herd of different animals including bears, elephants, tigers, and lions. All bears may be categorized as "1" on the nominal scale, all elephants as "2", tigers as "3", and lions as "4". The intervals do not signify anything in terms of magnitude or quantity. We cannot say that elephants are twice something as compared to bears. In other words, the difference in various points on the scale cannot be measured in amount, but only in kind.

Consequently, none of the mathematical operations, such as addition, multiplication, subtraction, division, and average can be applied to this scale.

ORDINAL SCALE

The ordinal scale improves on the nominal scale as it adds more value to the different categories represented. Categories can be ranked or logically ordered on the scale based on certain characteristics. In a nutshell, with an ordinal scale you can say whether a particular item possesses more or less of a characteristic as compared to other items.

To make more sense out of this scale, higher values are usually represented by higher numbers. An example of ordinal scale is as follows. You visit Orlando, Miami, and Tampa along with your family and at the end of your stay are asked to rate each of the hotels (on a scale of 1 to 5) with respect to certain characteristics. These could include quality of the room, quality of service and staff, facilities provided, proximity to Disneyland, and so on. You rate the hotel in Orlando as "4", the one in Miami as "2", and Tampa as "2". This becomes an ordinal scale where you can conclude that the hotel in Orlando is better than those in Miami and Tampa.

An important point to note is that the intervals on the ordinal scale may not be equal. Similarly, you cannot establish a ratio between two values. For instance, in the above example, it may not necessarily mean that the hotel in Orlando is twice as good as the other hotels.

A Brief History of Discovery and Development

Applications of scale can be seen in some of the very ancient paintings and maps. Researchers have found evidence that maps based on certain scales were being used more than 2,600 years ago. In 1963, an interesting painting—dating back to 6200 B.C. was discovered in the city of Ankara, Turkey. The painting depicted a miniature version of a city known as Catal Hyuk (part of modern Turkey) in great detail. The painting included streets and houses of this town. This is one of the first known examples of scale. There have also been discoveries of other maps in the regions of Egypt and Greece between the periods 2300 B.C. to 600 B.C.

Around the same period, scale became an integral component of architecture in Greece, Egypt, and Rome. Through the centuries, there has been evidence of scale being used to design and build monumental structures. Musical instruments that date back to the early 1500 A.D. were also designed using scale theory.

In the late A.D. 1700s, the French Academy of Sciences devised a more consistent and organized unit of measurement. This new unit, known as the "meter," was based on multiples of ten—a concept commonly used in modern scales. Eventually, scale also found use in other fields, including model railroading. By the beginning of the nineteenth century, people started using scale commonly for a number of purposes with respect to business, sports, and a host of sciences.

Real-life Applications

MAP SCALE

Maps, as stated earlier, are the most prominent examples of scale. Maps represent a much larger geographical area and can be of various types depending on the features they emphasize. The area represented by a

2 ----- 4 ----- 8 ----- 10 miles
------------- 10 miles

Figure 3: Examples of differences in linear and logarithmic scales.

map can be wide-ranging, from a small room to the entire universe. Most maps use both the metric as well as U.S. measurement units.

The relationship between a specific distance on the map and its actual distance is identified by scale (also known as map scale). The distance between two points on the map can thus be easily calculated. Map scale is generally indicated in three forms—verbal, representative fraction, and bar (or graphic). The verbal scale, which is the most basic form of representation, simply gives a written description of the map-to-actual distance relationship. For example, "One inch equals one mile." This would imply that a distance of one inch on the map is equivalent to one mile (63,360 inches) on the ground.

The representative fraction (RF) scale (sometimes referred to as ratio scale) is the most flexible of all the three scales. This scale indicates that the relationship between one unit on the map is equivalent to a specific number of units on the ground. The ratio scale is flexible because the unit of measurement can be assumed to be anything (centimeters, inches, etc.). This scale is usually expressed as a fraction (and thus the name). For example, 1:50,000 implies that 1 unit on the map is equal to 50,000 units on the ground. Again, this unit could be millimeter, centimeter, inch, and so on. Simply put, the map is 1/50,000 times the size of the actual area it represents.

The concept of RF scale is often used while designing scale models of automobiles, rail, and aircrafts. The size of most automobile scale models, for instance, is 1/32 or 1:32. In other words, one unit on the scale model is equivalent to thirty-two units of the actual automobile. RF Scales are also prominently used by geographers (people who experts of geography). Many people, however, also find RF scales confusing as they do not realize what unit of measurement to use.

The bar, or graphic scale, is the most widely used type of map scale. It is merely a single line marked with distances corresponding to the ground. Given below is an illustration of different kinds of bar scale.

The first illustration in Figure 3 indicates that the distance between two adjacent points is equal to two miles. Note that this is a type of interval scale. The best way to interpret such scales is by measuring the length of any interval with a foot-ruler, and then using this length as a reference for the entire map. For example, if the length between any two adjacent points is 0.5 in (inch) on the foot-rule, a distance of 0.5 in, anywhere on the map, would indicate 2 miles on the ground. Similarly, the second illustration denotes the length of the dotted line to be equivalent to 10 miles on the ground.

Based on their scale, maps can be categorized into two types—the large scale map, and the small scale map. The large scale map shows a smaller area but in greater detail, whereas a small scale map shows a larger area in less detail. A city map would be an example of a large scale map as compared to a world map (small scale).

ARCHITECTURE

We discussed earlier how scales were used in ancient architecture. Scales (and especially map scales) are extensively used by modern day architects and interior designers. Architects always draw plans (diagrams) before starting construction work on any structure, be it a building, a house, a football stadium, or even an entire city. Such diagrams are based on the concept of map scale.

These plans give a detailed view of the entire structure in terms of size and dimension. In other words, a plan would give an architect a much better sense of the final structure. For example, a plan for a house would specify the area (length, width, and height) of every room, including the living area, bedroom, and bathroom at every floor. It would also specify details of the garage, porch, and so on. Each of these is designed with respect to a specific scale corresponding to the actual house. The main purpose of this diagram is to ascertain whether all requirements are being satisfied within the given area. For example, are there enough bedrooms, is the garage large enough, and so on.

After conceptualizing the design on paper, 3-dimensional scale models are developed. These are miniature, yet detailed, prototypes of the actual structure. They are similar to the scale models for automobiles, discussed earlier. They are a certain proportion of the actual structure.

Interior designers also develop similar diagrams of a room before designing it to get a better idea of the space (area) available to them.

In a nutshell, scale models and diagrams allow architects to visualize a structure before it is built.

WEIGHING SCALE

We are all familiar with weighing scales and have used them frequently to measure our weight, or that of other things. As the name suggests, weighing scales work on the principle of scale. A weighing scale calculates the weight of an object and displays it on a scale. The units of the scale may be ounces (oz), pounds (lb), grams (g), or kilograms (kg). The weighing scale can be categorized as a ratio scale as it has an absolute zero and can go up to any weight depending on its type. For example, most home weighing scales indicate up to 300 lb.

Scale applications discussed till now mainly focused on the length (distance) property of an object. Weighing scales represent the mass (weight) of an object on an easy to comprehend scale. There are numerous types of weighing scales available apart from the common home scale, or bathroom scales as most people would call them. Such scales are widely used for medical, industrial (for example, weighing heavy equipment), and retail (for example, weighing food items or other groceries) purposes.

However, a majority of traditional weighing scales are now being replaced by digital scales. These simply display the weight in digital format (numbers) rather than show it against a scale. As a result, the use of scale as a weighing tool is decreasing.

THE CALENDAR

We look at the calendar every day. The calendar is merely a scale indicating the progression of time. It is one of the most universal examples of interval scale. Like an interval scale, the magnitude of intervals (years, months, days) between any two adjacent points on the calendar is the same. For example, the interval between January 1, 2001, and January 1, 2002, is the same as January 1, 2003, and January 1, 2004 (one year).

The years on a calendar (or even months, weeks, and days) can be meaningfully added or subtracted. The same is not true when you multiply or divide them. Moreover, as in the case of any interval scale, the calendar does not have an absolute zero point. The year 1 A.D. does not indicate the beginning of time. Time before this is specified as B.C.

Similarly, scale is also used in clocks.

ATMOSPHERIC PRESSURE USING BAROMETER

A number of modern-day instruments used for prediction and analysis of weather patterns include scales. One such instrument is the barometer. The barometer, like a thermometer, is an inverted glass tube dipped in mercury and sealed at the other end. It is used to measure atmospheric pressure, the weight due to the pressure (or force) of the atmosphere. As atmospheric pressure varies, the mercury in the barometer rises or dips accordingly.

The barometer consists of a scale (in inches) corresponding to the atmospheric pressure in millibars (unit for measuring atmospheric pressure). The height of the mercury on the scale would indicate the atmospheric pressure. For example, at sea level the atmospheric pressure is 1,013 millibars, and the corresponding mercury height would be 29.92 inches on the barometer scale. The atmospheric pressure at higher altitudes (height) is lower due to decreased air mass at and above the recorder.

In addition, the atmospheric pressure as indicated on the barometer scale can often be used to generally forecast weather conditions for the next twelve to sixteen hours. It is for this reason that weather experts use barometer readings in forecast reports.

An atmospheric pressure of around 1,015 millibars would indicate dry and calm weather. As the pressure increases, the temperature rises as well. In other words, higher the pressure, the sunnier are the conditions. Similarly, as the atmospheric pressure decreases, conditions usually become colder and wetter. A rapid fall in the atmospheric pressure would imply that a low pressure storm system might be approaching.

MEASURING WIND STRENGTH

Another type of instrument commonly used by weather experts is the Beaufort scale. The Beaufort scale, devised by Sir Francis Beaufort (a British admiral) in the early 1800s, is used to measure the speed of winds at sea. This instrument includes a scale of 0–12 (0–17 in some cases). Each number represents a certain strength of wind 10 m (meters) above the ground. The numbers also indicate the height of the waves in the sea, giving an idea of its state. Given below are a few observations and their implications on the weather.

A measurement of 2 (known as Force 2) would imply a light breeze blowing at a speed of 4–7 mph (miles per hour). The height of waves in the sea would be less than 0.1 meter (less than 4 inches), implying a relatively calm sea with small wavelets. A measurement of Force 6 on the scale would imply very strong breeze blowing at 25–31 mph. Such wind is capable of moving large tree branches and would make it difficult to control an umbrella if out in the open. The sea would understandably have large waves (up to almost 10 ft or 3 meters high), indicating rough conditions. Lastly, Force 12 on the scale would suggest the possibility of a hurricane. Winds would blow at enormous speeds of around 80 mph and waves in the sea can rise as high as 45 feet (approximately 14 m). Severe destruction can be caused in such cases. The devastating hurricanes that struck Florida in 2004 were measured at 12 on the Beaufort scale.

TECHNOLOGY AND IMAGING

Technology is continuously progressing. Software tools have become extremely complicated and can do

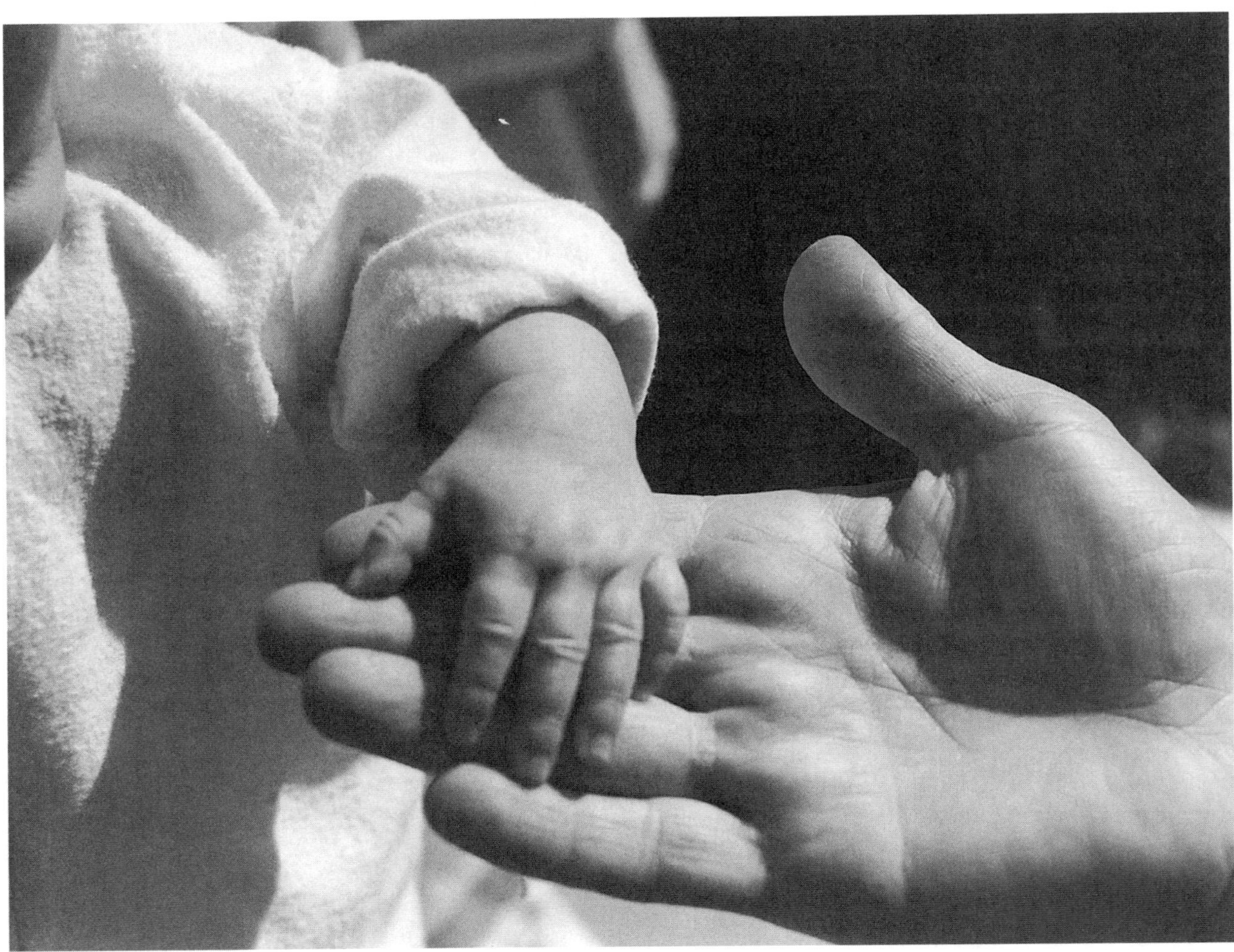

The difference in scale between an adult hand and that of an infant is clear. BETTMANN/CORBIS.

things we could not imagine a few years ago. Most software tools are based on mathematical concepts including scale. Take, for example, the car GPS (global positioning system) navigation tool. These systems help the driver in navigating from one location to another. All the driver has to do is enter the starting point as well as the destination, and the GPS would give detailed street-by-street directions on how to get to the destination. A map (much similar to a road map) is also shown on the GPS screen. The scales used by the GPS are similar to those used in printed maps.

Most architects now draw diagrams on the computer using specific software. The software tools make the architect's job easier and even faster. Scale diagrams can be effectively designed and printed. One such tool that is widely used by architects and interior designers around the world is AutoCAD.

Scale also forms an integral part of creating graphics and images, especially when using specific software tools. The building blocks of computer images are the pixels. A pixel is a specific number of blocks of color arranged in a grid. For example, a good quality 4 × 6 inch photograph would generally have 100 pixels per inch—400 × 600 pixels in total. The total number of pixels of any image is known as the resolution of that image.

The quality of an image is often measured by its resolution. In other words, the higher the resolution, the better the quality. What does scale have to do with the resolution of an image? To understand this relationship, take the above example again. If you have a 400 × 600 pixels computer image, you would be able to print a good quality 4 × 6 inch on paper. However, this does not mean that the size of the 400 × 600 pixels image on the computer is necessarily equal to the size of 4 × 6 inch image on paper. The size of the computer image (in inches) may be much smaller than the actual image printed out.

In fact, you can even print a much bigger image, say 8 × 10 inches with the same resolution (400 × 600 pixels). The quality of the printed image in this case, however would not be as good. In other words, the size of the printed image is not limited by the size of the computer image—the quality is. Using scale, you can define the relationship between pixels and inches to get a good quality image. In our example of the 4 × 6 inch image, the scale to get a good quality image can be defined as 100 pixels = 1 inch.

TOYS

Children love to play with toys. We always think of toys as a means of entertainment. However, toys also can be educational. Most toys, whether you buy them from the local Toys 'R' Us store or get them free with a kid's meal at McDonald's, are small scale representations of actual objects in the real world.

Take, for example, miniature dolls of characters from the Star Wars movies or cars from HotWheels.

THE RICHTER SCALE

The magnitude of earthquakes is measured using a numerical scale known as the Richter scale. Every earthquake, big or small, releases energy and produces shock waves of specific size within the Earth. The size of magnitude of these waves can be recorded on the Richter scale to get a better idea of how big or small the earthquake is. This scale is logarithmic. Thus, the increase of 1 unit of the scale would represent an increase in size of the shock waves (or magnitude) by 10 times. For example, an earthquake measuring 4 on the Richter scale is 10 times the magnitude of an earthquake measuring 3 on the scale.

The Richter scale starts with a unit of 1. It has no upper limit; however, it is important to note that theoretically it is not possible to have any earthquake bigger or equivalent to unit 10. An earthquake measured between 1 and 3 on the Richter scale would generally not be felt. On the other hand, an earthquake measured at 7 can be termed a massive one, capable of causing great damage up to a distance of 100 km (kilometers), or 62 mi (miles).

Tsunami, or seismic sea waves, are a series of strong ocean waves generated by the sudden displacement of large volumes of water. A very strong undersea earthquake caused a massive tsunami and killed hundreds of thousands of people in late 2004. The earthquake (ultimately measured at 9.0 magnitude on the Richter scale) created a series of waves that then radiated over thousands of square kilometers.

EXPANSE OF SCALE FROM THE SUB-ATOMIC TO THE UNIVERSE

Scale models can be used to represent parts of the solar system. Our solar system, and indeed the entire universe, are too big to comprehend. Without the use of small-scale models, it may not be possible to study them. Small-scale models are designed such that all parts of the model are in the same proportion in terms of size. In other words, the proportion of Earth's actual size to the actual size of the Sun is the same in the scale model.

To build a scale model, you must divide all sizes by a common factor. Scale models of different sizes can be built based on different common factors. However, the distances between planets and stars are so large that it may not be easy to find a common factor. For instance, the distance between Earth and the Sun is approximately ninety-three million miles, whereas the distance between Pluto and the Sun is three thousand seven hundred million miles. It is easier to make models based on distances (or sizes) measured in astronomical units (AU). 1 astronomical unit is equivalent to ninety three million miles (the same as the distance between Earth and the Sun). Consequently, creating a scale model based on the scale 1 AU = 30 inches would be much easier rather than dividing the distances by a common factor.

Over the years, scientists and researchers have used scale models to compare sizes of different objects ranging from the really large (the Sun, planets, and other stars) to those that have a more reasonable size (an elephant, or a human being) to the really minute objects that are not visible to the naked eye (an atom, an electron, or a body cell). The study of such "sub-atomic" objects is known as Nanotechnology ("nano" means really small). Simply put, the expanse of scale allows us to explore in detail, things from space and galaxies to nanotechnology.

MUSIC

Interestingly, music from the ancient period has been developed on various principles of mathematics. Scale is one of them. The most basic form of music is known as a "note." Each note corresponds to a certain frequency of sound. A series of such notes makes up music scales. The music scale is comprised of notes that are evenly spaced in terms of sound frequency. Thus, the music scale is based on the concept of interval scale.

Western music is made up of a number of major and minor scales. Each major and minor scale is comprised of seven notes. In other words, the frequency of various sounds is represented in the form of an interval scale to make a music scale.

THE METRIC SYSTEM OF MEASUREMENT

The metric system consists of units of measurement that are used to measure length, mass, or temperature. The most common units are meter (length), kilogram (mass) and Celsius (temperature). The meter, as we discussed earlier, was developed using the concept of scale. Moreover, length or mass can be measured with different units (within the metric system) related to each other by factors of 10—the basis of logarithmic scales. For example, 10 millimeters (mm) = 1 centimeter (cm), 100 centimeters = 1 meter (m), and 1,000 meters = 1 kilometer (km). This is true for units of mass as well.

The metric system is used in most countries. However, the United States and a handful of other countries still use the English system of measurement (foot, pound, Fahrenheit).

The use of the English system has always hurt the economy of the United States. It makes communication and trade with other countries difficult, and eventually affects the competitiveness of the United States. Due to this, most people within the Untied States have been encouraging the use of metric units. In fact, many organizations, including government agencies, have been using both systems of measurement.

SAMPLING

Sampling (taking small samples of a much larger quantity) is done extensively in the real world. A good example of sampling is blood tests. Doctors would take blood samples of a patient to determine his/her illness. To put it simple terms, to analyze a problem we do not study the whole object (blood, in our example) but we only take a small scale portion of it, commonly known as a sample.

Sampling is used in numerous other ways. Agriculture is another application. Farmers, around the world, grow different types of food at different places. What type of food (or crop) can be grown at a particular place depends on the nature of its soil. The soil mainly consists of four ingredients—water, air, minerals, and organic matter.

A specific crop would require the right amount of all of the above in order to grow well. Thus a farmer's job is not limited to only planting seeds and waiting for the crop to grow. The soil in different parts of his/her field must be tested (and recorded) to get a better idea of the contents of the soil. To do this, farmers take samples of soil from various parts, measure their content (using appropriate tools) and record them on a scale. The scale has an absolute zero (ratio scale). For instance, zero value of a mineral would indicate that there is no mineral in the soil sample. Similarly, if two samples (A & B) contain 2 gm (grams) and 1 gm of mineral respectively, we can say that sample A has twice the amount of minerals compared to sample B.

By recording data pertaining to the content of the soil, farmers can compare all the samples. They now know how much water and fertilizers to add and where. In other words, they can ensure that the soil contains the required amount of nutrients and water for the best possible growth of crops.

Where to Learn More

Books

Mills, Criss B. *Designing with Models: A Studio Guide to Making and Using Architectural Design Models.* Wiley, 2000.

Glazer, Evan M., and John W. McConnell. *Real Life Math: Everyday Use of Mathematical Concepts.* Greenwood Press, 2000.

Web sites

Discovery Magazine. "Size and Scale" <http://school.discovery.com/lessonplans/programs/sizeandscale> (March 14, 2005).

The IN-VSEE Project, Arizona State University. "Size & Scale: On the Relative Size of Things in Our Universe" <http://invsee.asu.edu/Modules/size&scale/unit1/unit1.htm> (March 14, 2005).

University of Washington Libraries. "How to use and understand Maps." Jan. 04, 2005. <http://www.lib.washington.edu/maps/use2.html> (Mar. 14, 2005).

Scientific Math

Overview

Numerous mathematical concepts are used for explaining complex real life situations using a scientific process. These concepts include trigonometry, calculus, rational exponents, statistical analysis, logarithms, and factoring. Mathematics, when used for scientific applications or processes, is referred to as scientific math. Such scientific mathematical concepts are widely used in real-life applications such as weather prediction, engineering, and astronomy. For example, scientific mathematical concepts can explain why bacteria multiply, thus providing a clear understanding of how to control them, and eventually benefiting from this process.

Scientific math is used in different aspects of daily life, including business-meteorology, aviation, biology, engineering, architecture, and basic sciences. The most common applications of scientific math are found in technology.

Fundamental Mathematical Concepts and Terms

The fundamental scientific math concepts cover an entire gamut of areas. Professionals such as engineers, scientists, accountants, and carpenters use scientific mathematics in different ways to manage and find solutions to improve their work. Thus, scientific math plays a vital role in various walks of life.

FUNCTIONS AND MEASUREMENTS

A function is a mathematical expression that specifies a relationship between two sets of numbers (perhaps representing physical characteristics of objects). In simple terms, it shows how a number belonging to one group can be related to a number belonging to some other group.

Take, for example, a tank of water. The total quantity of water in the tank can be expressed as a number. To estimate how much water the tank can hold, a function can be defined that presents a relationship between the shape of the tank and the volume of the tank (total quantity of water). A tank that is square would have a certain volume (quantity) of water, whereas a cylindrical tank would have a different volume. In a nutshell, the volume of the tank (and water) varies depending on the shape of the tank.

A variety of scientific processes can be explained using functions. For instance, meteorologists (people who study and predict weather) usually use relationships to predict the type of weather. There are certain factors that are vital in understanding weather patterns. Each of

U.S. Navy Blue Angels in formation. MUSEUM OF FLIGHT/CORBIS.

these factors is inter-related. Such relationships can be expressed easily in terms of functions. For example, there is a certain relationship between humidity and rain. If the weatherperson knows that the humidity is high, he or she may predict rain in the next few hours.

Functions can be simple or complex depending on the nature of the relationship. When a task becomes difficult and more factors are to be considered, a complex function may be used to explain the relationship. In the above example, if the tank is leaking, the function describing the relationship between the shape of the tank and the quantity of water would be far more complicated. This may involve the use of more elaborate mathematical concepts such as calculus and differential equations.

DISCRETE MATH

For every task, there can be two or more options. For example, a simple toss of a coin can result in a heads or tails presentation. The possibility of either a head or a tail showing up can be expressed by probability. Probability is the measurement of the likelihood of a particular event. For example, the probability of tails showing up, after a coin is tossed, is 50%. In other words, the probability of either the head or tail showing up is equal.

Workers in various disciplines, including engineers, mathematicians, marketers, government administrators, economists, biologists, or others who work with a vast collection of data, use probability and statistics. These concepts fall in an area of math commonly known as discrete mathematics. Like probability, most people also use statistics extensively in daily life activities. Statistics involves the analysis of recorded data. In other words, once a range of data is collected, it is organized and then studied to establish relationships and in turn, make more sense out of the data. For example, consider the population of a place. This data can then be organized by age or gender (male or female) and analyzed to understand various issues, such as determining the average age of the entire population, or the ratio of males to females, and so on.

Similarly, certain scientific predictions can be made from the above data using probability. For example, after determining the actual number of individuals who have received primary education from the total population of a city, the number of individuals opting for primary education can be calculated. This means that using the concepts of probability and viewing past data, one can determine future educational trends and whether the future population of this city will be educated.

It is important to note that probability is not an exact science. In other words, it allows for intelligent predictions, but can never determine an answer with absolute accuracy.

Biologists, scientists, and statisticians use the scientific concepts of discrete math (such as probability) to explain how to produce a better quality of product, such as a breakfast cereal. This process requires an understanding of how to make a better wheat plant by mixing and matching different types of wheat from across the country. Scientists use probability and statistical analysis to find matched types whose combination has the best attributes of wheat (say, a higher amount of vitamins). This combination can then be used to make breakfast cereal that has higher vitamin content.

In another example, an aeronautical engineer may study the impact of rain on the wings of a fighter plane by using another area of discrete mathematics called graphical analysis to create a model of rain drops. This eventually helps in designing a better (and more reliable) aircraft wing.

TRIGONOMETRY AND THE PYTHAGOREAN THEOREM

Mathematical principles of trigonometry, especially the Pythagorean Theorem, are used to find the height of a mountain, the distance of an airplane from the landing airstrip, the width of a river or valley, and much more without actually measuring these manually. The Pythagorean theorem was devised by the Greek mathematician Pythagoras (569 B.C.–475 B.C.). It presents a relationship between the short sides of a right triangle and its long side. The Pythagoras equation is given as $a^2 + b^2 = c^2$ where a and b are the lengths of the short side, and c is the length of the long side of the triangle. (See Figure 1.)

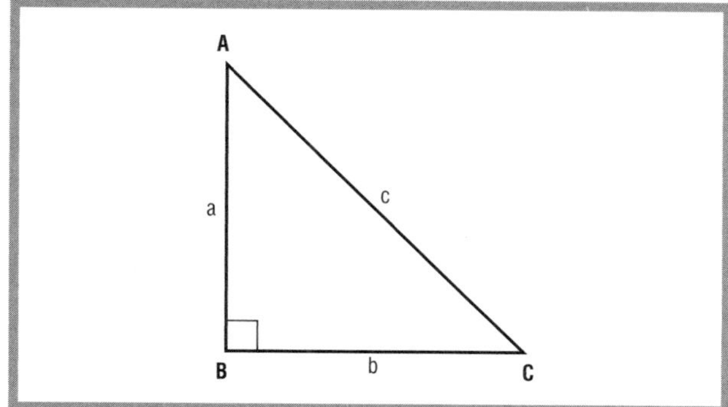

Figure 1.

Procedures that employ the Pythagorean theorem are greatly useful and far easier than manual measurement. For example, it would always be simpler to estimate the depth of a valley if it is extremely uneven and unsafe, rather than treading the terrain physically. In such cases, by taking simple measurements, the entire depth of the valley can be calculated using trigonometric concepts.

In addition, trigonometry is used for a variety of other applications. For instance, certain trigonometric concepts are used to survey distances to plan the design and development of a six-lane highway. A carpenter uses the same scientific concepts to design and build furniture. The weather department would use the Pythagorean theorem to calculate considerable data vital for predicting weather.

LOGARITHMS

Logarithms (and scales based on logarithms) are often used scientifically to simplify several processes. In mathematical terms, a logarithm is defined as the power to which a base must be raised to equal a given number. Scales that have units as log of a value instead of the value itself are known as Logarithmic scales. Simply put, the logarithmic scale represents the logarithm of the quantity rather than the quantity itself.

Logs can be expressed as exponents—$10^1 = 10$, $10^2 = 100$, $10^3 = 1{,}000$, or $10^6 = 1{,}000{,}000$ (a million). The superscripts are actually the logarithms of the final result. In other words, the log of 10 is 1, the log of 100 is 2, the log of 1,000 is 3, and so on.

A log scale generally has units such as 10, 100, 1,000, and so on, instead of 1, 2, 3, etc. The natural steps on a logarithmic scale increase in a multiplicative fashion rather than an additive or linear fashion. For example, in this case the natural steps increase by a multiple of ten ($10 \times 10 = 100$, $100 \times 10 = 1{,}000$, and so on).

Subsequently, the total length of any given road, the distance from New York to San Francisco, and even the distance from Earth to the planet Pluto can be easily compared using the log scale. In short, log scales facilitate comparison of wide ranging data. This is critical for most applications.

A geologist observing the vibrations of the earth records an extremely wide range of data (small vibrations to huge vibrations). There is a high variation in the observations, and hence it becomes much easier to present such data in terms of logarithms (and log scales).

MATRICES AND ARRAYS

A matrix is a square or a rectangular array of numbers. The numbers are represented as rows and columns. Matrices (plural of matrix) can be 2-dimensional, 3-dimensional (for example, cubical matrices), or higher-dimensional. Simply put, a 2-dimensional matrix can record different values for two characteristics of an object, whereas a 3-dimensional matrix can record different values for three characteristics of an object.

Matrices are used in several real world applications. A medical radiologist takes a patient's CT scan to show the presence of a tumor in the brain so that a surgeon can remove it. The scan of the brain uses the mathematical principle of matrix to create a 3-dimensional image. Put simply, sophisticated technology uses a matrix to convert a 2-dimensional image into a 3-dimensional one. This greatly enhances the quality of the image and makes it easier for the surgeon to pinpoint the exact location of the tumor.

Engineers use similar principles and concepts to explain the structural nature of a material. For example, using matrices, they can create a 3-dimensional view of the wings of a space shuttle. This allows them to study the characteristics (structure) of the wings in order to make them stronger and more reliable.

EQUATIONS AND GRAPHS

Scientists, engineers, and researchers use mathematical concepts of variables, expressions, or equations to show a relationship among different objects and entities. An equation is an expression made of two or more members related by an equality sign. Each member of the equation can either be a fixed number or a variable. The Pythagorean theorem described earlier is also an equation (as given below): $a^2 + b^2 = c^2$. Here, the members a, b, and c are variables representing the sides of a right angle triangle.

Equations can be used to represent different objects (and their characteristics) in the form of tables and graphs. In other words, an object can be mathematically represented in the form of an equation, or a graph, without using actual values and numbers.

The purpose of such visual representation is to understand an object (or abstract equation) better. Equations and graphs help in identifying a wide variety of patterns for the object. Moreover, complex calculations can be performed easily. In short, they help create a model to better understand a situation and to solve a problem. In addition, they also facilitate comparison between two or more objects.

A Brief History of Discovery and Development

Babylonians four thousand years ago knew of Pythagoras' theorem. As stated earlier, the Greek mathematician Pythagoras also developed this equation describes the relationship between all three sides of a right triangle. This equation has, ever since, been used extensively in architecture and carpentry. Subsequently, it was also employed in a number of other fields including the science of flight and measurements of height for various purposes.

Although ancient global civilizations knew of the importance of mathematics for various applications, it was in the eighteenth century that the Swiss mathematician Leonhard Euler (1707–1783) invented two new branches of mathematics, namely calculus of variations, and differential geometry. With these, people started recognizing the importance and value of mathematics for scientific processes.

Euler was also instrumental in pushing forward with research in number theory, which was eventually used in several scientific applications. Towards the end of the eighteenth century, Italian-born French mathematician Joseph-Louis Lagrange (1736–1813) worked on the theory of functions and equations.

Real-life Applications

WIND CHILL IN COLD WEATHER

Wind chill is a vital aspect of weather. Meteorologists provide details on wind chill along with the temperature in the winter. The reason for doing so is that a high wind chill makes a place feel colder, even though the temperature remains unchanged. In other words, a temperature of around 40°F (around 2°C) and a zero wind chill would indicate cold weather. However, the same temperature with a high wind chill would make the effects of the cold seem more efficient, although the temperature is still the same.

In simple terms, wind chill gives a measure of the discomfort caused due to a combination of the speed of wind and the air temperature at a particular time. The speed of wind does not increase or decrease temperature. Nevertheless, a high wind speed (in cold weather) always makes a person feel colder than it actually is. This is one reason why the windy city of Chicago (with normally high wind chills) has harsh winters. High wind chill can also be very dangerous, as it may result in acquiring frostbite and other problems related to extremely cold weather more quickly.

A mathematical relationship exists between the wind speed, and air temperature. It is commonly known as the wind chill index, and was devised by the National Weather Service. The equation Wind Chill Index (°F) = $35.74 + 0.6251T - 35.75(v0.16) + 0.4275T\ (v\ 0.16)$ relates the above-mentioned factors where v is the wind speed in miles per hour and T is the temperature in Fahrenheit. The unit of the wind chill index is °F. The wind chill index gives a "truer" indication of the temperature. For example, if the air temperature is 40°F and a wind is blowing at 10 miles per hour (around 16 kilometers per hour), the wind chill index is 34°F (calculated using the above formula). This implies that although the actual air temperature is 40°F, it feels like 34°F due to the wind.

Meteorologists record the necessary data and use the above equation to calculate the wind chill index. The equation makes it simpler to compute wind chill as the factor values are easily determined.

Wind chill indices are of great importance to the armed forces. Often, military personnel are posted in places with adverse weather conditions (for example, Siberia). Wind chill becomes critical in such cases. In addition, wind chill indices are also used to study the effect of wind and cold weather on other life forms.

WEATHER PREDICTION

Meteorologists predict weather patterns over the next few days or even few weeks. Weather prediction is based on a number of factors. These include the geographical location of a place, its weather in the last few days, existing temperature, humidity, formation of clouds, wind systems, pollution, historical weather trends, air pressure, direction of the wind, and many more factors.

Meteorologists use the mathematical concept of relationship to explain the cause and effect of each of these factors (in the form of a mathematical equation). Each of these relations is then studied to develop an overall prediction. Based on these findings, weather is usually expressed in six different ways—as the temperature (in °F or °C), the humidity of the air, the type and amount of cloudiness, the types and amount of precipitation, the atmospheric pressure, and the speed and direction of the wind. These weather attributes describe future weather patterns. For example, high amounts of precipitation and cloudiness would indicate rain in the next few hours.

The weather at a particular place can also be predicted by the scientific relationship between a number of other factors, such as the amount of solar radiation, the geographic location (latitude and longitude), position of the sun, and the seasonal variation in the altitude of the sun. There are other factors including the type of rainfall, the type of clouds, and ocean currents.

In a nutshell, a large collection of factors affect weather and their relationship can be expressed in the form of an equation. These equations are also used in a host of other applications. For example, astronomers and scientists use them to predict and study weather patterns of other planets. The National Aeronautics and Space Administration (NASA) uses these findings to model spacecrafts. Similarly, climatic conditions in places having extreme weather (such as many parts of Antarctica) can be understood on the basis of data collected by studying similar relationships. Explorers can then use these weather predictions to their advantage. Meteorologists also use similar equations to understand the impact of weather on crops.

Conversions

The units of measurement around the world are based on two systems, the Metric System, and the English System. Most the countries, apart from the United States, employ the Metric system of measurement. The Metric system of measurement includes units such as centimeter, meter, kilometer, gram, kilogram, °C, and so on. The English system of measurement includes units such as foot, mile, pint, gallon, quart, °F, and so on.

There is a definite relationship between a unit of measurement in the Metric system and its counterpart in the English system. This relationship can be expressed in the form of an equation. Such equations are often used in daily life activities for converting one unit to another.

For example, the relationship between one mile and one kilometer can be shown as: 1 mile = 1.61 kilometers. Similarly, other units can be expressed in terms of mathematical equations.

Conversions based on equations and relationships are possible for any unit of measurement, even within the same measurement system. Some of the real world examples based on equations discussed earlier are also types of conversions. In other words, any equation always results in the conversion of one entity (or unit) into another.

Conversion is also used extensively in scientific processes. For example, the amount of electricity produced from thermal energy can be expressed as an equation. The calorific value of fuel can be converted into heat, and then into power using simple equations. The relationship between water, its solid form (ice), and gas form can be shown by an equation. This would state at what temperature water gets converted into ice or gas.

ESTIMATING DATA USED FOR ASSESSING WEATHER

Meteorologists must gather information about the environment to predict weather. To collect and measure data critical for determining weather patterns (temperature, pressure, precipitation, solar energy, wind speeds, and so on) they use specific tools. All these tools are set on a balloon, which is raised into the atmosphere at a particular altitude (height). The balloon is attached to a cable. Using this process, meteorologists can record data at different altitudes; the balloon is raised to a certain height and brought down to note down the readings. It is raised again to a different height, and the same data is again recorded. This process is repeated for varying heights as it helps meteorologists predict weather better. Moreover, it is also repeated on a regular basis every day. Changes in weather patterns at certain altitudes can assist in forewarning people about difficult conditions.

The height at which data is gathered is crucial. This would be equal to the length of the cable. However, measuring the length manually would be an extremely difficult and time-consuming task (as the balloon is often raised to very high altitudes). This is where the Pythagorean theorem comes in. Using the Pythagorean theorem, the length of the cable can be easily calculated. In fact, it enables meteorologists to pre-measure lengths allowing them to simply place the balloon at different altitudes. Once the necessary data is gathered, other scientific mathematical concepts are used to forecast weather patterns.

BRIDGING CHASMS

Engineers and architects construct dams and highways in difficult terrains such as those across river gorges, and through mountain ranges. These terrains make it difficult to measure distances manually. With the help of certain instruments and mathematical concepts of trigonometry, especially the Pythagorean theorem, the distance or span of a bridge in such geographical locations can be calculated easily. (This is another application of math in scientific processes and hence Pythagoras' theorem could be considered as a scientific mathematical term.)

For example, if a bridge has to be built across a river, the engineer could use the principle of the Pythagorean theorem to calculate the width of the river (rather than measuring it manually). The theorem states that the sum of two smaller sides (squared) of a right triangle is equal to the square of the biggest side. One way of measuring the length of the river is as follows. A pole of known height is placed on one side of the river (perpendicular to the river). One side of a string is then attached to the top of the pole, whereas the other side is tied to the other side of the river (exactly at the end of the river). The string is then taken out and its length is measured. Using the length of the string and the length (height) of the pole, the length of the third side (which is the length of the river) can be calculated.

In hazardous conditions, an engineer can use this simple scientific concept of math to measure the river's width. Similar processes are also used in other areas. For example, the Pythagorean theorem is also used in space explorations. By simply studying the length of shadows, the depth of craters and height of mountains on the Moon (or a planet) can be estimated.

AVIATION AND FLIGHTS

A supersonic fighter jet is a complex machine to launch, especially from a small runway, such as the deck of an aircraft carrier. Controlling the fighter jet requires great skill and knowledge of how to fly the machine from a restricted runway. There are certain factors that control the take-off and landing of the fighter jet. The pilot applies scientific math concepts to launch an F/A-18 Hornet from an aircraft carrier. The amount of lift or force required to fly the F/A-18 Hornet can be expressed as a mathematical relationship dependent on factors such as the air density, the wind velocity, and the surface area of the wings. In order to allow the aircraft to take off, lift force must overcome gravity and equal the weight of the aircraft. This entire process can be shown as a simple math equation.

The benefit of applying this equation is that it allows the pilot to concentrate only on some key indicators of the equation to fly the plane from the deck of an aircraft carrier. The equation provides the pilot with critical data, including the ideal speed of the plane, to get the right lift for take-off or landing. The equation also shows the pilot how much time he or she has for safe landing of the plane on a shortened runway, in case there is low fuel and a large payload.

In another similar example, an athlete uses a similar relationship to assess the length and height of his or her long jump and the high jump or pole vault jump, respectively. Like fighter jets, the athlete also has a short run up but must jump as long (or high) as possible. Similar equations can thus be of great help.

Equations are also used to determine take-off and landing maneuvers for larger airplanes. For example, the pilot of a large Boeing or Airbus jet has to maneuver the plane, and approach the runway in a precise manner (for safe landing). Planning the approach ensures a smooth landing within the "touchdown zone" of the runway (this is an area on the runway that ensures that after touch down the airplane has can be smoothly and steadily brought to a halt). Pilots must sometimes execute visual approaches that vary in size, shape, and angle based on a variety of factors, such as other aircrafts on the runway, obstructions, noise abatement, and prevailing weather conditions. In other words, all these factors contribute to the safe landing of an airplane.

Pilots use mathematical concepts of relationship and equations (similar to those discussed in the case of fighter jets) in working out the approach strategy for landing the aircraft. Despite airplanes being equipped with modern technology instruments, a pilot must know and understand the relationship between various factors, to determine the distance and angle of descent, required to land the plane. As discussed earlier, such relationships can be expressed in the form of equations.

Using these landing equations, the pilot can figure the total distance required to land the plane from a particular height. This simple scientific computation enables the pilot to land safely, despite the distractions caused by differing conditions at various airports. In other words, the equation shows the effect on the landing caused by

Mathematics of Flight

Have you ever wondered how people measure the height at which an airplane (or a bird) is flying? Airplanes have advanced tools that constantly measure their altitude (height). However, these tools are based on simple mathematical principles, the same principles that can be used to measure the height of any flying object. The concept that is used here is again Pythagoras' theorem.

Airplanes continually record the distance they have traveled since take-off. With the help of instruments such as radar, it is also possible to pinpoint the exact location of the airplane, corresponding to the ground. Tools that measure altitude would then take this data and calculate the height using the Pythagorean theorem. Similarly, it is also possible to determine the exact location of an airplane if the height and total distance traveled are known, especially if for some reason the plane cannot be detected on the radar.

Mathematical principles are also used for other aspects of flight. Aeronautical engineers use concepts such as functions and equations, to find out how rain affects the wings of airplanes and eventually its flight. As raindrops are not all the same size, varying from tiny droplets to large blobs, each raindrop affects the wings differently. The concept of relationship is used here as well. The relation between the size of the raindrop and how it affects the wings (in terms of damage) can be expressed as a mathematical equation. For different sizes of raindrops, their corresponding impact on the wings is recorded (during an average rainy day, or even a storm). Using the equation, the impact for any size of the raindrop can be estimated (which would not be possible manually).

An entire range of data is then represented in the form of a graph. The purpose of doing so is similar to that discussed in the example on bacteria. After plotting all recorded values on the graph, a line can be drawn that represents the pattern in which raindrops affect airplane wings. The line can then be further extended to assess the impact of different sizes of raindrops, the ones that are not measured. Wings are central to the flight of an airplane (or any other flying object). Such equations and graphical representations allow engineers to assess the damage that can be caused during rain and thunderstorms, and in turn, build far more stable and reliable wings.

Furthermore, to study the effects of vibration on astronauts during a space shuttle launch, space engineers employ methods based on logarithms. Before launch, the vibrations felt by an astronaut inside the space shuttle are negligible (similar to those anyone would feel on the ground). However, as the space shuttle is about to be launched, the vibrations increase enormously. The magnitude of vibrations at different times during the space shuttle launch is expressed in terms of a logarithmic scale. This suggests that the vibrations increase in magnitude in multiples of ten. Another reason for using a log scale is that the magnitude of vibrations ranges drastically, from small tremors to large shuddering shakes. Consequently, this cannot be expressed on a linear scale (or by a linear equation).

a change in any one (or more) of the factors. For different airports, the magnitude (value) of the factors forming the equation may be different. For example, weather conditions at airports would vary. This would change the landing process (in terms of the distance required and the angle of descent). Thus, the equation would ensure that the pilot knows exactly what is the new distance, and angle of descent for that particular airport (for a smooth landing).

In addition, the pilot may not always have the opportunity to bring in the plane from a specific height every time. In such cases, the angle of descent would have to be modified. For these scenarios, pilots also create a graph (or chart) based on the landing equation that shows the relation between altitudes and the corresponding angle of approach. In simple terms, the angle of descent for different heights is known from this graph.

Such equations are also used in a variety of other applications. A baseball batter would use it to assess the force he requires (and the swing angle) to hit the ball for a home run. A trapeze artist uses it to define his or her swing while performing. Racecar drivers use similar equations to control the speed of their cars at sharp turns on a racing circuit. The mathematical concepts for all these remain the same.

SIMPLE CARPENTRY

Architects, designers, and carpenters need to understand dimensions of the structures they work on. Any

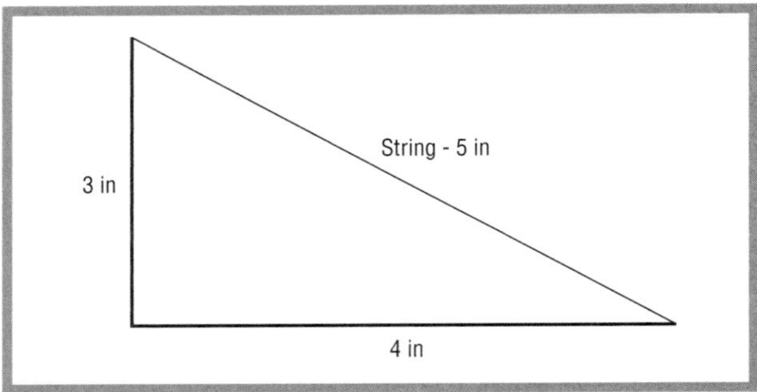

Figure 2.

construction is unique in its shape. To work out the size, shape, and dimensions of any new structure, be it a building or a simple wooden table, the Pythagorean theorem (and other trigonometric concepts) is used extensively. Although the use of mathematics in architecture is commonly known as architectural math, some of these concepts are also being used in a scientific manner, and hence can be termed as scientific math.

Continuing with the discussion on Pythagorean theorem, take for example the process of designing and creating an entertainment center for the living room. A carpenter uses the Pythagorean theorem to make sure the corners of the cabinets are at a perfect right angle. To do so, the carpenter would cut two pieces of wood (forming two sides of the cabinet). Consider that the length of one side is 3 inches, and the other is 4 inches. To join both these pieces such that they form perfect 90° angle, the carpenter would cut a string 5 inches long. As per the Pythagorean theorem, if the string fits perfectly between the free ends of both the pieces, the angle between them is an exact 90° angle. (See Figure 2.)

The same process would be repeated for the remaining two sides and a square cabinet is created.

The benefit of using the Pythagorean theorem is that the carpenter does not have to always manually measure the angle between sides, using complex tools. He or she can do this by simply measuring the corresponding lengths and sizes, a process that is far more convenient. Pythagoras' theorem can be similarly used on larger scales as well. An architect (or engineer) designing a highly technical structure, or even a model, uses this same mathematical concept.

The benefit of using trigonometry in carpentry or architecture is that the relationships between shapes and sizes hold true in most conditions. In other words, the relationship between two sides of a cabinet as defined by the Pythagorean theorem would be the same even for a much larger or more complex structure. Besides, in some cases, an architect may not need sophisticated tools. Knowledge of mathematical concepts and simple tools (such as a string in our case) can do the trick. Ancient structures around the world were built using similar methods, as they did not have advanced tools at the time.

MEDICAL IMAGING

In earlier days, in order to diagnose internal problems of a patient, doctors could only rely on x-rays that created 2-dimensional images. This would make complicated operations such as surgery rather difficult. Medical imaging, over the years, has progressed immensely. Newer technologies that produce 3-dimensional images have become extremely common. These technologies that greatly facilitate complex operations are based on mathematical concepts.

A radiologist undertakes imaging of the human body to find any growth, say in the brain, using computer tomography (CT) or nuclear magnetic resonance (NMR). An explanation of these technologies is not within the scope of this article. For our understanding, these are imaging methods that use mathematical principles of algebra and matrices to create 3-dimensional images.

CT measures the length of x-ray beams passing through a part of a body, from hundreds of different angles. Subsequently, based on the evidence of these measurements, computer software is able to reconstruct 3-dimensional pictures of the body's interior. In doing so, the software uses matrices to define dimensions of small portions from the body. In other words, the body (or a part of it) is considered as a number of smaller parts. The dimension of each part is defined using a 3-dimensional matrix. The 3-dimensional matrices for all parts are then joined together to get a complete image. Sonography is another technology that is based on similar concepts.

As stated earlier, the benefits of 3-dimensional imaging are numerous. Doctors can pinpoint the exact location of a problem area and perform surgery with greater effectiveness. Simply put, doctors can see the inside of any part of the body (as if they are the actual thing itself) and diagnose health-related problems far more efficiently. Three-dimensional imaging is also used in a range of other applications. This includes architecture, aviation, automobile engineering, computer games, and much more.

ROCKET LAUNCH

Rocket scientists are always looking for cheaper and more effective ways of launching a rocket. The space agency NASA launches its space shuttles from Florida, and with a reason. According to the Coriolis force, a scientific

Bacterial Division and Replication

Medical scientists, biologists, and health officials constantly study the impact of various diseases on humans and other living beings in order to seek better cures. To study these diseases, these scientists must first understand how bacteria multiplies and at what speed. Bacteria are small cells (living organisms) that cause many diseases. Once inside another body, bacteria multiply quite rapidly. Each single-celled bacterium divides to form two or more bacteria cells. Each of these then split into two or more bacteria cells. The process, known as binary cell fission, is efficient at causing tremendous bacterial growth.

The time that it takes one bacterium to split into two (or more) cells is known as generation time. Generation time varies greatly among different species of bacteria. For example, certain bacteria such as *Escherichia coli*, which causes severe diarrhea, takes only about twenty minutes, whereas *Mycobacterium tuberculosis*, the bacterium responsible for tuberculosis, would take as much as twenty-four hours.

Cell fission follows certain mathematical principles. Consequently, the entire process can be expressed through algebra and basic calculus. Scientists use these mathematical concepts to develop a model for understanding the behavior of bacterial division and replication. For example, they would be able to figure out up to what number different bacteria can grow within one hour, and eventually an entire day. Most bacteria grow proportionately. In other words, the rate of replication is proportional to the population of existing bacteria. This has been established by studying the growth of bacteria in different environments, at different intervals of time.

Based on this fact, relationship models (or equations) are developed between existing bacteria population and time. Such relationship models ensure a better understanding of bacterial growth, which is extremely vital to the progress of medical science. Furthermore, scientists also study the impact of the bacterial growth through a visual graph, known as the exponential graph. The population growth trend for bacteria is an exponential curve. Each generation doubles in number. For instance, the process starts with one bacterium that replicates itself to two bacteria, two grow into four, eight, and so on. The numbers (quantity of bacteria cells) can be plotted on a graph against time. A single straight line that connects most (if not all) of these points would represent the growth trend of the bacteria. The purpose of such models is to estimate the growth after a certain time, by way of extrapolation. In other words, the line on the graph can be drawn further to estimate the growth of bacteria after subsequent intervals of time. Thus, instead of actually measuring the growth every hour through experiments, scientists can simply predict it.

Similar models can also explain the growth of a particular disease among living beings. Most contagious diseases (diseases that are spread through contact) affect living beings exponentially, at least early on in an epidemic. For example, initially one person may be infected, and after a certain time two would be infected, and so on. Exponential graphs can be used here to predict the number of people affected after a certain period of time. This helps government officials control the problem, especially in cases of an epidemic.

Similarly, the utility department can use such relationships towards controlling water borne diseases. Similar principles are also employed in handling radioactive materials in medicine, studying the impact of their exposure in space, or in case of a nuclear accident.

principle (based on mathematics) developed by French Mathematician Gustave Coriolis (1792–1843), as Earth's axis of rotation at the poles is nearly vertical to the horizon, any still object in the sky would spin horizontally. In contrast, at the equator Earth's axis are nearly horizontal. Subsequently, still objects in the sky would move vertically. Thus, objects here would get a vertical boost.

In other words, if a space shuttle was launched nearer the equator it would get a vertical boost. Consequently, because of this boost, less fuel is utilized. In fact, NASA saves considerably on the cost of a launch dollars because of this strategy. In a nutshell, the science behind rocket launch is based on a principle that presents a mathematical relationship between Earth's axis of rotation at a particular place, and the corresponding direction and speed at which an independent object moves at that place.

The same principle can be used to explain why most airplanes that fly around the world fly near the poles rather than the equator. At the poles, an object spins horizontally. In other words, its horizontal speed would be higher. Subsequently, the speed of airplanes near the poles would be higher.

Key Terms

Equation: A mathematical statement including an equals sign.

Function: A mathematical relationship between two sets of real numbers. These sets of numbers are related to each other by a rule that assigns each value from one set to exactly one value in the other set. The standard notation for a function, $y = f(x)$, developed in the 18th century, is read "*y* equals *f* of *x*." Other representations of functions include graphs and tables. Functions are classified by the types of rules which govern their relationships.

Logarithm: The power to which a base number, usually 10, has to be raised to in order to produce a specific number.

Matrix: A rectangular array of variables or numbers, often shown with square brackets enclosing the array. Here "rectangular" means composed of columns of equal length, not two-dimensional. A matrix equation can represent a system of linear equations.

SHIPS

The design of ships and submarines involves extensive use of a scientific principle known as the Archimedes principle. According to this principle, floating objects (or even objects that are fully or partially submerged in a fluid) displace a certain amount of fluid. Due to this, the object feels a certain up-thrust. The magnitude of the up-thrust is equal to the amount of fluid displaced. Note that this principle is a mathematical equation based on the relationship between amount of fluid displaced, and the up-thrust experienced by the object. Ship architects and engineers use this principle to assess how a ship would lie in the water before it is launched. In other words, the Archimedes principle helps in designing the ship such that its movement and position is ideal once it is in water. The same principle is also used for submarines.

GENETICS AND MATHEMATICS

Doctors and scientists perform studies to predict what characteristics a child inherits from his or her father and mother. Such scientific studies are based on the concept of probability. Consider that the father and mother have brown eyes. However, it is possible that the child does not have brown eyes. The reason for this is that every characteristic of the human body (in this case color of the eyes) has two genes (cells that possess the characteristic). One gene is dominant, whereas the other is recessive. In this case, both father and mother would have one gene that is responsible for the brown color. Simultaneously, they would also have another gene (responsible for the eye color) that would have some other color. A child is likely to inherit the dominant genes, which may be the non-brown ones, and hence not have brown eyes at all.

Using the principles of probability, scientists can figure out the characteristics a child is most likely to have. The scientist must also have complete information on the genes of the parents. Thus, characteristics such as the color of the skin, color of the eyes, facial features, build and physique, and much more can be predicted. It is important to note that scientists and doctors can only state characteristics that a child is most likely to inherit.

Scientists also use these principles to create genetically modified plants and animals. For example, scientists can genetically modify a cow so that she gives birth to a calf that ends up giving more milk. Similarly, different breeds of animals are also genetically modified so that they give birth to offspring of mixed breeds.

EARTHQUAKES AND LOGARITHMS

A common example of a logarithm scale (scale based on log values) is the Richter scale for measuring magnitude of earthquakes. During an earthquake, an enormous amount of energy (in the form of heat) is released from the surface. The magnitude of the earthquake depends on the amount of energy released. The magnitude is shown on the Richter scale.

The relationship between each step on the scale (magnitude of the earthquake) and the corresponding amount of energy released by the earthquake can be explained by an algebraic equation. The Richter scale is a log scale; the difference in magnitude (in terms of the energy released) between two consecutive steps on the scale is ten-fold. For example, the amount of energy released by a 6.0-magnitude earthquake on the Richter scale has ten times more than the energy released by a 5.0-magitude earthquake on the same scale.

Potential Applications

GENETICS

The use of probability in genetics is fast increasing. Genetic technology is being used in agriculture and forestry to improve plants, increase disease resistance and the yield of crops, and adapt non-native crops to a new environment for specific benefits. This has improved the quality and quantity of food production in many parts of the world.

Animal breeders have been using genetic principles for many years to develop characteristics within a species that they feel are desirable. There are areas where genetics can be used; all of these are based on the principles of probability. However, the maximum benefit would be in the field of medicine. Genetic modification of certain bacteria has helped find better cures for many diseases, and holds true potential.

Where to Learn More

Books

Hazewinkel, Michael. *Encyclopedia of Mathematics.* Dordrecht, Netherlands: Kluwer Academic Publishers, 2000.

Owen, George E. *Fundamentals of Scientific Mathematics.* Mineola, NY: Dover Publications, 2003.

Web sites

"Physical Effects of the Earth's Rotation (Coriolis Effects)" <http://cseligman.com/text/planets/coriolis.htm> (April 9, 2005).

"Scientific Mathematics/Mathematics as a Science" <http://huizen.dto.tudelft.nl/deBruijn/grondig/science.htm> (April 9, 2005).

Scientific Notation

Overview

When dealing with very small or very large numbers, such as within the scientific fields of biology, chemistry, engineering, mathematics, and physics, professional men and women use scientific notation as an efficient way to read and write such numbers. The method of scientific notation uses the significant digits of a number and multiplies it by specific integral powers of ten. This method of notation is an easy way to read and write numbers so that the resulting representation makes more sense. It is also quicker to perform various mathematical operations such as addition and multiplication when large and small numbers are changed to scientific notation.

Scientific notation is used when working with the numerous sizes and time frames often found in the sciences, such as the very large distances between stars and the very small diameters of atoms and molecules.

Fundamental Mathematical Concepts and Terms

When using scientific notation, the expression of a number n is represented as $n = a \times 10^p$, where the variable a (generally representing a number between 1 and 10) is multiplied by an integer power (p) of 10. A power of 10^p is 10 multiplied by itself a specified number of times (p). For example, the number 1 would be written as 1×10^0 where $10^0 = 1$; 10 would be written as 1×10^2 where $10 \times 10 = 100$; and 1,000 would be written as 1×10^3 where $10 \times 10 \times 10 = 1{,}000$. As a specific example, when $n = 71{,}000$, the number n can be written in scientific notation as 7.1×10^4, where $a = 7.1$ and $p = 4$.

Writing numbers in scientific notation allows scientists to reduce, and often times eliminate, many zeros while indicating that zeros are still significant. The number 71,000, as shown above, is the same as 7.1 multiplied by 10,000 (10^4) and is written 7.1×10^4 in scientific notation. In scientific notation, numbers that are smaller than one will contain negative exponents. The number 0.00523, for example, is the same as 5.23 times 0.001 (10^{-3}) and is written 5.23×10^{-3}. (The term 10^{-3} means that 1 is divided by $(10 \times 10 \times 10)$.)

To convert a large or small number to scientific notation, move the decimal point in the number to the right of the first nonzero digit. For example, within the number 45,630,000.00, move the decimal point seven places to the left so that it is positioned to the right of 4, the first nonzero digit. (Remember that a decimal point is implied at the end of every whole number, even when it is not written; thus, $33 = 33.0$.) Then, indicate the movement by multiplying by a power of 10 that shows the exact number of positions moved. In this case, since the decimal point was moved seven positions to the left, 4.653 would be multiplied by 10^7, or 4.653×10^7. If the decimal point is moved to the left (as in the above example), the exponent p in $a \times 10^p$ is positive (+). If the decimal point is moved rightward, then the exponent p is negative (−). In this second case, the number 0.0000376 would be written as 3.76×10^{-5}.

The process is reversed when changing a number from scientific notation to regular notation. That is, move the decimal point the same number of places as the value of the exponent and then move the decimal point to the right if the exponent is positive or to the left if it is negative. Finally, add zeros if necessary.

Scientific notation is also important because it provides a clear indication of the number of significant digits within a number or calculation. For example, if a truck weighs 4,007 pounds (accurate to the nearest pound) then both 4,007 and 4.007×10^3 give an accurate measurement. However, if a truck's weight is 4,000 pounds then it is not (necessarily) apparent that this weight is accurate to the nearest single pound because it might be rounded to the nearest ten pounds. Scientific notation, on the other hand, shows that all four digits are significant. That is, when the truck weight is shown as 4.000×10^3 pounds, the extra (significant) zeros to the right of the decimal point show that the precision of the measurement is down to the single pound.

A Brief History of Discovery and Development

The publishers of the Oxford English Dictionary are interested when a new word or term shows up in print for the first time. The first recorded use of the term scientific notation appeared in the third edition of the *New International Dictionary of the English Language*, which was published in 1961. Because scientific notation did not appear in the second edition of that dictionary, which was published in 1934, language experts widely assume that the term was probably invented sometime during the decades of the 1940s or 1950s, and it is also assumed that the term gained widespread usage during the 1960s.

In 1963, the term was used inside the article "Digital Computer Technology and Design" that was part of the Oxford English Dictionary. Within this article, scientific notation referred to any number of the form: a first number times a second number raised to a third number. Since scientific notation came from a computer science reference book, the term is likely to have been regularly used by the pioneering computer users who were already buying electronic calculators and experimenting with simple computers. It is assumed that computer enthusiasts wanted a specific way to describe how a number is stored in a computer because at that time there was a big difference in how integers and fractional numbers were stored. (By the way, it is assumed that mathematicians, physicists, and engineers did not invent the term because they were already using the term exponential notation as an alternate form for scientific notation.) In 1973, scientific notation was defined in an introductory textbook on computer science and two years later appeared in the *Physics Bulletin* as a feature contained on calculators. By this time, the term scientific notation had spread from the computer science community out into the community of physicists and other physical scientists.

The modern meaning of the term scientific notation has changed from its original meaning. In the 1960s the meaning of scientific notation referred to any number of the form "first number times second number raised to third number." In modern usage of scientific notation, the second number is always 10, while the more general term exponential notation is used when this second number is any numerical value.

Real-life Applications

MATHEMATICAL OPERATIONS

One of the most obvious reasons to use scientific notation is when adding, subtracting, multiplying, and dividing very large and very small numbers. Adding two or more numbers with scientific notation involves converting all of the numbers to the same power of 10 and then adding the digit terms of the numbers. (When measurements are added or subtracted, the accuracy of the answer is no greater than the least accurate measurement.) For example, adding 5.045×10^{-6} and 2.65×10^{-4} involves: $5.045 \times 10^{-6} + 265 \times 10^{-6} = 270.045 \times 10^{-6} = 2.70 \times 10^{-4}$. When subtracting two or more numbers with scientific notation, convert all of the numbers to the same power of 10 (as with addition) and then subtract the digit terms of the numbers. For example, $7.99 \times 10^5 - 4.534 \times 10^3 = 7.99 \times 10^5 - 0.04534 \times 10^5 = 7.94466 \times 10^5 = 7.94 \times 10^5$.

Multiplying two numbers with scientific notation involves using the rules of exponents: $10^m \times 10^n = 10^{m+n}$. (When measurements are multiplied or divided, the answer contains no more significant figures than the least accurate measurement.) As an example, multiplying

46,850 and 0.0000417 with the use of scientific notation results in $(4.685 \times 10^3)(4.17 \times 10^{-5}) = (4.685)(4.17) \times 10^3 \times 10^{-5} = 19.53645 \times 10^{3} + (^{-5}) = 19.54 \times 10^{-2} = 1.95 \times 10^{-1}$. Dividing (/) one number such as 650,000 by another number 25,000,000 results in: $6.5 \times 10^5 / 2.5 \times 10^7 = 6.5 / 2.5 \times 10^5 \times 10^{-7} = 2.6 \times 10^{-2}$. Generally, from between one-hundredth (1/100) to 100, scientific notation is not usually needed, but on either side of this range, it is useful to apply scientific notation.

CHEMISTRY

In chemistry—which involves the study of the interactions between matter and energy and the composition of matter itself—the calculation of the parts of matter such as the dimensions of atoms uses very small numbers and thus needs the application of scientific notation. For example, the weight of a single atom of hydrogen is better expressed as 1.7×10^{-24} grams, rather than 0.0000000000000000000000017 grams.

One of the fundamental laws of chemistry is Avogadro's Law, which expresses the fact that under a closed, theoretical environment of pressure and temperature, identical volumes of gases contain an equal amount of molecules. The law helped in the early development of chemistry, but the number itself (Avogadro's number) was not calculated until the last half of the nineteenth century due to limitations in technology. Avogadro's number is usually stated as the number of molecules existing in one mole, or gram molecular weight, of a substance; and is formally defined as the number of atoms in 12 grams of the element carbon-12. The number of molecules in one mole has been scientifically determined to be approximately 6.0221367×10^{23} molecules, a number that is obviously easier to read under scientific notation than with many zeros.

ELECTRICAL CIRCUITS

Physics is the study of all the physical events that take place in the universe. Physicists use mathematical equations to describe and predict physical events, and because there is so much variety in the universe, physicists encounter very large and very small numbers within their observations, theoretical calculations, and experimentation. Because of that fact, all the many divisions within physics, such as astronomy electromagnetism, mechanics, optics, quantum mechanics, and thermodynamics, use scientific notation.

In electromagnetism, for example, a scientist might consider the number of electrons passing through a point in an electrical circuit of one ampere every second. Scientific notation would be helpful during such calculations because one ampere contains about 6,250,000,000,000,000,000 electrons per second; that is, approximately 6.25×10^{18} electrons per second. In this case, the advantages of scientific notation are obvious: the number is not as cumbersome when written on paper, and the significant digits are easy to identify. Both advantages are important in nearly all physics calculations.

LIGHT YEARS, THE SPEED OF LIGHT, AND ASTRONOMY

Astronomy is the study of the universe, including all materials such as celestial bodies, dust, and gases within it. Work within astronomy includes theories and observations about the solar system, the stars, the galaxies, and the general structure of space itself. Much of the research performed in astronomy involves very large and very small numbers so that scientific notation is regularly used as an important way to handle such numbers. For example, one of the largest objects visible to the naked eye is the Andromeda galaxy, with a diameter of over 1.0×10^5 (100,000) light years. A light year is the distance that light travels in one year, or about 5.9×10^{12} miles (about 9.5 trillion kilometers). The Andromeda galaxy appears large even though it is over 2.0×10^6 (two million) light years away from the Earth.

Astronomers use scientific notation when measuring the distance between stars because these distances can be as small as a few light years to hundreds of light years or more—but in any case are very large numbers. The speed of light is approximately 3×10^8 meters per second (or about 1.86×10^5 miles per second). Even the simple calculation concerning the time it takes light to travel from the Sun to the Earth involves two very large numbers. This calculation can be simplified with the use of scientific notation by knowing the distance between the Earth and Sun is about 1.5×10^{11} meters (m) and the speed of light to be 3×10^8 meters per second (m/s). Dividing the first number by the second results in $(1.5 \text{ m}) / (3 \text{ m/s}) \times 10^{11} \times 10^{-8} = 0.5 \times 10^3 = 500$ seconds, or about 8.3 minutes.

Astronomers also use scientific notation when examining electromagnetic radiation from all of the wavelengths emitted by celestial bodies. The most obvious radiation with respect to humans is visible light, which is the only radiation that humans can see with their eyes. However, visible light is only a small part of the electromagnetic spectrum that includes radio waves, microwaves, infrared light, visible light, ultraviolet light, x-rays, and gamma rays. Radio waves, which include waves used to provide sound in AM and FM radio, are the longest type of radiation, with a wavelength of about 1×10^8 meters. The shortest type of radiation is gamma waves, with a wavelength of about 1×10^{-16} meters. Gamma waves are emitted by such objects as radioactive isotopes and in some nuclear reactions, both created by mankind and occurring within the center of stars.

Professional astronomers use powerful telescopes, computers, and instruments while performing their jobs. Most work includes, first, observing astronomical bodies by using telescopes and instruments to collect relevant information. Astronomers, secondly, analyze the resulting images and data. Computational astronomy is one way that astronomers use computers and scientific notation to simulate and analyze astronomical events. Examples of events that are simulated by computers include the massive explosions of stars as they end their lives to make way for supernovas and the creation of the earliest galaxies within the universe. Thirdly, they compare their results with existing theories to determine whether their observations coincide with what theories predict, or whether the theories can be improved or, in some cases, replaced. Some astronomers work only on observation and analysis, and others work primarily on developing new theories, but in all cases the men and women within the field of astronomy use scientific notation in order to do their very difficult and complicated work.

COSMOLOGY

Cosmology, a branch of astronomy, is the study of the origins of the universe. It includes the Big Bang theory, which is the currently accepted explanation of the beginning of the universe. The theory proposes that the universe was once extremely dense and hot. Then a cosmic explosion called the big bang happened about 1.37×10^{10}, or approximately 13.7 billion, years ago, and the universe has ever since been expanding and cooling. Cosmologists best understand the universe from about one hundredth of a second after the big bang through to the present day. However, particle cosmologists attempt to describe the state of the universe that occurred only 1.0×10^{-11} seconds after the big bang—information that is hard to verify. For that reason, sophisticated computer models along with scientific notation are used in order to make predictions about unknown characteristics from the few facts known—a process called extrapolation.

Besides working on the early beginnings of the universe, some cosmologists work with quantum cosmology in order to study the origin of the universe itself. Because of the tiny and huge numbers involved, cosmologists use scientific notation within their research. This study is an attempt to characterize processes at the earliest times of the universe, that of the Planck epoch at 1.0×10^{-43} seconds after the big bang. It is widely accepted that from the instant of the big bang to about 10^{-32} seconds later, the universe expanded much more rapidly than it did later—to about 10^{50} times its original size. At the Planck epoch, the universe was extremely hot in temperature. In fact, cosmologists do not even talk in terms of familiar temperature units such as degrees Kelvin, Celsius, or Fahrenheit, but use gigaelectron volts (GeV) when dealing with such very hot temperatures. At the Planck epoch, the temperature of the universe is believed to be 10^{19} GeV, which is equivalent to about 1.0×10^{32} degrees Kelvin.

ENGINEERING

When engineers use scientific notation they call it engineering notation because the powers of ten are limited to multiples of three. For instance, electronic multimeters are set up in ranges that accommodate engineering notation. A reading of 3.06×10^{-5} amperes would not be valid (because 5 is not a multiple of 3) but with the use of engineering notation the value would be converted to 30.6×10^{-6} amperes (amps) and represented as 30.6 microamps, where micro stands for one millionth of an ampere. The prefixes associated with engineering notation include (in the positive) 10^3 = kilo, 10^6 = mega, 10^9 = giga, 10^{12} = tera; and (in the negative) 10^{-3} = milli, 10^{-6} = micro, 10^{-9} = nano, 10^{-12} = pico.

COMPUTER SCIENCE

Computer science involves the engineering, experimentation, and theory that goes into the design, production, and use of computers. Writing out very large and very small numbers can be tedious and cause mistakes, which is one reason why in early computers these large and small numbers were often written out with scientific notation. However, when inputting such large and small numbers into a computer with scientific notation, another problem arises. A number such as 1.4×10^5 was not easy to input into early computers because the times ($\times$) symbol is different from the letter "x," and most of the computers of those early years did not have a way to indicate superscripts. So, when computer languages were first developed, an alternative way of writing scientific notation was developed, the exponential notation. The "$\times$ 10" (or times 10) was replaced with the capital letter "E" and the exponent itself was written without the superscript. Thus, the value of 1.4×10^5 was written as: 1.3E5 (some other equivalent representations include +1.4E + 05, 1.4E + 05, and 1.4000E05). Because computers in the twenty-first century are used in every conceivable field of science, business, and everyday life, the inputting of very large and very small numbers is now an easy task.

MEDICINE

The process of diagnosing, treating, and preventing illnesses, diseases, and injuries is called medicine. Although practicing doctors and other similar health care professionals rarely use scientific notation as part of their daily routine, medical researchers and scientists often use scientific notation in their search for new medical knowledge and technology in such areas as drugs, medical treatments, and equipment and devices.

Controlled clinical trials are a method used by medical professionals to decide whether new drugs and treatments are safe. In a controlled clinical trial, a group of patients, normally called the treatment group, receives a new drug or treatment. Another group, referred to as the control group, is given a placebo (an inactive drug) or a currently accepted method of treatment. Over an appropriate period of time, researchers compare the two groups as to their overall reactions. The resulting data is collected and analyzed with statistical techniques, which includes scientific notation, to determine if the new treatment is better than standard treatments or no treatment at all. For instance, in one study, volunteers might receive a one 1-milliliter (1×10^{-3} – mL) injection of an experimental vaccine at a dosage of 1×10^{10} units. Due to the extremely small amount of vaccine given to the volunteers, scientific notation would be used to analyze the resulting data, and to ultimately determine the safety of the new vaccine. Amounts of such medical materials as bodily fluids, drugs, DNA samples, and plasma and blood all potentially need to be measured in terms of scientific notations in order to be efficiently researched and analyzed by the medical community.

ENVIRONMENTAL SCIENCE

The study of the environment involves dealing with all of the external factors such as other living organisms and nonliving factors like ocean currents, rainfall, and temperature affecting an organism. Environmental scientists study the long-term consequences of human actions on the Earth's environment and other smaller environments. During their studies, these scientists are confronted with many large and small numbers. For instance, there are 3.34×10^{22} (or 33,400,000,000,000,000,000, 000) molecules in one gram of water, the life providing material of all organisms on Earth.

Environmental scientists are likely to measure the average volume of river water flowing into a particular ocean, which may commonly reach values of 1×10^{9} cubic meters per year. For example, at New Orleans, Louisiana, the average flow rate of the Mississippi River, one of the principal freight transportation arteries in North America, is 6×10^{5} cubic feet per second, which relates into 1.9×10^{13} cubic feet per year (5.4×10^{11} cubic meters per year). Because the Mississippi River is so important to the health of the United States, it is necessary for environmental scientists to study the river's overall condition. Because so many of the river's statistics are large numbers, scientific notation is regularly used to analyze the very large numbers that describe the Mississippi River with regards to transportation, farming, fishing, and the general environmental conditions of the areas surrounding the river.

GEOLOGIC TIME SCALE AND GEOLOGY

The study of the history, features, and the processes acting upon the planet Earth is called geology. A specific type of calendar is used by geologists in order to find out (for example) how long ago a dinosaur lived or why a volcano was formed. Such a calendar, which is able to go back millions of years into Earth history, is called the geologic time scale. It begins when the Earth was first formed, about 4.6×10^{9}, or 4.6 billion, years ago, and continues up to the present. Instead of months and days, the geologic time scale divides Earth's history into: (1) eons (the longest unit of geologic time comprising several eras), (2) eras (the second longest unit of geologic time comprising several periods), and (3) periods (a third longest unit of geologic time, shorter than an eon or an era).

Scientific notation is critical to the proper use of the geologic time scale because of the large numbers involved. For instance, the Hadean eon occurred about 4.6×10^{9} years ago, at the time when the Earth was formed, while the Proterozoic eon occurred about 2.2×10^{9} years ago, at the time when the mechanics of plate tectonics began to slow down and operate much like it does today. During the Jurassic period, the second division of the Mesozoic era which occurred about 2.06×10^{8} years ago, reptiles were the dominant form of animal life, having adapted to life in the air, in the sea, and on the land.

FORENSIC SCIENCE

Forensic science—the application of science to law—uses advanced technologies to uncover scientific evidence in a variety of fields. There are many subspecialties within the field of forensic science including anthropology, biology, chemistry, pathology, odontology, toxicology, psychiatry, and physics. Forensic scientists in each subspecialty use scientific notation in their own way to perform the science necessary within their specific jobs. In fact, the amount of digital evidence that forensic scientists collect each year and store in data storage devices is greatly increasing. Many digital evidence computer programs are searching terabytes of data each year, where one terabyte is one thousand billion, or 1×10^{12}, characters.

Recent technological developments easily permit scientists to analyze the deoxyribonucleic acid (DNA) of forensic evidence in order to determine whether it came from a victim or a suspected criminal. During the last quarter of the twentieth century, its use in forensic science has dramatically helped to solve ever increasingly complicated crimes. In fact, the high-technology process known as polymerase chain reaction (PCR) is an impressive technique that quickly multiples very small samples of DNA into much larger samples and results in the use of scientific notation when very small numbers are converted into very large numbers. Repeated cycles of replication (multiplication) involve the heating and cooling of a DNA sample within a solution of heat-resistant enzymes. This action results in a particular DNA sequence being multiplied at a rapid rate. Within several hours, a tiny sample, for instance 1 nanogram (or 1×10^{-9} grams) of DNA, would have been increased by about a million times, to give a milligram of sample material, enough material for many DNA tests by numerous laboratories.

ELECTRONICS

Electronic engineering, the largest field within engineering, is concerned with the application, design, development, and manufacture of devices and systems that use electrical power. While working in the field, electronic engineers encounter many very large and very small numbers that would be quite inconvenient to write out with traditional notation. For example, a capacitor might have a value of 0.000001 farad or a resistor a value of 150,000 ohms. Because these numbers are inconvenient to write, it becomes much easier to use 1×10^{6} Farad or 1 micro-Farad and 1.5×10^{5} ohms or 150 kilo-ohms. Another familiar measure used in the electronics field is the coulomb, which stands for the quantity of electrical charge. One coulomb equals the amount of electrical charge carried by 6.25×10^{18} electrons.

ABSOLUTE DATING

Anthropology is the study of all aspects involving the ways and means that humans live. Anthropologists study such topics as: what people think about, how they react to their environments, and the reasons that humans evolved over time. Archaeologists use specialized methods and tools for the excavation and recording of recovered remains of ancient peoples and their artifacts. They use a variety of dating methods—all using scientific notation to achieve reliable results—which involve various scientific analyses to uncover the characteristics of materials buried for thousands, even millions of years. One of these dating methods is called absolute dating, which determines the age of a material with respect to a particular time scale that involves very large numbers.

EARTH SCIENCE

Earth science, as the name suggests, is the study of the Earth. Because Earth science deals with many very large and very small numbers, it is essential that earth scientists use a form of shorthand to represent such numbers. As a result, scientific notation is used, for example, to represent very large numbers such as the mass of the Earth as 6.000×10^{24} kilograms, rather than as 6,000, 000,000,000,000,000,000,000 kilograms, and the average circumference of the Earth as 4.0074×10^{7} meters, rather than 40,074,000 meters. Scientific notation is also used to represent very small numbers within earth science such the concentration of gold in seawater as 5×10^{-8} grams per liter rather than the unwieldy 0.00000005 grams per liter. In each case, it would be both confusing and requiring of a great deal of space to continually use the longer version.

Real-Life Math and Audio Engineers

Audio engineers make audio amplifiers such as those found commonly in radios and television sets. The word audio refers to signals with information in the frequency range that is audible by humans, which involves very large numbers calculable with scientific notation. An audio amplifier consists of an electrical circuit manufactured to increase the current, power, or voltage of an applied signal, which is then converted to sound. For example, electromagnetic signals between 300 hertz and 3,000 hertz are called audio-band electromagnetic signals. Generally, audio signals operate at frequencies below 20,000 hertz, or 20 kilohertz-where 1 kilohertz equals 1,000 (10^{3}) cycles per second-but can operate up to 100 kilohertz (or 100,000 (10^{5}) cycles per second).

Key Terms

Anthropology: The study of humankind.

Exponent: Also referred to as a power, a symbol written above and to the right of a quantity to indicate how many times the quantity is multiplied by itself.

Potential applications

PROTEINS AND BIOLOGY

Biology is the study of life, and its data is usually acquired through measurements of very small and very large values of mass, volume, length, temperature, pressure, and pH, which again necessitates the need for scientific notation. For example, the basic length scale used to describe any type of molecule is a nanometer, where 1 nanometer $= 1 \times 10^{-9}$ meters. Currently, biological data is subdivided into sections based on the cell, the molecule, the organism, and the population. Using an example from molecules, when compared to a water molecule, a protein molecule is gigantic with a typical mass of about 1×10^{-22} kilograms. Made up of thousands of atoms (mostly atoms of carbon, hydrogen, oxygen, and nitrogen), proteins serve numerous purposes such as a constituent of bones and tendons; an ingredient within red blood cells; a part of oxygen in the lungs; a material in the hair and skin, and an aid in the digestion of food. Discovering how atoms are arranged in a protein molecule is one of the most challenging research projects in the biological sciences. With an estimated 30,000 different proteins in the human body, only about two percent have been adequately described, which provides ample need for future research into the discovery of these descriptions, and along with it the need for calculations with scientific notation.

NANOTECHNOLOGY

Nanotechnology is a relatively new science that involves the creation and use of materials and devices at extremely small sizes. These materials and devices are generally in the (nanoscale) range of 1 to 100 nanometers, where one nanometer is equal to one-billionth of a meter (0.000000001, or 1×10^{-9}, meter), which is about 50,000 times smaller than the diameter of a single length of human hair. Scientists refer to the materials at the nanolevel as nanomaterials or nanocrystals. The transmission electron microscope, a pioneering nanotechnology invention, is already a popular instrument for visualizing individual atoms within semiconductor nanocrystals. This instrument and other such breakthroughs have already applied nanotechnology, but future research and development will hold the key to nanotechnology's major impacts in such fields as energy conservation, medicine, environmental protection, electronics, computers, and world defense.

Where to Learn More

Books

Bluman, Allan G. *Mathematics in Our World.* Boston, MA: McGraw-Hill-Higher Education, 2005.

Florian, Cajori. *A History of Mathematical Notations.* New York: Dover Publications, 1993.

Tussy, Alan S., and R. David Gustafson. *Basic Mathematics for College Students,* 2nd Ed. Pacific Grove, CA: Brooks/Cole Thomson Learning, 2002.

Web sites

Bagenal, Fran. Atlas Project, University of Colorado at Boulder. "The Scientific Notation." Mathematical Tools for Astronomy.<http://dosxx.colorado.edu/~atlas/math/math1.html> (March 15, 2005).

Charity, Mitchell N. "10^X: An Exponential Notation Meta Page (and scientific notation too)." Vendian Systems. <http://www.vendian.org/mncharity/export1/exponential_notation_meta/> (March 15, 2005).

Feldmeier, John. "Units and Unit Conversion." Case Western Reserve University. <http://burro.astr.cwru.edu/johnf/scales.rev.txt> (March 15, 2005).

Molecular Expressions. "Secret Worlds: The Universe Within." Optics and You (Interactive Java Tutorials). January 17, 2005. <http://micro.magnet.fsu.edu/primer/java/scienceopticsu/powersof10/index.html >(March 15, 2005).

University of Washington. "Scientific Methods: Scientific Notation." Department of Astronomy. February 3, 2000. <http://www.astro.washington.edu/labs/clearinghouse/labs/Scimeth/mr-scnot.html> (March 15, 2005).

Sequences, Sets, and Series

Overview

Sets, series, and sequences are all interrelated. They are each helpful in dealing with groups of numbers, especially large groups of numbers. Sets can contain useful data and can help mathematicians understand these groups of numbers with greater clarity. Sequences always form patterns. These patterns have many practical applications in other areas of mathematics and also in life. From these sequences emerge ratios and formulae that have many applications. One such sequence is the Fibonacci sequence, which is directly related to the golden ratio. This ratio was often used in ancient architecture and is still used today.

Fundamental Mathematical Concepts and Terms

A series is a natural extension of a sequence. However, series can stretch further. The formulae that can emerge from a series have applications in probability and many other areas of mathematics.

SETS

A set is simply a collection of things. These things can be of any genre: numbers, coins, animals, trees, etc. However, in mathematics, a set is usually a set of numbers. The things that make up a set are called elements.

A set contains elements of a specific characteristic. There is usually a reason that elements in a set are part of the set. There is some commonality between the elements. In this way, sets, and sequences are connected.

SEQUENCES

A sequence is an ordered set of mathematical terms. It is usually formed by a specific rule. A sequence can also be called a progression. There are two main types of sequences: arithmetic sequences and geometric sequences.

An arithmetic sequence occurs when the difference between successive terms is the same. For example: 2, 4, 6, 8, 10. The difference between each term and the one before it is 2. Therefore this is an arithmetic sequence. In general terms this can be stated as the nth term minus the $(n-1)^{th}$ term is equal to a constant. The formula for finding the nth term of an arithmetic sequence is: $t_n = a + (n-1)d$.

A geometric sequence occurs when the ratio between successive terms is equal. For example: 3, 9, 27, 81, 243 . . . , where a is the first number in the sequence, d is

Fibonacci Sequence

The Fibonacci sequence is one of the best-known sequences. It was written down, and its properties examined, by Fibonacci in 1202. Fibonacci, also known as Leonardo of Pisa, was an Italian mathematician. The sequence is formed by adding the previous two terms to obtain the next term: $t_n = t_{n-1} + t_{n-2}$. In other words, $t_1 + t_2 = t_3$. The sequence is: 1, 1, 2, 3, 5, 8, 13, 21, 34, 55 . . . *ad infinitum* (continuing forever).

The Fibonacci is directly related to the golden ratio. The ratio between successive terms approaches the golden ratio as the number of terms approaches infinity. The golden ratio is usually written as the Greek letter *phi* (ϕ). The exact value of the golden ratio is

$$\frac{1+\sqrt{5}}{2}$$

The golden ratio is used in ancient architecture. The proportions of the length to the height on the front of ancient Greek and Roman buildings are often the golden ratio. This ratio is proven to be the most aesthetically pleasing ratio. The golden ratio is also found in nature. The nautilus shell spiral can be created by drawing a curve through successive golden rectangles. The pinecone is another example of this same spiral.

the difference between each term in the series and the next, and n goes on as 1, 2, 3, ... In this case the ratio between a term and the term preceding it is 3. In other words the 4th term divided by the 3rd term gives 3. In general, this can be stated as the nth term divided by the (n−1)th term is equal to a constant. The formula for finding the nth term of a geometric sequence is $t_n = ar^{n-1}$, where a is the first term, r is the factor by which each term differs from the one before, and n goes as 1, 2, 3, ...

SERIES

A series is a sequence that is derived from the sum to n terms of another sequence. It can also be defined as the sum of a specific number of terms in a sequence. For the sequence $t_1, t_2, t_3, t_4 \ldots t_n$ the corresponding series would be: $t_1 + t_2 + t_3 + t_4 + \ldots + t_n$. Sequences and series are related, but different.

An arithmetic series is formed from an arithmetic sequence and a geometric series is formed from a geometric sequence.

The sum of an arithmetic series can be found using the formula $\frac{1}{2}n[2a+(n-1)d]$.

The sum of a geometric series can be found using the formula:

$$S_n = \frac{a(1-r^n)}{(1-r)}$$

and the sum to infinity:

$$S_\infty = \frac{a}{1-r}$$

Just as a sequence can be finite or infinite, a series can also be finite or infinite. An infinite series can be convergent or divergent, also just as a sequence.

A Brief History of Discovery and Development

Zeno of Elea, who lived about 490–425 B.C. was the first mathematician to write about infinite series and sequences and their sums. Archimedes discovered a way to show that infinite sequences could have finite results. Chinese mathematicians used methods that have led to an understanding of long-term behavior and limits of infinite sequences. Mathematicians have used sequences over the years to develop new methods in calculus. More recently sequences have found applications in computing.

Real-life Applications

OPERATING ON SETS

A set P consisting of the numbers 1, 2, 4, 8, and 16 is written in set notation as: P = {1, 2, 4, 8, 16}.

To state that 4 is an element of P, the following notation is used: $4 \in P$. $4 \notin P$ means that "4 is not an element of P."

If set A consists of 2 and 4, i.e., A = {2, 4}, then it is a subset of P. This is written in set notation as $A \subseteq P$. Alternatively it can be written as $A \subset P$. However, this means that A is a proper subset of P, which means that it is a subset of P but is not equal to P. To write that A is not a subset of P the following notation is used: $A \not\subseteq P$.

Including all numbers in an infinite set would be impossible. Simple set notation makes this an easy task.

"M is a set containing all integers greater than one" is written as: M = x ∈ Integers(x > 1). This reads as "M is a set of all integers x such that x is greater than 1." The vertical bar means "such that." It states a condition. Commas separate multiple conditions.

Occasionally a set will have no elements. This set is called the null, or empty, set. It is represented by the symbol: ∅. For example: A = { } = ∅.

Sets can also be part of a group. That is, data in a set can be part of a larger group of data. There is set notation that deals with multiple sets. A Venn diagram most easily represents these because a Venn diagram is a method of representing multiple sets graphically. A∪B means "A union B";. This is a way of writing the elements that are in both sets combined. (See Figure 1.) A Venn diagram is a method of representing multiple sets graphically. A∩B means "A intersection B". This is a way of writing the elements in both set A and set B—in other words, the elements common to both set A and B. (See Figure 2.) These types and/or principles are used in database searches, the most common being an Internet search. A search may be conducted for something containing the word cat AND dog—mathematically this would be cat ∩ dog.

Sets are frequently used in everyday applications. Their most common application is in classification schemes, whether it be for clothes, food, animals, or socks. Catalogues are another example of non-mathematical sets. Sets are also used in data analysis in genetics. Chromosomes are sorted and arranged in sets according to shape and specific lengths of chromosome arms and other factors.

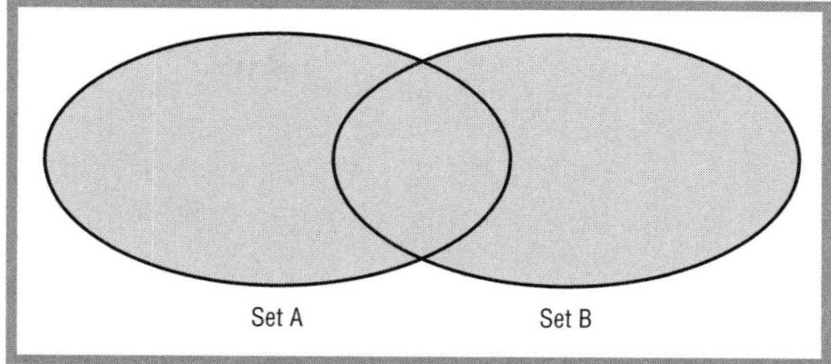

Figure 1: The elements in set A and B.

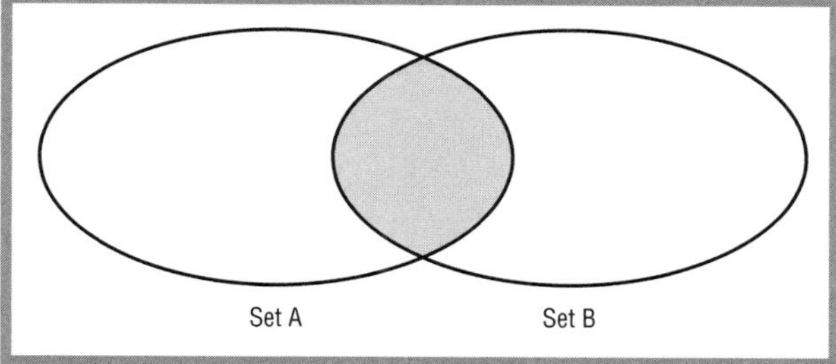

Figure 2: Venn diagram of the intersection of sets A and B (intersection is shaded area).

USING SEQUENCES

There are finite sequences and infinite sequences. A finite sequence has a limited number of terms. An infinite sequence has an unlimited number of terms. An infinite sequence can be a divergent or convergent sequence. A divergent sequence is truly unlimited. A convergent sequence approaches a limit as the number of terms approaches infinity. Convergent sequences are usually geometric sequences. This is because if the rule for finding the nth term is t_n; = $ar^{1/n}$ then as n approaches infinity the nth term approaches 0. This is because $r^{1/n}$; = $1/r^n$.

Sequences always have a specific pattern to them. However, occasionally there is a sequence where a pattern exists that is unknown but it would be beneficial to understand the pattern. An example of this is the stock market. Researchers have been searching for many years to find a pattern behind the stock market, with varying degrees of success.

ORDERING THINGS

Associated with sequences is the notion of a specific order. Placing objects or numbers in an order can give them meaning, such as placing soccer teams on a ladder with the leader at the top and the team with the least number of points at the bottom. This is a simplified sequence.

GENETICS

Sequences are also used commonly in the field of genetics. Specific genes are sequenced (e.g., their base sequence is identified) to determine exactly what gene is associated with a specific physiological function, characteristic, or disease. Sequences of bases determine what gene is formed and what the gene does.

SERIES

A convergent series converges, or comes to, a finite sum. The series 0.5 + 0.25 + 0.125 . . . , is a convergent series because even if extended to infinity its sum is finite. Using the above formula it is easy to see that the sum to infinity of this series is 1, and so it is a convergent series. Another way to think of a convergent series is to think of a radio signal that is attenuated by half it's strength each time it pulses (a sequence used in some timing devices). At each step the signal strength decreases but will mathematically never reach zero. Mechanically the signal reaches a functional zero when the signal strength is too weak to be measured. This is the concept of a sum approaching a specific figure as the number of terms approaches infinity.

Potential Applications

Series can be used to predict regular repeating events, for example, earthquakes and the weather. The data mathematicians collect can be analyzed as a sequence or a series and thus future events predicted in an accurate manner. Sequences can be used in speech recognition. The sound waves produced can be converted to a sequence, similar to a sine wave. Each pitch has its own specific sine wave, which in turn has a specific sequence of y values and x values and these can be converted to a sequence. Sequences also have potential uses in the communications industry, most specifically in signals analysis and wireless connections.

Where to Learn More

Books

Bowerman, Bruce L. *Forecasting and Time Series: An Applied Approach,* 3rd ed. Belmont, CA: Duxbury Press, 1993.

Engquist, Bjorn. *Mathematics Unlimited: 2001 and Beyond.* Berlin/New York: Springer, 2001.

Han, Te Sun. *Mathematics of Information and Coding.* Providence: American Mathematical Society, 2001.

Nelson, David. *The Penguin Dictionary of Matematics,* 2nd ed. England: Penguin, 1998.

Web sites

Drexel University. “Golden Ratio, Fibonacci Sequence.” Math Forum. <http://www.mathforum.org> (October 19, 2002).

Rider, J.W. “Mathematical Set Notation.” <http://www.jwrider.com/lib/setnotation.htm> (January 21, 2005).

Taylor, S. “Applications of Sequences and series.” Algebra Lab: Mainland High School. <http://www.damtp.cam.ac.uk/user/gr/public/gal_milky.htm> (February 10, 2005).

Thomas' Calculus, 10th ed. “History of Series and Sequences” <http://occawlonline.pearsoned.com/bookbind/pubbooks/thomas_awl/chapter1/medialib/custom3/topics/sequences.htm> (February 10, 2005).

United States Naval Academy. “Set Notation.” <http://mathweb.mathsci.usna.edu/faculty/traveswn/sm230old/Set%Notation.htm> (January 21, 2005).

University of Cambridge. “Arithmetic Sequence.” <http://thesaurus.maths.org> (January 21, 2005).

University of Cambridge. “Arithmetic Series.” <http://thesaurus.maths.org> (January 21, 2005).

University of Cambridge. “Geometric Sequence.” <http://thesaurus.maths.org> (January 21, 2005).

University of Cambridge. “Series.” <http://thesaurus.maths.org> (January 21, 2005).

Weisstein, E.W. “Sequence.” Mathworld: Wolfram Resource. <http://mathworld.wolfram.com/Sequence.html> (January 21, 2005).

Sports Math

Overview

Sport, at its best, is a perfect marriage of emotion and execution. The exhilaration of competition and the endorphin-fueled rush of physical fitness are not capable of being measured with mathematical precision. However, the results achieved in sports of every type, whether the endeavor is an individual pursuit or a team game, are invariably assessed in two ways: using the basic math of counting to keep score and assessing the subjective perspectives of the participants about how they regarded the performance.

When the subjective, emotional components of the particular sport are stripped away, the desire to improve performance will often become the focus of an athlete or a team. The desire to improve technique and to achieve better results in competition has spurred the development of numerous mathematical approaches within every sport to measure and to compare aspects of performance.

Math principles are fundamental to sport. They manifest themselves on a number of different levels, from simple counting and keeping track of a score or time, to mathematics as a tool of human discovery as to how a particular sport can be played better.

Sports math can be grouped into three general categories. The first is rules math, in which mathematics is the regulatory basis for the sport. The second grouping is math as an interpretive or demonstration tool in which the understanding and the illustration of aspects of athletic performance is assisted by the application of mathematical concepts, or in which math is used to indicate or predict future performance. The third grouping is performance math, in which mathematics alone, or in conjunction with other science concepts, particularly those of physics, is used to assist in the improvement of athletic performance.

Fundamental Mathematical Concepts and Terms

Counting is fundamental to the appreciation of sport, both in competition and in training. This concept is present in an elemental fashion in sport through the recording of scores, the measuring of distances, and the keeping of accurate time, both as a standard of achievement as well as a competition limit.

Counting in a more sophisticated form is found in most sports through the compilation and use of statistics. Statistics is defined as the branch of the science of mathematics related to the collecting, classification, and use of numbers, often in large quantities. Statistics in sport are often expressed as percentages.

A percentage is a fraction with a denominator of 100. A percentage may be expressed using the word "percent," as in 25 percent, or using the % symbol, as in 25%. Percentages are the natural mathematical extension of three other familiar concepts: fractions, ratios, and proportions. A fraction is a number written as one whole number divided by another; for example, one half is expressed as ½. A ratio is the relationship between two magnitudes. For example, if the payroll, meaning the total of the salaries paid to all members of a professional baseball team, was $40 million 10 years ago and $80 million today, the relationship between the two payroll figures may be expressed as a ratio of 1 to 2. A proportion is a pair of ratios expressed as a mathematical equation. For example, if a college basketball team has a roster of 15 players, and three of the players are left-handed, the ratio of the team members who are left-handed will be expressed as 3/15, or 1/5.

All percentages are an expression of a relationship based on 100. As is set out in the various sports applications that follow, every fraction, ratio, and proportion may be expressed as a percentage. Percentages may also be expressed where required using decimals, as in the figure 66.92%.

Mathematics is the language of physics. Physics and its particular applications to sport, both in the understanding of the mechanics of human movement as well as in equipment construction, is made clearer in its application through statements expressed as mathematical equations.

A Brief History of Discovery and Development

RULES MATH

How math came to occupy its place as the prime regulatory tool in sport is best understood by the following simple progression: Physical activity, followed by specific activities, followed by informal competition, followed by structured competition, followed by codified rules of competition (time, space, distance), followed by scorekeeping (simple math), followed by performance analysis (advanced math, statistical measures, and mechanics of sport).

At its root, sport is competition, from the informal challenge that individuals make to themselves as they run to keep fit, simply testing themselves personally, to the organized event against a rival or a team of rivals. Competition, to be organized and certain, requires a framework, a structure within which the event can occur with certainty for every participant that everyone is competing in exactly the same fashion. The structure must have limits of time to give the competition a fixed duration, and space or distance to provide boundaries within which the sport can occur.

The evolution of many sports has been accompanied by a progression in the rigidity of rules concerning time and space. Lacrosse, as originally played by Native Americans and referred to as "the little brother of war," was played with teams of hundreds of men, goals set miles apart, in contests lasting as long as three days, and no particular rules of engagement. Timekeeping and precise boundaries were unimportant to the competitors. As another example, ice hockey was originally played on large frozen ponds or rivers, with no lines or markings to regulate play. Modern sport places a greater premium on precise and effective counting of time, space, and distance in the creation of a venue for a competition.

The origins of sports math are best understood in the context of the math applications at the heart of traditional athletics: the track and field competition. Modern track and field events are modeled to a large degree upon the motto of the ancient Olympics of Greece: higher, faster, stronger. Traditional athletics, whether the high jump, the shot put, or foot races ranging from the sprints to the marathon, required the barest of mathematical measures: a defined, accurately determined distance to run, a precisely set object to jump, or an accurately weighed object to throw as well as the measurement of the throw itself.

A timing device to measure performance was a latecomer to sport. By the mid-1800s, the time it took a runner to run a particular distance became important as standards of athletic achievement had become an important public fact. Handheld precision stopwatches were the timing standard in track and field competitions until the 1960s.

The results in these athletic disciplines are calculated in simple mathematical terms, including the fastest time, the furthest distance, the greatest height. Advances in technology have taken the mathematics involved to even greater degrees of precision. For instance, a 100-meter race at the St. Louis Olympics in 1904 took place on a cinder track, where the distance was measured by a steel tape and the races timed by handheld stopwatches. Modern events are run on tracks measured by way of sophisticated electronic means and timing is similarly accurate to thousands of one second. However, the basic mathematics involved in determining the "higher, faster, stronger" concept is not changed.

The geometry of basketball. ELLEN H. WALLOP/CORBIS.

Math as the rules of a sport extends to team competitions. Fundamental to all team games are the following mathematical concepts:

- keeping score
- keeping time
- keeping track of the players
- ensuring conformity with rules concerning size (i.e., a regulation basketball court is 94 feet long; a regulation American football field is 120 yards long; an ice hockey goal is 6 feet wide by 4 feet high)
- keeping a record of the statistics common to the game

Real-life Applications

MATH TO UNDERSTAND SPORTS PERFORMANCE

Beyond the result achieved in any competition, numbers are used throughout sport to explain and to better understand performance. Statistics are widely relied upon in virtually every team sport as a means of enhancing the understanding of both individual competitions as well as entire seasons, for both the participants and the public at large. The interpretation of sports performance through mathematics will often provide an understanding of a result that the scoreboard does not reflect.

Statistics in sport must be regarded with caution. Media commentators often speak with apparent authority about a sport through their reliance on the numbers that are associated with competition. An understanding of the mathematics involved in these interpretive aspects of sports math is critical to separating the statistical "wheat from the chaff."

There are levels of interpretation that mathematics can provide. Some simple statistics provide a peephole on performance, others a picture window. The math involved is not the entire story, but merely an insight into the actual result achieved. The better the relationship between the math and the subject sport, the more likely the analysis is more insightful.

The following statistics are examples of how sport observers, especially in the media, often present arguments to support a contention that a certain player is outstanding:

- An NBA basketball player averages 22.0 points per game over the course of a season. Is such a player an elite performer?
- A major league baseball pitcher wins 18 games in a season, and he has an earned run average calculated at 4.5 runs per game. Is this pitcher an elite player?
- An NHL hockey goaltender has a save percentage of 0.93 (for every 100 shots on his goal, he saves 93). Is this athlete an elite level player?

In each of these three examples, the answer to the question of whether each player is an elite must be a resounding "maybe" or "perhaps." The NBA player scoring 22.0 points per game may be a terrific all-round player, with his strong scoring average one facet of his skills, or he may be a weak defensive player who is a liability at his end of the floor. The pitcher might be truly dominant, or he may be one of those athletes with impressive overall statistics, who performs poorly in important situations in critical games. The NHL goaltender could be a steady performer, or might be inconsistent, getting a shut-out one game and allowing six goals the next. Without a more in-depth and focused use of mathematical principles, these simple numbers are not the basis for great insight when taken by themselves.

In most team sports, such as baseball or basketball, individual statistics are an indicator, and not a determinant, of team success. The final score in a team contest is the only absolute measure of the team's success on a given day.

Sport statistics are almost universally expressed as either a percentage, or as a decimal. In sports, the score in a game or the result achieved by an athlete in an individual competition represents "what." Various kinds of mathematical and statistical applications can tell a great deal about the related questions of "why" and "how." Math is variously an interpreter of past performance and a predictor of future performance. The game of baseball provides some useful insights in this regard.

BASEBALL

Baseball is a sport saturated in statistics. Its fans will often exaggerate when explaining the nuances of baseball to more casual observers that this sport is the only major team sport played without a clock, which is said to give it a subtlety that can be captured by a wide variety of statistical measures. Further, because baseball has inherent repetitions of a number of actions within the game over a season (a major league player may face a pitcher more than 500 times in a season, make hundreds of throws, and run the bases hundreds of times), each player and each team perform in ways that can be readily converted into a statistical measure.

Some baseball statistics, such as batting average and earned-run average, are ingrained in the public consciousness as key indicators of performance. Over the past 30 years, the desire to delve further into the interpretation of baseball performance led a number of statisticians to develop the analysis into a field now known as sabermetrics. This analysis extends the mathematics of baseball from the relatively simple formulae of batting average and earned-run average to detailed calculations used to both rank player ability and to predict future performance.

Baseball batting averages are expressed as percentages. A player's batting average is the number of base hits made by the player divided by the number of at bats (an "at bat" in baseball is defined by the number of times the player comes to bat, less all walks, errors, and times hit by pitch ball). For example, Player A has 140 hits, 30 walks, was hit three times by a pitch, and reached first base five times as a result of errors in a season. He took 220 total bases. He has 600 at bats. His batting average is 140 / 600 = 0.233. In baseball terms, Player A is said to be a "233 hitter."

The on-base percentage statistic determines how effective a player is at getting on base, by all possible means. The on-base percentage is Number of times to first base / Number of plate appearances = 140 + 30 + 3 + 5 / 600 + 30 + 3 + 5 = 178 / 638 = 0.279. One would conclude from this calculation that Player A's simple batting average is deceiving; Player A is more effective, by the ratio of 0.279 to 0.233, at getting on base by any means than the simple batting average statistic reveals. As the object of baseball is to score more runs than one's opponent, getting on base, and being in a position to score a run, is a more accurate statistical measure of a player's worth to a team than the number of hits the player may collect.

For many years, baseball placed a premium on the total runs scored by a player as a measure of effectiveness. Sabermetrics took this analysis in a more detailed direction, that of "runs created" by an individual player. As in the on-base percentage example, the object of baseball is to score more runs than the opponent. If determining how effective a player is at getting on base is a more useful indicator of a player's contributions than simple batting average, determining how effective a player is at creating runs will be an even more useful statistic to analyze the abilities of a given player. The player who can create the most runs, in whatever fashion, will likely be the most valuable offensive player on a team.

Total bases are the number of bases that the batter's hits amounted to over the season, with a single being one base, a double two bases, a triple three bases and a home run all four bases. In an individual game, for example, where a player hit two doubles and a home run, the player would have eight total bases. Therefore, Runs Created is (Hits + Walks) × Total Bases / (At Bats + Walks). With this equation, Player A's statistics read (140 + 30) × 220 / (600 + 30) = 37,400 / 630 = 59.37.

Using this analysis, Player A created approximately 59 runs for his team over the season. If the team scored 700 runs, and if 12 players had been used regularly as hitters in the team's line up over the course of the season, Player A is proven to be a slightly above average contributor to his team's offensive production (700 team runs divided by 12 players means that the average regular hitter would be expected to create 58.3 runs). To complete the progression from the simple batting average calculation to the runs created determination, a relatively modest .233 batting average translates in the example to a slightly better than average contributor to the offence of this team.

As will be further illustrated, and as the above baseball examples confirm, the rough rule of thumb regarding sports statistics is that the more intricate and involved the desired analysis of athletic performance, the more typically involved the mathematics will be.

NORTH AMERICAN FOOTBALL

Football is a game of territorial conquest, and the measurement of the amount of territory gained by competing teams has been a focus of performance analysis in this sport. Like the evolution in the extent and the sophistication of mathematical applications in baseball, North American football has grown from a game where bare statistical measures of leading scorers and yards gained by individual players have given way to involved analyses of every aspect of the game. Thirty years ago, a quarterback, the single most important player on a team, would generally be assessed on the following set of statistical indicators:

- Completion percentage: The number of passes completed divided by the total number of passes attempted; percentages in the range of 55–60% were typically considered good.
- Number of touchdowns versus the number of interceptions: A good quarterback typically would throw for more touchdowns than interceptions.
- Yards gained through total number of passes: In a territorial game, quarterbacks who passed for a greater number of yards were generally more valuable.

With these statistics, it was possible to have an understanding of the performance of an individual quarterback, but the statistics did not give a complete picture. For example, a quarterback on a team that threw the football less frequently might appear to be an inferior player when compared with a player whose team threw the ball a great deal. Using years of data, analysts were able to incorporate known and reliable statistics to create useful mathematical tools with which to assess performance, as well as to establish a standard that would permit comparison between individual players.

The desire to better understand quarterback performance lead football analysts to develop the Quarterback Rating Index, which is calculated as follows: (1) Total pass completions divided by total pass attempts; (2) Subtract 0.3 from (1); (3) Divide by 0.2 and record the total (the result cannot exceed 2.375 or it may be less than 0), and this gives Subtotal one; (4) Total passing yards divide by total pass attempts; (5) Subtract 3 from (4); (6) Divide by 4 and record the total (the result cannot exceed 2.375 or may it be less than 0), and this gives Subtotal two; (7) Total number of touchdown passes divided by total pass attempts; (8) Divide result by 0.05 and record the total (the result cannot exceed 2.375 or it may be less than 0) and this gives Subtotal three; (9) Total number of interceptions divided by total pass attempts; (10) Subtract 0.095 from (9); (11) Divide the result in (10) by 0.04 and record the total (the result cannot exceed 2.375 or it may be less than 0), and this gives Subtotal four; (12) Add the four subtotals recorded; (13) Multiply by 100; (14) Divide by 6. This final total is the Quarterback Rating.

This more complicated Quarterback Rating Index is seen as more reliable because it incorporates every aspect of the quarterback's ability to pass into the equation. It takes the analysis beyond interpretation into an understanding of the individual player's performance.

However, as with any statistic that is not the final score or result, even the complicated ratings of this Index are not a compete picture. If a quarterback throws for three touchdowns and many yards after his team is hopelessly behind, the individual rating may be enhanced, but the team performance not assisted. An interception thrown in the first quarter of the game is given the same weight as an interception thrown in the last minute of a tied game that is returned by the other team for a winning score. As with all forms of statistical analysis, the factors not calculated in the statistical equation must be assessed as well.

BASKETBALL

Basketball is a simpler, more free flowing game than either baseball or North American football, and its statistical base has long been the individual statistics of players, totaled for team assessment. In basketball, statistics are seen as a measure of tendency, as opposed to the interpretation of individual performance. For all of the importance attached to scoring averages of players, basketball experts key on two chief statistics: rebounding (broken into offense and defense) and free-throw shooting.

Studies have illustrated that where a team has more rebounds than an opponent, that team will be expected to win 70% of its games. When that same team is more effective from the free-throw line, the success rate for the team is between 85% and 90%. These statistics bear out the nature of the game itself, i.e., rebounding advantages mean that a team is controlling the ball on defense and likely getting more than one shot on any given sequence on offence. Good free-throw shooting means that the team is getting in a position on offence to take shots, drawing fouls from the opponent, an indication that the team is better controlling the ball and the game than its opponent.

These statistics also underscore the fact that each sport has subtleties that are inherent in the interpretation of the statistics gathered, and each sport has its own measure of what constitutes success. In baseball, a professional hitter who has a batting average of .350 is very likely a successful offensive player; an NBA basketball player with a free-throw shooting percentage of 35% would be considered a very poor free-throw shooter; an NFL quarterback with a passing completion rate of 35% would not succeed as a player at that level.

PREDICTING THE FUTURE: CALLING THE COIN TOSS

In many respects, the coin toss is a metaphor for life itself: if one's call is wrong on one occasion, one will likely get another chance at some later time. The coin toss in North American football is one of the great rituals of the game; the winner of the coin toss has the right to elect to receive the opening kickoff and thus take initial possession of the ball. The loser of the coin toss can select which end of the field they will defend. In a game that is, at its core, one of territorial conquest, success at the opening coin toss in a football game is important, especially given the variables of wind, field condition, and weather where the game is being played outdoors. The direction and the outcome of the game may well be influenced by the result of the coin toss.

In the American professional game, the coin toss takes on special significance if the game is tied at the end of regulation time. To determine which team will take possession of the ball at the commencement of sudden death overtime, where the first team to score in any fashion wins, the referee will toss the coin and the winner of the coin toss will inevitably elect to receive the kickoff.

The question is, Is there a mathematical predictor for how the coin toss should be played by either team? The probability of a coin being heads or tails is 1:1, or an even chance, every time. No matter that, for example, the previous five coin tosses may have been heads, on the sixth and subsequent tosses the probability of heads versus tails remains even. In other words, each coin toss must be approached as a unique, free-standing event: any history will be irrelevant.

The probability of heads (or tails) is expressed as Number of Favorable Outcomes / Number of Possible Outcomes = $\frac{1}{2}$. The probability of two heads (or tails) being tossed in a row is $\frac{1}{2} \times \frac{1}{2} = \frac{1}{4}$. The probability of four heads (or tails) being tossed in a row is $\frac{1}{2} \times \frac{1}{2} \times \frac{1}{2} \times \frac{1}{2} = 1/16$.

PASCAL'S TRIANGLE AND PREDICTING A COIN TOSS

The French mathematician Blaise Pascal (1623–1662) is regarded as the developer of the device known as Pascal's triangle (although history confirms that Chinese mathematicians developed a very similar construct 500 years before Pascal). Pascal's triangle has a number of algebraic applications. (See Figure 1.)

To read this table in terms of coin toss probabilities, suppose that the number of tosses is one. On the corresponding line, the triangle provides the numerators for two possible outcomes, heads or tails. As the total number of outcomes is two, the denominator for this calculation shall always be two, meaning that the probability of heads or tails is $\frac{1}{2}$.

When the coin is flipped twice, there are three possible outcomes: (1) heads once, tails once, (2) twice heads, and (3) twice tails. The extreme possibilities, two heads or two tails, are represented by a "1" on each side of the triangle. The one head, one tail result (which may occur in two different orders), is represented by "2." The extreme possibilities are therefore $\frac{1}{4}$, and the probability of one head, one tail is 2/4, or the expected 50%.

If the third line of the Pascal's triangle in Figure 1 is examined, the total number of possibilities is $2 \times 2 \times 2 \times 2$, or 16. By using the triangle as a calculator, the probabilities can be determined as All heads, 1/16; One head, Three tails, 4/16; Two heads, Two tails, 6/16; Three heads, One tail, 4/16; All tails, 1/16. (As the triangle is symmetrical, it does not matter which side is called heads or tails.)

There has been interesting research carried out recently that suggests that the side of the coin facing up in a coin toss is slightly more likely to be the side turned up on the flip. The premise of such research appears to be that as coins are tossed by real people, who are subject to bias, and the person tossing the coin sees the head or tail facing upwards, they may tend to unconsciously catch the coin with that face up. The probability is said to rise to 0.51 in favor of the exposed side as opposed to the accepted 0.50. This theory will no doubt be the subject of further study.

Number of Coin Tosses	Numerator for Probabilities	Denominator for Probabilities
1	1 1	2
2	1 2 1	4
3	1 3 3 1	8
4	1 4 6 4 1	16
5	1 5 10 10 5 1	32
6	1 6 15 20 15 6 1	64
7	1 7 21 35 35 21 7 1	128
8	1 8 28 56 70 56 28 8 1	256

Figure 1: Pascal's triangle as a probability tool.

FOOTBALL TACTICS—MATH AS A DECISION-MAKING TOOL

In many team games, math is a tool for strategy decisions, many of which are rooted in the concepts of probability. The basic question that is often addressed is, in a given situation, what play or strategy affords the best chance of success?

American football is a game of field position and territorial advantage. As a general rule, the team with the consistent best position on the field will be in the best position to score and therefore win. A common tactical decision in American football related to field position is, given where the team has the ball on the field, whether the should team punt the ball away to the other team, attempt to gain a first down and keep possession of the ball and ultimately score, or kick a field goal.

Based on the data gathered from more than 700 NFL football games, the following statistical analysis can be made: Team A has a fourth down on the Team B 2-yard line. Team A is assessing its options: punt, attempt to score a touchdown, or attempt a field goal. A punt by Team A is not a sensible option, as the ball would be kicked through the 10-yard end zone and would be placed at the Team B 20-yard line, where Team B would take over on offense. The two realistic options for Team A are the attempt to score a touchdown or to kick a field goal.

At the NFL level, the statistical data confirms that an attempt to score a touchdown from the 2-yard line, coupled with a virtually automatic extra-point conversion, has a probability of success of 40%. Therefore, the value of that choice can be calculated as 6-point touchdown $\times$ 0.40 = 2.4 points; 1-point convert $\times$ 0.40 = 0.4 points. Therefore, the total value of attempt = 2.8 points.

An attempt at a 3-point field goal from the 2-yard line is as likely to be successful as the 1-point conversion after a touchdown (the success rate is slightly under 99%). The value of the field goal choice is 3-point field goal $\times$ 0.99 = 2.97, which is in essence 3 points.

Using the value of each choice with the probabilities for each calculated, it would seem that Team A would have a slightly better option with the field goal over the touchdown attempt (3 points versus 2.8 points). However, as stated above, American football is a game that turns, to a large degree, on field position; the decision as to attempt a field goal versus the try for a touchdown must also be assessed considering the field position consequences that flow from each choice.

By rule, after a successful field goal, Team A would kick off to Team B. On average, statistics confirm that an NFL kickoff will be returned to the receiving team's 27-yard line. A first down and 10 yards to go situation at that position for Team B is worth 0.6 points to Team B, based on the probabilities of scoring from that position.

If Team A were to attempt to score a touchdown from the Team B 2-yard line and fail, the ball would be turned over to Team B at the same 2-yard line. With Team B 98 yards from the Team A's goal line, this poor field position statistically is worth −1.6 points to Team B. With the field position components factored into the calculation weighing the attempt at a touchdown versus a field goal, the value of each choice can be calculated as Value of Touchdown Attempt = 2.8 points; Field position (−1.6 points to Team B) = 1.6 points to Team A; Total Touchdown Attempt Value = 4.4 points; Value of Field Goal Attempt = 3.0 points; Field Position (+0.6 points to Team B) = −0.6 points to Team A; Total Field Goal Attempt Value = 2.4 points.

With the field position information now factored in, it is apparent that what was a slightly preferable course of action, the field goal attempt, is now a significantly lesser option when compared to the touchdown attempt.

As with any mathematical model employed to predict an event or to select the optimum course of action, there will be variables that cannot be reduced to a number or an equation. In the above example, factors, such as how much time is left in the game, field conditions, or an injury to a key player on either side, would potentially alter the decision that the probability calculation otherwise directs as the best choice. Math is rarely determinative with respect to decisions in team sports, but it is often very illuminating.

UNDERSTANDING THE SPORTS MEDIA EXPERT

It is standard in the television coverage of virtually every sport, from professional competitions to the various events in the Olympics, for the broadcasting network to engage the services of an expert to assist in the presentation of the event. A typical broadcast will have a commentator giving the audience a play-by-play of the action being telecast, and the expert, often referred to as a color commentator, provides insight and analysis about the event, both as it is unfolding and at various stoppages in play.

In addition to what the expert may be saying to the audience, it is common for the presentation to display statistical summaries in relation to either the individual players, the teams, or the season to date. Where the casual fan is seeking information to enhance their enjoyment of the broadcast and the game itself, the numbers cited by the experts are often not helpful, but can actually be confusing. It is important to approach these statistics-filled commentaries with caution.

For example, in professional baseball, basketball, and hockey, it is common for a playoff series to be played as a best-of-seven-games event. In a baseball World Series, where one team is ahead of its rival three games to none, there will inevitably be an expert commentator who might suggest that "never in World Series history has a team come back from three games down to take the Series." This statement might be true, a powerful sounding pronouncement, which can be examined more closely using math principles to test its weight.

Between 1920 and 2004, there have been only 20 World Series where a team led 3-0. Assuming for a moment that all other factors are equal, and that the two teams are relatively evenly matched, the odds of a team winning four straight games can be calculated using the same probability as the coin toss analysis or Pascal's Triangle ($\frac{1}{2} \times \frac{1}{2} \times \frac{1}{2} \times \frac{1}{2} = 1/16$). Based on this calculation, one would therefore expect a four-game comeback to take a series as very rare. Mathematically, one would not expect this result with high probability in 20 World Series. Again, variables such as talent disparity, injury, the tendencies of teams in certain stadiums, and similar factors will play their role. The basic math, however, underscores that the expert's breathless pronouncement about the difficulty of a comeback is an overstatement.

Another common example of a statistic used in a superficial fashion is the emphasis placed by the expert on an aspect of the game that is not essential to a team's success, yet stated in an authoritative fashion. It is common in the course of an NBA basketball season to hear references to a certain player being the best slam dunker on his team or in the league as a whole. A dunk, or slam dunk, is the delivery of the basketball by a shooter through the cylinder to score with the ball being propelled down after the player has jumped high enough for the shooting hand and ball to be above the rim.

For example, a statement such as "X is the most prolific dunker in the NBA" might be made by the expert analyst. While there is no question that the dunk is in many situations an emphatic and athletic maneuver, from a mathematical perspective, assessing the relevance of this statistic to team success, this expert statement is of little value, because the dunk is worth the same as any other 2-point field goal attempt in basketball, therefore the manner in which the basket is made has no greater effect on the scoreboard. Further, the dunk is less important than the 3-point shot (a 50% greater value per successful attempt). In a team game, the analysis as to how the ball got to X for the dunk is more important than the dunk itself—how did X become open to make the dunk, what passes or other maneuvers were made by X or his teammates. In basketball, rebounding the other team's miss is a typical way that the ball begins its path to the other team's goal. Therefore, one would expect rebounding to be far more important in an analysis than is a dunking statistic.

In the simple analysis above, one would conclude that rebounding is far more important than dunking. In fact, based on statistics gathered using NCAA men's and women's basketball data over a five-year period, the simple analysis is confirmed. Where a team out-rebounds its opponent, it will win 72% of its games. Where that team also shoots more free throws than its opponent, its success rate climbs to almost 90%. Dunking, dramatic as it may be, is not a significant factor in team success.

Every sport has its statistics that, when employed in commentary, may impress but not necessarily inform the audience. In NHL hockey, frequent references will be made during telecasts as to how hard a particular player can shoot the puck. There is no question that from a physical standpoint, it is an impressive feat for a player to be able to deliver a shot towards the opposing goal at speeds in excess of 100 miles per hour. However, much like the dunking example in basketball, this statistic does not really contribute to the understanding of the game, especially if the information is taken in isolation.

The object of the game is to put the puck in the opposition goal, it is shooting accuracy that will be at a premium.

RATINGS PERCENTAGE INDEX (RPI)

Determining a winner in a team sport competition on one level is an easy task, i.e., the score at the end of the game is the sole indicator of success. In a series or in a season of competition, the winner is readily determined by the season standings, and so long as all teams in the competition have competed the same number of times against all others in the league, the final standing will be the determinate as to the champion.

Standings, and what are referred to as win/loss records, are a less useful standard where there are multiple teams and all do not compete against all of the others in a given season. For example, in American college sports, there may be as many as 350 teams in a particular division. While each team may play in a conference of between eight and 15 other programs, assessing a team using its won/loss record alone among conference rivals is straightforward, but to compare the team regionally or nationally from its conference play is problematic. For example, if a basketball team won 27 games and lost one, as opposed to a team they did not play who compiled a record of 16 wins and 12 losses, the team with the best record is not necessarily the better team, if its opponents were weaker than those of the 16 win team.

When an issue such as the determination of a national champion in a particular sport is at stake, consideration is given as to how teams that did not compete against one another in the regular season might be compared and ranked to create a fair championship that included all deserving teams. The Ratings Percentage Index (RPI) was created to achieve this objective.

The RPI is calculated in different ways for different sports, so as to reflect the nuances of that particular competition, but the RPI is an algorithm that takes into account the common features of a team's season record, the record of its opponents, and the record of its opponents' opponents. The theory in the construction of the RPI is simple: wins and losses must be assessed on a qualitative basis as much as they are counted on a quantitative basis. To calculate the RPI, Team A RPI = 25% (Team A record) + 50% (Team A's opponent's record) + 25% (Team A's opponent's record).

There are a number of variables that the RPI does not address. The following examples show the RPI as a potential rectifier of disparity that appears from a review of comparison teams' win/loss records alone.

Team A and Team B are both NCAA Division 1 women's basketball teams. Each is being considered for a place in the elite National Championship tournament. Team A is from California and Team B is from Connecticut, and the teams play in different conferences. Team A and Team B did not play one another during the course of the regular season.

Team A had a very successful season, compiling a record of 26 wins and five losses, for a wining percentage of 83.8%. Team B struggled for large parts of the season, achieving a record of 16 wins and 14 losses, 53.3%. Team A was therefore more than 30% more successful at winning games than Team B.

For example, Team A winning percentage (83.8%); Opponent's winning percentage (45%); and Opponents/Opponents (58%), the calculation is RPI Team A = $(83.8 \times 0.25) + (45.0 \times 0.50) + (58.0 \times 0.25) = 20.95 + 22.50 + 14.50 = 57.95$ (round to 58.0).

Continuing the example, Team B winning percentage (53.3%); Opponent's winning percentage (75%); and Opponents/Opponents (60%), the calculation is RPI Team B = $(53.3 \times 0.25) + (75.0 \times 0.25) + (60.0 \times 0.25) = 13.33 + 37.50 + 15.00 = 65.83$ (round to 66.0).

It is evident from this analysis that while Team A had a far more successful season in terms of winning games, Team B played a much more difficult schedule. The RPI would lead to the conclusion that if one were to choose between these teams, Team B is likely the better team. However, while the RPI is a more involved calculation than simple wins and losses, it has significant variables that cannot be reduced to mathematical equation. Those variables include injuries to key players, whether the wins were early or later in the season, or the margin of victory (the RPI calculation treats as equal a win by 25 points and a win in overtime by a single point, the margin is not relevant to the RPI).

As the RPI is used as a statistical measure more frequently, individual sports can customize the calculation to reflect a feature in its game. For example, in 2005, the NCAA adjusted its RPI for basketball championship calculations. The NCAA concluded that as a home court was a significant advantage to the home team in college basketball, a visiting team deserved extra credit for a victory in a hostile environment. The RPI was thus adjusted so that road win = 1.4 wins; road loss = 0.6 losses; home win = 0.6 wins; home loss = 1.4 losses; neutral site = 1.0.

NCAA hockey adopted a similar approach to fine-tune its RPI, with a road win worth 1.5 wins, a neutral site win valued at 1.3 wins, and a home win worth 1.0.

The RPI will never be determinative of ability; arguably, the only reliable measure of that standard in team sports is head-to-head competition. Used in conjunction with other tools, the RPI provides insight to complicated ranking and seeding issues.

MATHEMATICS AND THE JUDGING OF SPORTS

In most athletic disciplines, mathematics will illuminate aspects of performance. As noted in the discussion of team sport statistics, the more involved the mathematics, often the greater degree of insight into present and future performance.

Mathematics is a rather poor tool when used to explain sports that are subjectively assessed, such as figure skating, diving, synchronized swimming, and other similar disciplines. However, it is important to understand what is being sought to be achieved in the scoring of these disciplines, and to be careful in attempting to interpret any result beyond the obvious ranking of the participants.

Figure skating has always posed particular difficulty for judges: How can a subjective opinion, however knowledgeable, concerning elements of beauty, presentation, and grace be reduced to a score, a hard, certain mathematical proposition? Figure skating has had a number of judging scandals, usually turning on improper collaboration among judges to guarantee that certain participants would achieve certain scores. In 2004 the international figure skating adopted a scoring system referred to as a "code of points." The general proposition of this system is that every aspect of each skater's performance, every jump, spin, turn, and movement will be graded individually. The judges, typically numbering eight at an international event, would then add or subtract points for the skater's execution of each part.

This grade of execution applies to five overall components; the maximum score available in each component is 10.0. Each judge submits their individual score for each skater. The highest and the lowest scores from the judging panel are discarded, with the maximum score attainable of 50.0.

Similar to the manner in which team sports telecasts communicate statistics that are purportedly used to describe performance, but without proper reference to the fundamentals of the sport itself, figure skating judging and the mathematical results generated are difficult to understand. The mathematics here is an imperfect attempt to put a hard number to a purely subjective discipline.

Unless one has an in-depth knowledge of the judging criterion, the viewer is left with a number that is disconnected from performance. If a swimmer races 100 meters in a pool in 55 seconds, that result is observable. If a football player scores a touchdown by running with the ball, that result is observable. The math underlying the scoring in the judged sports is only an indicator to the most expert in that discipline; the more casual fan must treat the numbers generated as a simple ranking.

MATH AND THE SCIENCE OF SPORT

The aim of mathematical applications in sport is not to change the sport in question, but to better understand it. One must be able to understand the essence of performance. As already discussed, mathematics is a very helpful interpretive tool in sport. Math as the language of science can be applied in virtually every sport to understand how the game is played. However, the following examples of math explain how a particular sporting activity is performed, or how it might be performed better.

BASEBALL

The home run is arguably the most dramatic play in baseball, the product of a one-on-one confrontation between hitter and batter. Science will assist in the understanding of a number of features concerning how far a baseball can be hit. Assuming that the bat is constructed of wood and is 32 ounces in weight, that a ball is pitched at 85 miles per hour (an approximate average speed for a pitch thrown by an American major league pitcher), and that the ball was struck squarely on the "sweet spot" (the optimum part of the bat for striking the ball, given that the end of the bat is moving more quickly than the handle). To send the ball 400 feet, the bat must strike the ball at a speed of 70 miles per hour.

To take the analysis one stage further, the difference between the properties of an aluminum bat and a wooden bat can be examined. This analysis will turn on a calculation known as determining the coefficient of restitution (COR), a determination of how "springy" the surface of each bat is, which will impact upon how much energy will be lost in the transfer from the pitched ball to the bat when it is struck.

Due to the nature of each bat's construction, a 32-ounce aluminum bat will have a barrel of 2.75 inches (the maximum size permitted by major league baseball); the wooden bat will have 2.50 inches, as it is not possible to have the weight distributed more to the barrel of the bat, with the thin handle, as the bat tends to shatter on impact with a ball.

The aluminum bat has a barrel circumference that is 1.21 times bigger than the wooden bat (the ratio of 23.74 to 19.63), which translates into a surface available to strike a ball that is approximately 10% larger than the wooden bat. Precise scientific trials using standard baseballs show that 25% of the energy created in the collision between an aluminum bat and ball will be restored to the ball, as opposed to 20% of the similar energy generated between a wooden bat and a ball. The COR for the wooden bat is 0.45, the COR for the aluminum bat is 0.50. Using this general relationship (and assuming that other variables such as elevation, wind speed, and direction and the angle at which the bat strikes the ball are not factored), if a ball were hit 380 feet with a wooden bat, one would expect an aluminum bat to generate a hit of approximately 410 feet.

It is this type of analysis that has persuaded the authorities to forbid the use of aluminum bats in North American major league competition. The physics of the aluminum bat would not only make home runs an easier proposition, the speed of the ball created on impact with the aluminum bat would create a greater risk of danger for those fielders closest to the batter, including the pitcher, the first baseman, and the third baseman. As noted, the coefficient of restitution is a measure of the springiness of a surface, expressed as the ratio of the speed of the object before and after collision. If a rubber ball were thrown against a wall at 50 miles per hour, and it returned at 30 miles per hour, COR = 30/50 = 0.6.

The following will illustrate why the pitcher and his first and third base teammates are at greater risk from a ball hit with an aluminum bat. The regulated distance from the batter to the pitcher is 60 feet, 6 inches. However, on the delivery of a pitch, the pitcher must maintain contact with the point of measurement (the pitching rubber) only as the ball is being delivered, and as the pitcher throws the ball, the natural pitching motion will carry the pitcher one stride closer to the batter, plus any extra distance that is created by his follow-through. It is safe to assume that the pitcher will be no further than 54 feet from the batter as the ball is released.

Pitches at the major league level vary in speed, but a typical pitch will travel approximately 85 miles per hour (assuming a constant speed, although in practice the pitch will be faster on delivery and slower as it has traveled to the batter's box). The pitch thrown at 85 miles per hour is traveling at a speed of 124.67 feet per second. To travel the 60.5 foot distance to home plate, the ball will reach home plate in approximately 0.49 seconds. Studies using wooden bats confirm that the approximate time for a hard-line drive delivered directly at a major league pitcher is approximately 0.40 seconds.

Studies at the American college level, where aluminum bats are legal for use (with players presumably not as strong or as skilled as major league players), confirm the following: At speeds off the bat at between 103 and 113 miles per hour, batted balls were reaching the pitcher 54 feet away at 0.357–0.315 seconds. It would be reasonable to conclude that a major league batter would deliver the ball harder and therefore in a shorter time than the college sample.

As it is generally accepted that human reaction time is rarely faster than 0.20 seconds to an event, even where the event is anticipated (as with a pitcher who might expect a ball to be struck at him), not only would balls travel further when hit by the aluminum bat, the difference in the available reaction time to the pitcher and the time for the ball to reach the pitcher from the bat when struck would decrease to as little as 0.10 seconds in major league play.

CYCLING—GEAR RATIOS AND HOW THEY WORK

Cycling, at both the recreational level as well as international competition level, requires an understanding of a number of physical principles. The gears used by cyclists are an example. The "penny farthing" bicycles of the period from 1870 to approximately 1900 were constructed with a huge front wheel and a tiny rear wheel, with the pedals connected to the front wheel only. The penny farthing did not have gears connecting the large front wheel and the small rear wheel, and it depended upon the fact that one pedal by the rider would create one rotation of the large front wheel. The penny farthing construction is identical in principal to that of a child's tricycle. As an example, if a tricycle front wheel, to which the pedals are directly connected, is 16 inches in diameter, that wheel will have a circumference 50 inches. One revolution of the front wheel means that the tricycle travels 50 inches. If the child pedaling the tricycle pedals at a speed of 60 revolutions per minute, or one revolution per second, the tricycle will be traveling at 50 inches per second, which is a speed of 2.8 miles per hour. Greater speed can only be achieved through greater revolutions per minute.

If an adult wished to ride a tricycle at a speed of 15 miles per hour, a typical speed at which to ride a bicycle in a recreational fashion, the tricycle front wheel would have to be very large or the cyclist would not be able to generate enough revolutions per minute to get distance. If the adult pedaled at 60 revolutions per minute, to achieve a speed of 15 miles per hour, the front wheel would necessarily be 84 inches, or 7 feet, in diameter. Gearing was necessary to make cycling a more efficient form of movement.

The concept of gearing for a bicycle was first theorized by Leonardo da Vinci (1452–1519) in the 1500s, but the concept was not developed for commercial application until Frenchman Paul de Vivie, alias "Velocio," (1853–1930) built the first functional derailleur, the device that permits the gears on a multi-speed bicycle to be changed by the rider as the bicycle is ridden. The gearing on a bicycle permits a cyclist to use energy more efficiently to climb hills, and to maximize speed on a downhill.

A typical bicycle has wheels that are 26 inches in diameter. Gears are typically measured by the number of teeth each gear has on its circumference. For example, if a front gear ring on a bicycle has 54 teeth, and a rear gear wheel has 27 teeth, every time the front wheel rotates, the rear wheel will rotate twice, creating a 2:1 gear ratio.

The lowest gear ratio on a bicycle might be a front chain wheel with 22 teeth and a rear chain wheel with 30 teeth. This creates a lower gear ratio of 0.73:1. For each pedal stroke, the rear wheel will turn 0.73 times, meaning that the bicycle will move forward approximately 60 inches (approximately 3.4 miles per hour if the bicycle is pedaled at a 60 revolutions per minute rate). The highest gear ratio on the bicycle might be a front chain wheel with 44 teeth and a rear chain wheel with 11 teeth, creating a 4:1 gear ratio. As the bicycle wheels are 26 inches in diameter, the bicycle will move forward 326 inches with each pedal stroke. If the cyclist pedaled at a rate of 60 revolutions per minute, the bicycle will be traveling at a speed of 18.5 miles per hour. If the pedaling rate were doubled to 120 revolutions per minute, the bicycle would travel at a speed of 37 miles per hour. Gearing permits the bicycle in this example to travel at speeds ranging from 3.4–37 miles per hour, to climb steep hills or to travel along level terrain.

The gearing of the bicycle must be considered along with the cadence or the rate at which the bicycle is pedaled. A smooth, even cadence means that the bicycle will be pedaled efficiently, as the energy expended by a cyclist is delivered most efficiently at certain cadences. For example, on the very steep terrain where mountain bikes are effective, a cyclist may climb a hill at as little as 50 revolutions per minute, whereas a road racer will typically operate at cadences of between 80 and 120 revolutions per minute.

SOCCER—FREE KICKS AND THE TRAJECTORY OF THE BALL

The free kick, bent with skill around a wall of defenders, is one of the most dramatic aspects of soccer. The success of this tactic is dependent upon a host of variables, including the distance from the goal, the height of the opposing players' wall, the height of the goal, the force of the kick, the spin imparted to the ball, the air flow around the ball in flight, and lesser variables, such as air temperature, humidity, and the friction of the grass when the ball is struck.

Using a famous free kick goal from the England vs. Greece 2001 World Cup qualification game as an example, English midfielder David Beckham scored a goal analyzed as: The ball was kicked at 36 m/sec. (80 miles per hour); the ball was kicked from a distance 27 meters from the goal; the ball moved laterally in flight 3 meters (from Beckham's right to left, facing the goal); the ball cleared the defender's wall by 0.5 meters; as it entered the goal, the ball was traveling at a speed of 19 m/sec (42 miles per hour).

The analysis assists in understanding how players taking a free kick can best strike the ball for a similar effect. The high speed of the initial kick results in the ball having very little drag as it moves through the air. However, as it passes the opposition wall, the ball begins to slow, entering a smooth airflow (laminar) phase of its travel. Greater drag on the ball now occurs, when coupled with forces generated by the spin imparted on the ball by the foot of the player (Magnus force), the ball will appear to bend and dip, fooling the goalkeeper.

There have also been various studies conducted in recent years in world soccer due to the rise in the importance of the penalty shot. The penalty shot awarded for a foul committed against an offensive player in the defender's penalty area has long been a feature of soccer. Only since 1982 has the penalty shootout been the sanctioned method of deciding an international soccer game. The ball is placed at a spot 36 feet from the goal; the goalkeeper may not move until the ball is struck. The international soccer goal is 8 feet high and 24 feet wide. In an analysis of penalty kicks taken in the World Cup between 1982 and 1998, it was determined that 211 such kicks were taken, with 161 successfully made: Success rate = 161/211 = 76.3%.

Further analysis revealed that the goalkeeper during the penalty dove to the side to which the ball was directed (the correct side) 63% of the time. Also, 41% of all successful goals were scored within 6.2 feet of the goalkeeper's initial position. By studying the tendencies of an opponent (that is, whether the kicker tends to kick the ball in a particular direction, or to a particular part of the net on penalties), a soccer goalkeeper can increase the chance of saving a penalty.

GOLF TECHNOLOGY

In recent years, there has been considerable public debate about the technology of golf clubs, and the ability of golfers to hit a golf ball farther than ever. One key area of debate has centered on the construction of the driver, the club used to generate the greatest distance. World golf regulatory bodies have imposed rules with respect to the construction of the driver, based on the principles of coefficient of restitution (COR). As noted in the baseball math segment, COR is the ratio of the speed of an object measured before and after a collision with a fixed object, such as a wall. It is impossible to have an object speed after such a collision that is greater than the object speed prior to collision. The higher the COR (that is, the closer the COR is to 1.0), the faster the object is expected to move after collision. If the object, in this case a golf ball, were struck by a very bouncy, trampoline-type surface on the face of the golf club, one would expect it to travel farther than if struck by a denser, less elastic material. Huge sums are spent each year by golf equipment manufacturers to design surfaces for clubs that create high COR values.

The current COR for a driver legal for use in international golf is 0.83, meaning that an object striking the material used on the driver's surface at 100 miles per hour would expect to rebound at a speed of 83 miles per hour.

FOOTBALL—HOW FAR WAS THE PASS THROWN?

There are innumerable circumstances where mathematical principles can be used to assist in assessing performance. For example, to determine how far a quarterback actually threw a pass, consider the following: The quarterback takes the snap from the center and drops back to pass. The pass is delivered and is caught by a receiver 20 yards from the quarterback's position. The ball was thrown at an angle of 30° from the line of the hash mark on the field. What was the actual down-field gain for this pass and catch? The distance thrown down field is represented by d: $d = 20 \times \sin 30° = 20(0.5)$; $d =$ 10 yards.

MONEY IN SPORT—CAPOLOGY 101

To even the casual observer of the modern sporting world, the media coverage of teams and competitions is seemingly fixated with the financial aspects of sport. Player contracts, television contracts, the sale of professional franchises for huge sums of money, ticket prices to attend events, all these issues have captivated the sporting public to an ever-increasing degree.

The salary cap is a financial tool in place in a number of professional sports. By definition, a salary cap is a prescribed limit placed upon how much money individual athletes may earn from their playing contract, and the salary cap is also a limit as to how much a team may collectively pay its roster of players. NFL football and NBA basketball are the two best North American examples of leagues with a salary cap in place.

As with other examples of mathematics in sport, the expression of the salary cap in a sport as a finite number may appear simple; the calculation and the impact of the salary cap on different aspects of team organization and player transactions is often very complicated. The salary cap and its rules as employed in various sports have created a species of sports administrator commonly referred to as the team "capologist," an expert with respect to the interpretation of salary cap rules made by the league in question. The capologist will assist the team management in determining whether, from a financial perspective, certain types of player transactions comply with the salary cap rules.

The salary cap is generally intended to create two important results for a professional team. One, the owners of the team will have a measure of cost certainty, in that they will know that in the given season, the team's player payroll will not (or should not) exceed the cap limit. For example, if an NBA basketball team is said to have a salary cap of $82 million, the team payroll, in theory, may not exceed this amount, and the team must budget accordingly. Two, the level of the salary cap will impact decisions that the team may wish to make concerning trades and other acquisitions of players. As noted, the salary cap may be set out in a finite number.

The NFL salary cap structure is a complicated calculation, taking into account numerous factors. For the sake of illustrating the function of the salary cap and its impact upon team personnel decisions, the example is the amount of the salary cap. This figure is calculated using 64% of the team's "defined gross revenue" calculated from the previous year. For the purpose of this calculation, such revenues will include the team's share of the league television revenues, stadium ticket sales, merchandise sales, and related revenues generated by the games themselves for each team.

Therefore, in an imaginary NFL season, if a team had defined gross revenues of $120 million, the salary cap = $120,000,000 × 0.064 = 76,800,000; the salary cap for the next season would be $76.8 million.

By salary cap rules, the top 51 players' contracts on a team are included for the purposes of the salary cap. The amount of the contract is defined by both its face value, for example, if a player has a contract worth $10 million over a four-year period, as well as any bonuses that the individual contract may provide. All bonuses are prorated for the purpose of a salary cap calculation over the four years of the sample contract.

In the sample, if the player had a $10 million contract over four years, and a $1 million bonus he received upon signing the contract, for the purposes of the salary cap, the contract is expressed as $10 million / four years = $2,500,000/year (salary component); $1 million / four years = $250,000/year (bonus component); net salary = $2,750,000/year. The player salary is treated as $2.75 million against the salary cap total of $76.8 million. This calculation will be made with respect to all 51 current contracts.

Assume that the total player salaries are $71.2 million. The total available monies with which to sign other players to contracts is $5.6 million for the coming season, subject to releasing or otherwise terminating any existing contracts to create a greater cushion under the salary cap limit of $76.8 million.

How does the salary cap work if a team wishes to acquire a player beyond their means? In the example, the available money for player acquisitions is $5.6 million. The team finished the previous season at eight wins and eight losses, and it did not qualify for the postseason playoffs. The head coach and the general manager believe that a certain wide receiver, who is not under contract to any team and is therefore a free agent, would be a player who might take the team that extra step needed to make the league playoffs in the coming year.

This wide receiver is an elite player, and he is expected to command a salary of $10 million per season, and he will command a contract of four seasons. Can the sample team with only $5.6 million left in its salary cap sign this player? The capology options include:

- No bid for this player: The current roster, subject to other contingencies such as injury, remains intact. (In a salary cap, where a player has been injured, they remain in receipt of their salary for the life of the contract, all counted in some fashion against the salary cap.)
- Sign the elite player at $10 million per season for four seasons. To get "under" the salary cap in this example, the team would be required to cut other players whose salaries total $4.4 million for the coming season ($10 million in new salary, less the available $5.6 million). The team in this scenario would be required to assess whether the benefit to the team in terms of performance was worth the loss of other players; further, the variable of injury for the new player would be considered.
- Sign the elite player, but structure the $10 million salary in year one of the four years as follows: Agree that the contract will be a $20 million bonus, and $20 million in salary over the following three years. The bonus is prorated over four years, meaning only $5 million would count against the salary cap this coming season. As $5.6 million is available as room under the team's cap, the bonus/deferred salary structure works, at least for the first year. The team will have to assess how it deals with this contract in each successive year, as it will be required to count this player's salary contract in year two as Bonus = $5 million (25% calculated over four-year period) and Salary = $20 million obligation now payable over three years.

This math application in essence borrows from the team's future to pay for the present needs of the team. In the realm of the salary cap, the best interests on the team on the field and the best financial interest of the team do not always exist in harmony.

The more involved the mathematical equations dealing with salary cap, the less important are the players themselves. Further, it is a reasonable presumption that the greater the room available to a professional sports franchise in its salary cap, the greater potential profits to the ownership of the franchise.

Some salary caps have a punitive component for those teams that breach the salary cap rule; these penalties are often referred to as a luxury tax. The premise behind these measures is that the richer franchises that exceed the salary cap limits will pay monies back into the general funds of the league, which are then distributed among the franchisees that abided by the salary cap rules.

In the NBA, the tax on the individual player salary that broke the cap ceiling is 10%. The team is also obligated in general terms to pay a 10% team tax on its payroll that is in excess of the cap. There are a multitude of exemptions and qualifications; the bottom line for the owner is, are they prepared to exceed the salary cap and pay the penalties imposed if they get a team that might win a championship?

MATH AND SPORTS WAGERING

Team sports wagering has grown from its clandestine roots in taverns and clubs to a multi-billion dollar enterprise that includes private bookmakers and state-run sport bets. All forms of sport gambling have a mathematical basis, rooted in the concepts of probability and understanding the statistics relied upon by odds makers to establish betting systems. There are a number of different types of wagers available, each generally involving a different math principle:

- Straight bet: This is a wager placed on the final outcome of an event. For example, if a team is chosen as the winner and does win, the successful bettor gets a return on their money 1:1. If $100 were wagered on the team, the winner recoups his initial bet, plus $100.
- Odds: As with the straight bet, the wager is with respect to the final outcome, with the odds, or the

Key Terms

Average: A number that expresses a set of numbers as a single quantity that is the sum of the numbers divided by the number of numbers in the set.

Odds: A shorthand method for expressing probabilities of particular events. The probability of one particular event occurring out of six possible events would be 1 in 6, also expressed as 1:6 or in fractional form as 1/6.

Percentage: From the Latin term *per centum* meaning per hundred, a special type of ratio in which the second value is 100; used to represent the amount present with respect to the whole. Expressed as a percentage, the ratio times 100 (e.g., 78/100 = .78 and so .78 × 100 = 78%).

Statistics: Branch of mathematics devoted to the collection, compilation, display, and interpretation of numerical data. In general, the field can be divided into two major subgroups, descriptive statistics and inferential statistics. The former subject deals primarily with the accumulation and presentation of numerical data, while the latter focuses on predictions.

probability, of the event added to the wager. For example, as in the earlier example, if the team were not likely to beat the opponent, the odds of such an event occurring might be as remote as 10:1 against, meaning that it is stated to be 10 times more likely that the team will lose than win. If $100 were wagered on 10 to 1 odds, and the team were successful, the successful bettor would again recoup the initial $100 wagered, plus 10 × 100, or $1,000.

- Point spread (also referred to as the line and other terms): This variation in sports betting is very popular in sports such as football and basketball. The nature of the point spread in any given game is typically calculated by professional gambling organizations, and published in major media. The bettor does not wager necessarily on the best team, but the wager is with respect to the difference in points between the team's scores at the end of the game. For example, Team A and Team B are NFL football teams scheduled to play on a Sunday afternoon. The professional gambling organization reviews the teams' records, injury situation, home field advantage, and the play of each team to date, and determines that "Team A is a 5-point favorite," which means that the gambling organization believes that Team A will beat Team B by 5 points or more. The organization will then take bets on the outcome of this game using that 5 points, referred to as the spread, as its betting standard for that game. The results in this type of bet for a bettor placing $100 on Team A are that Team A must win by 5 points or more. If Team A wins by 5 points exactly, the result is referred to as a "push": the bettor gets his $100 back, less the fee charged by the gambling house, 10%. Another result is for a bettor who places $100 on Team B. Because Team A is favored by 5 points, this bet will succeed if either Team B wins altogether, or Team B loses by 5 points or less. As with the straight bets, these wagers pay on a 1:1 ratio, less the 10% customarily charged by the betting establishment.
- Over/under: This bet and its variations are based upon the total number of points scored in a game, including any overtime played, by both teams; the win or loss of the game itself is not relevant. For example, in a basketball game, the wagering line would be established as 176 points, wagers invited as being over and under the mark. If a wager is successful in predicting whether the teams were a total over or under the line, the return is again a 1:1 ratio to the money wagered.
- Parlay: This form of wagering permits the bettor to gamble on two or more games in one wager. The better must be correct in all of the individual wagers to claim the entire bet. The reward multiplies in parlay betting, as does the risk of missing out on one wager in the sequence:In three-game parlay, Game has 12.7:1 odds; Game 2 has 3.3:1 odds; Game 3 has 1.9:1 odds. On a $5.00 wager on this three-game parlay, the return if each team selected were successful would be 2.7 × 3.3 × 1.9 = 16.93; $5 × 16.93 = $86.45. As is illustrated, a return of almost 17 times the initial $5 wager would be a successful gambler's reward in this scenario; a loss of any of the three games would mean the bettor would lose the entire parlay.
- Future event: It is common for both North American and world sporting events to be the subject of odds

posted by various professional gambling agencies. For example, in the lead-up to the World Cup of Soccer, every team will be the subject of odds of winning the quadrennial championship; a perennial soccer power like Brazil might be listed at 3 to 1 odds, while a traditionally less successful nation, such as Saudi Arabia or Japan, will be listed at more dramatic numbers such as 350 to 1. Wagers are typically binding at the odds quoted, no matter what might happen to the subject team in the period between the date of the wager and the date of the event. For example, if Brazil's best scorer and best goaltender were injured, the actual odds quoted for Brazil might be quite higher at the start of the championships; the wager would remain payable at the initial 3 to 1 odds.

Where to Learn More

Books

Adair, Robert K. *The Physics of Baseball,* 3rd ed. New York: Perennial, 2002.

Holland, Bart K. *What are the Chances? Voodoo Deaths, Office Gossip and Other Adventures in Probability.* Baltimore, MD: Johns Hopkins University Press, 2002.

James, Bill. *Baseball Abstract,* Revised ed. New York: The Free Press, 2001.

Periodicals

Klarneich, Erica. "Toss Out the Toss Up: Bias in Heads or Tails," *Science News,* February 28, 2004.

Postrel, Virginia. "Strategies on Fourth Down, From a Mathematical Point of View," *New York Times,* September 9, 2002.

Square and Cube Roots

Overview

Finding the square and cube roots of a number are amongst the oldest and most basic mathematical operations. A number, when multiplied by itself, equals a number called its square. For example, nine is the square of three. The square root of a number is the number that when multiplied by itself, equals the original number. For example, three is the square root of nine. The cube root is the same concept, but the cube root must be multiplied three times to yield the original number. These two concepts get their names from the relationship they have with the area of a square and the volume of a cube.

In our three dimensional world, lines that have one dimension, squares that have two dimensions, and cubes that have three dimensions form the basic shapes that mankind uses to build models of the world. The square and cube of a number, and their inverses the square and cube roots, allow us to relate the length of a line to the area of a two-dimensional square or the volume of three-dimensional cube respectively.

Examples of the square and cube roots will be found in any area of design where a model of an object will need to be conceptualized before the object can be built, for example in the architect's plans for a new house or the maps for the construction of roads, or the blueprints of an aircraft. During the design phase, whenever areas and volumes need to be manipulated, the square and cube roots would be used to calculate these quantities.

Fundamental Mathematical Concepts and Terms

The definition of the square root is a number that when multiplied by itself, will yield the original number. As an example, again consider the value 9. It has a square root of 3, so $3 \times 3 = 9$. The value 9 is called the square of 3. The cube root is similar, but now the value that has to be multiplied is multiplied by itself three times, for example, the cube root of 8 is 2, so $2 \times 2 \times 2 = 8$ and the value 8 is called the cube of 2.

The names square and cube root come from their relation with these shapes. Consider a square, where each side has an equal length; if you know the area of the square, the square root will give you the length of one side. Since all the sides are an equal length, you have found the length of them all. The area may be some square land where you want to know how much fencing is needed to mark the edge of your land. If the area is 100 square meters then the length of one edge is 10 meters. As there are four edges to the square, you will need to buy 40 meters of fencing.

The cubed root comes from the same idea. Imagine a wooden cube, where each edge is again exactly the same length. If we know the volume of this cube, the cube root will give us the length of one of the edges; since it is a cube, we know the length of all the edges. For example, an architect has calculated that his building will need a foundation with 1000 cubic meters of cement to hold the weight of the structure safely. The cube root of 1000 is 10, so the builders will know that by marking a 10 by 10 meter square out on the floor and digging down 10 meters this hole will be the right size for the cement.

NAMES AND CONVENTIONS

In mathematical text the radical symbol is used to indicate a root of a number. The square root is written as $\sqrt{9} = 3$.

To indicate roots or higher than the square root, for example the cubed root, the number of the root is entered into the top left part of this symbol. For example the cubed root is written as $\sqrt[3]{8} = 2$.

This notation was developed over a period of about 100 years. The right hand slash and line above the numbers first appeared in 1525 in the first German algebra book, *Die Coss*, by Christoff Rudolff (1499–1545). It is thought that the notation of adding the number 3 for a cube and numbers for higher roots as a symbol to the top left of the radical was first suggested by the Western philosopher, physicist, and mathematician René Descartes (1596–1650). The addition of the "vee" to the left side of the symbol is thought to have been developed in 1629 by Albert Girard (1595–1632), a French mathematician who had some of the first thoughts on the fundamental theorem of algebra.

The name root comes from a relationship with a family of equations called polynomials, these equations contain all the powers of a variable x in an infinite series and have the form, $y = a + bx + cx^2 + dx^3 + ex^4 \ldots$ and so on, forever. All the letters on the right hand side of the equals sign, apart from the x, can have any values we want. Setting a value to zero will eliminate that term in the series.

A Brief History of Discovery and Development

In ancient times numbers held a deep religious and spiritual significance. Mathematics was heavily based on geometry, philosophy, and religion. Early thinkers about the nature of geometry saw lines and other geometrical shapes as the fundamental and logical building blocks of the heavens and Earth. The idea that nature could always be expressed with lines and shapes lead to the development of Pythagoras' famous proof for triangles, a relation that uses the square root to calculate the final answer.

Pythagoras of Samos (c. 500 B.C.), was an extremely important figure in the history of mathematics. Pythagoras was an ancient Greek scholar who traveled extensively throughout his life. He founded a school of thought that had many followers. The society was extremely secretive but was based on philosophy and mathematics. The school admitted women as well as men to follow a strict lifestyle of thought and practice of mathematics.

Pythagoras' proof is for a triangle with one right angle and it relates the length of the longest side to the lengths of the other two sides. In the modern era, the proof is included in school textbooks and so it is hard for us to understand the deep impact on their way of life that this new method of logical thinking had on our ancestors. The proof—and knowledge of mathematics in general—were venerated as sacred secrets.

Today, Pythagoras' proof is learned as a formula with symbols, but this system of thinking would not have been known to its founder. Moreover, the proof that Pythagoras found was based purely on geometry. Legend has it that a philosopher of Pythagoras's society, called Hippasus, made the discovery at sea that if the two shorter sides of the triangle are set to 1 unit of length, then the result for the longer sided is an irrational number when the square root is taken. This special number could never be drawn with geometry and the legend goes that the other Pythagoreans were so shocked at this discovery that they threw him overboard to drown him and so keep his discovery a secret.

There is another important property of taking roots of numbers that was not understood until English physicist and mathematician Sir Isaac Newton's (1642–1727) time: the concept of taking the root of a negative number. If you try this on a calculator it will most likely give you an error. However, it was shown that it is possible to extend our number system to deal with taking the root of a negative number if we add a new number, given the symbol, i, in mathematics. This opened a whole new world of algebra that mathematicians call complex numbers and allows solutions to be found for problems that had previously been thought impossible.

From a practical viewpoint, this development affected almost every area of modern physics, which relies on complex numbers in some form or another. Some examples of their usage are found in electromagnetism, which gave us television, radio, and quantum mechanics, which gave us, among many other things, computers and modern medical imaging techniques.

PYTHAGOREAN THEOREM

Using just pure geometry, Pythagoras is famous for proving that, for a right angled triangle, the square of the lengths of the longest side, called the hypotenuse, is equal to the sum of the squares of the other two sides. This rather long sentence is much easier to follow if it is written as an equation: $h^2 = a^2 + b^2$.

In this equation, the letter h is the length of the hypotenuse and a and b are the lengths of the other two sides. As this equation has only squared terms, we must take the square root if we want to find the actual length of h.

For example, in a rectangular room, how long would a wire have to be if it was to be run in a straight line, across the floor, from the back, left hand corner, to the front, right hand corner? The room is full of furniture and it would be impossible to just measure the distance with a tape measure. However, we notice that the walls and the wire form a triangle pattern. Each wall is at right angles and lengths of the walls form the shorter two sides of the right angle triangle. The wire, running across the room, forms that longer side, the hypotenuse.

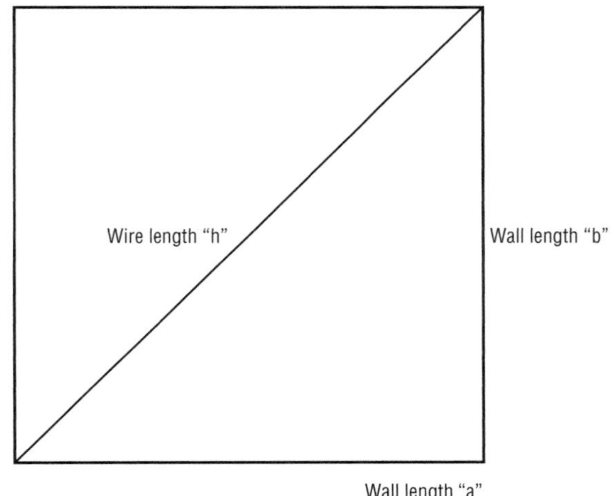

$$h^2 = a^2 + b^2$$
$$h = \sqrt{h^2} = \sqrt{a^2 + b^2}$$

One wall is 3 meters long, and the other is 7 meters, so: $h^2 = 3 \times 3 + 4 \times 4 = 25$. So the length of the wire is given by the square root of 25 as 5 meters long.

HIPPASUS' FATAL DISCOVERY

How long is the wire in the previous example if we have a room where each wall is just 1 meter long? $h^2 = 1 \times 1 + 1 \times 1 = 2$. Now take the square root of 2 to find 1.4142136.

In fact the digits of this number go on forever. It is a member of the family of numbers called irrational numbers. These numbers have the property that the fractional part of the digits continue forever and never repeat the same pattern. From the practical perspective of installing our wire, this is no problem as we would simply round up the length. However, in the exact world of mathematics the consequences are much more dramatic. Due to the fractional part having an infinite nature, it cannot be expressed as a ratio of integer values (a fraction).

What is even stranger is that we have made this length in something that is a perfectly reasonable and real geometric shape, a square box with sides equal to 1 meter. In this case, what exactly does the length of the line from one corner to the opposite corner of the box "mean"? Something that at first glance would seem child's play to measure is soon found to be impossible. No matter what we do, the length, given by the square root of 2, will always be wrong to some degree if we try to give it an exact value. In the legend of the death of Hippasus at the hands of his fellow Pythagoreans, it was the discovery of this anomaly that shattered the idea that the Heavens and Earth could be expressed totally and completely by lengths and their ratios.

Real-life Applications

ARCHITECTURE

The knowledge that some lengths are related with squared ratios has been known since Egyptian times, even though they would not have known the proof. Examples of this include the lengths 3, 4, 5, which are related by Pythagoras' theorem and are thought to be found in the construction of the Egyptian pyramids.

Today, squared and cubed roots are used in construction and design. If you were to design a car you might wish to change the volume of the driver's compartment. A modern three-dimensional (3D) design would be stored, as a wire frame model, in the memory of a computer. A computer program will divide the 3D space into thousands of tiny cubes, a job that is easy for a computer to do. Next, a program is run that counts the number of cubes within the driver's compartment and returns a value. The total volume is equal to the number of cubes found in the compartment, multiplied by the

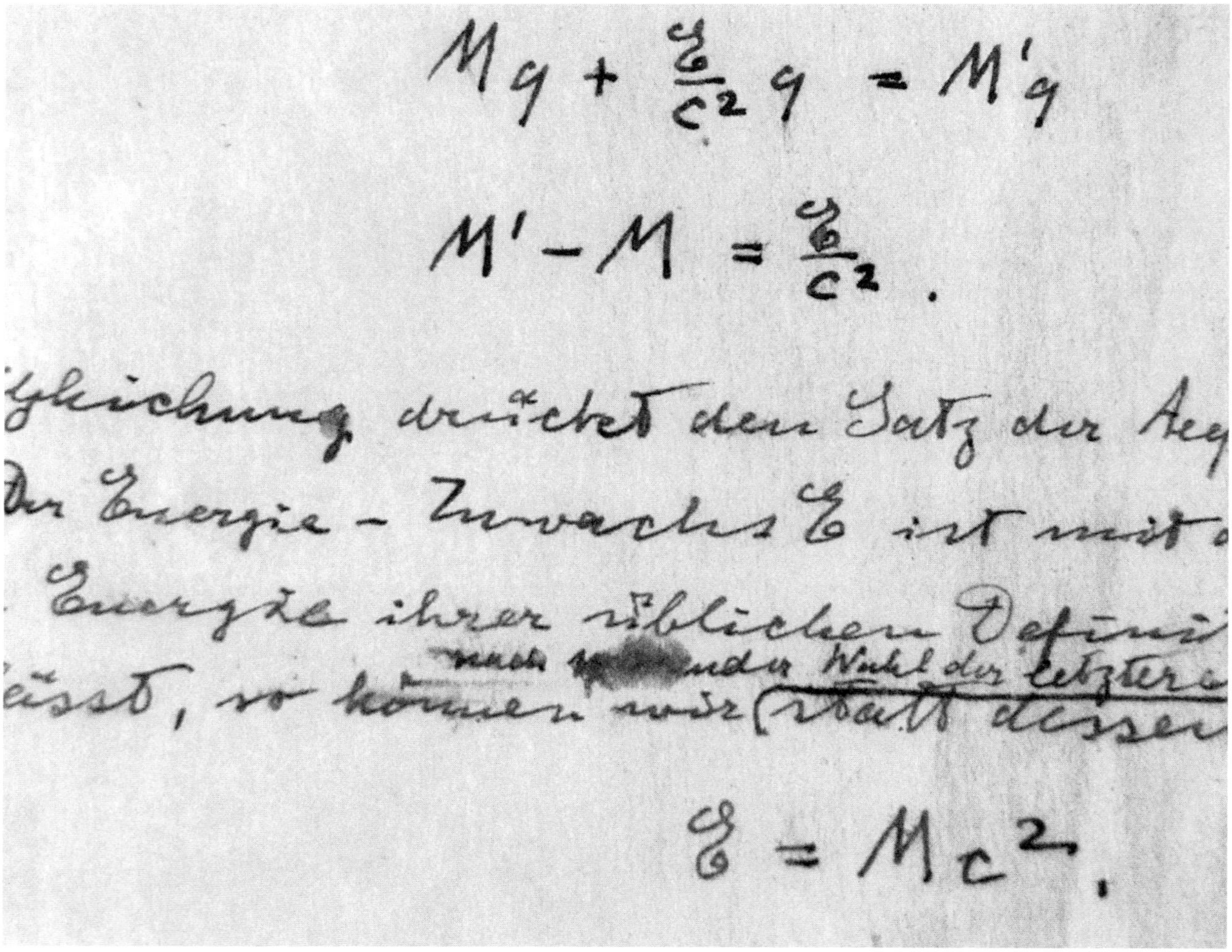

This paper, written in 1946, was written by Albert Einstein. He explains how he derived the formula E = Mc^2, a consequence of his Special Theory of Relativity, first published in 1905. The formula specifies that c (the speed of light) is squared. AP/WIDE WORLD PHOTOS. REPRODUCED BY PERMISSION.

volume of one cube. The one cube is called the unit cube and has real dimension; this allows us to make modifications to the actual size of the 3D wire frame without altering the wire frame itself.

To change the volume of the compartment, you change the volume of the unit cube. The amount that you would need to scale the sides of the unit cube is found by taking the cubed root of the original volume

NAVIGATION

The use of Pythagoras' theorem allows distance to be calculated on maps using coordinate systems. A coordinate system is a grid-like structure that is used as reference for points on the map's surface. Lines between one point and another form vectors and the calculation of lengths of vectors requires the use of square roots. Vectors can also be used to map velocity, a combination of speed and direction. These systems are used on land by the military, at sea by the navy and shipping firms, and in the air by aircraft, to plan and negotiate the terrain they are moving over. As an example, if two ships are moving perpendicular to each other, i.e, at 90 degrees to each other, and one ship is traveling at 3 knots and the other at 4 knots, using Pythagoras' relation, the navigators on the deck of each ship would measure the speed of the other as moving away from them at 5 knots.

SPORT

Football pitches, tennis courts, race tracks, and swimming pools are some examples of areas used by professional

Key Terms

Cubed root: The relation of the volume of a cube to one of its edges.

Root: The solutions of a polynomial equation, of which the square and cube root are special cases.

sports people that need to be accurately measured if the events are to be considered fair. The areas to be surveyed and locations of the various markings must be set down. The process of surveying these areas requires the use of roots in the calculations of various lengths for the markings

STOCK MARKETS

Many of the transactions used in stock markets use statistics to estimate the market trends and the best times to buy and sell stocks and shares. These calculations will often use something called the standard deviation, a measure of the spread of random events, and will give the traders some idea of the accuracy of their estimates. This calculation will require the use of roots.

Another occurrence of the root comes when the errors of predictive models are calculated. Models used to predict the stock market or anything else will have some sort of error depending on the accuracy of the data fed into it. If the error is much smaller than the size of the result, then the result can be trusted.

For example, if your model suggests that you buy gold next Wednesday, within an error of one hour, this is fine, but if the error is ten years then the it would be foolish to trust the result. As there may be many sources of error they will all have to be accounted for they need to be combined to give a final overall error. This technique is well defined in statistics, which requires the use of the square root.

Potential Applications

GLOBAL ECONOMICS

As global finance becomes more sophisticated, mathematicians and economists investigate the patterns of these transactions and look for relationships that will indicate the growth and decline of large groups of companies or even countries. It has only been recently, with the large scale computing and the application of a number of areas of science to economics that such models have come into use.

Successful interpretation of these trends, and new ideas and concepts in understanding the trends, are vital to the future development and stability of corporations and governments. This science, macroeconomics, is statistical in nature and allows predictions of important economic indicators such as inflation, interest rates, and the prices of materials. The use of squared and cubed roots in making these judgments incorporates fundamental formulas of probability and statistics that rely on square and cube roots.

Where to Learn More

Web sites

Wolfram. MathWorld. <http://mathworld.wolfram.com/> (February 1, 2005).

Statistics

Overview

Statistics is the branch of applied mathematics concerned with characterization of populations by the collection and analysis of data. Its applications are broad and diverse. Politicians rely on statistical polls to learn how their constituents feel about issues; medical researchers analyze the statistics of clinical trials to decide if new medicines will be safe for the general public; and insurance companies collect statistics about automobile accidents and natural disasters to help them set rates. Baseball fans immerse themselves in statistics that range from slugging percentages to earned run averages. Nervous travelers comfort themselves by reminding themselves that, statistically speaking, it is safer to travel in a commercial airliner than in an automobile. Students preparing for college fret over grade point averages and standardized test score percentiles. In short, almost every facet of daily life involves statistics to one degree or another.

Fundamental Mathematical Concepts and Terms

POPULATIONS AND SAMPLES

A statistic is a numerical measure that characterizes some aspect of a population or group of values known as random variables. They are random variables because the outcome of any single measurement, trial, or experiment involving them cannot be known ahead of time. The weight of men and women, for example, is a random variable because it is impossible to pick a person at random and know his or her weight before he or she steps on a scale. Random variables are discrete if they can take on only a finite number of values (for example, the result of a coin toss or the number of floods occurring in a century) and continuous if they can take on an infinite number of values (for example, length or height).

In some cases the populations are finite, for example the students in a classroom or the citizens of a country. While it may be impractical to do so if the population is large, a statistician can in theory measure each member of a finite population. For example, it is possible to measure the height of every student attending a particular school because the population is finite. In other cases, especially those related to the outcome of scientific experiments or measurements, the populations are infinite and it is impossible to measure every possible value. An oceanographer who wants to determine the salt content of sea water using an electronic probe is faced with an infinite population because there are an infinite number of places where he or she could place the probe.

In many practical situations, the underlying objective of statistics is to make inferences about the characteristics of a large finite or infinite population by carefully selecting and measuring a small sample or subset of the population. A political pollster, for example, may infer the likely outcome of a national election by asking a sample of a few hundred carefully chosen voters which candidate they prefer. An environmental scientist may collect only a few dozen samples in order to determine whether the soil or water beneath an abandoned factory is contaminated. In both cases it would have been impractical or impossible to analyze each member of the population, especially because the number of possible samples that could be collected is infinite. So, representative samples are chosen and statistics are calculated to draw conclusions about the population. Statistics that are calculated from measurements of an entire finite population are known as population statistics, whereas those that are based on a sample of either a finite or infinite population are known as sample statistics.

Because sample statistics are used to make inferences about populations, it is essential that the samples are representative of the population. If the objective of a study is to calculate average income, then it would be misrepresentative to poll only shoppers at a yacht brokerage because people who can afford yachts probably have incomes that are higher than average. By the same token, it would be just as misrepresentative to ask people waiting in line to file unemployment claims, because their incomes may generally be lower than average. Therefore, real world applications of statistics demand that considerable attention be given to experimental designs and sampling strategies if the statistical results are to be reliable.

One way to obtain a representative sample is to select members of the population at random. In simple random sampling, each member of the population has an equal chance of being selected or measured and there is no predefined sampling pattern. Random sampling is often accomplished using a computer program that generates random numbers or by referring to published random number tables. It is impossible to generate truly random numbers using a computer program, because the program itself must have some underlying structure or pattern. Mathematicians have been able to develop methods or algorithms, however, which generate nearly random numbers that suffice for most practical applications. To select a random sample of 100 people attending a sporting event, a statistician might assign a number to each seat in the stadium or arena. Then, he or she would generate 100 random integers and the people in the seats corresponding to those 100 numbers would comprise the random sample. Likewise, a scientist interested in measuring the soil nutrients in a farmer's field might divide the field using a grid of north-south and east-west imaginary lines. If the objective were to sample the soil at 20 random locations, the scientist would then use 40 random numbers to generate 20 pairs of north-south and east-west coordinates. One sample would be taken at each of the 20 locations specified by the coordinates.

Although simple random sampling works well for homogeneous populations, it may not produce truly random samples of heterogeneous populations that consist of distinct sub-populations or categories. In such cases, stratified random sampling provides more representative samples. The first step in stratified random sampling is to define the sub-populations. In a political poll, the sub-populations might be registered Democrats, Republicans, and Independents. In a marketing survey, the sub-populations might be defined in terms of age, sex, and income. Each sub-population is randomly sampled and the results are weighted so that they are proportionate to the relative size of each sub-population. Thus, stratified random sampling provides results that characterize each sub-population and the population in general, which the contribution of each sub-population proportional to its size.

PROBABILITY

It is possible to use basic statistical results without reference to the concept of probability. A diehard baseball fan, for example, can compare Babe Ruth's lifetime batting average of 0.342 to Hank Aaron's lifetime batting average of 0.305 and argue passionately that Ruth was the better hitter of the two. Batting averages are statistics, one is clearly larger than the other, and there is no need to worry about the nature of probability.

Unlike simple comparisons of batting averages, real life applications of statistics are in most cases closely tied to the concept of probability. The type of probability that is most often taught in basic statistics courses is known as relative frequency probability (or just frequency probability), and those who advocate this definition are known as frequentists. Relative frequency probability is defined as the number of times an event has occurred divided by the number of trials conducted or observations made, where the number of trials or observations is large. Flip a coin many times and the results should be very close to 500 heads and 500 tails, so the relative frequency is 500 ÷ 1,000 = 0.5, or 50%. All other things being equal, therefore, the probability of obtaining a head with the next toss is 50%. A slightly more complicated example might involve the measurement of a quantity that has an infinite number of possible outcomes, for example weight. If each

This tablet displays ancient Sumerian measurements and statistics (ca. 2400 B.C.). BETTMANN/CORBIS.

of 1,000 students in a high school were weighed, and 100 of them weighed between 140 and 150 pounds, then the relative frequency of a weight in that interval would be 100/1,000 = 0.1, or 10%. Therefore, the probability that a student selected at random would weigh between 140 and 150 pounds is 0.1. The determination of values of a random variable, in this case the weights of students in a school, by repeated measurement produces an empirical probability distribution.

Mathematicians have devised a number of theoretical probability distributions that play an important role in statistics, the best known of which is the normal, or Gaussian, distribution. Named after the mathematician Karl Friedrich Gauss (1777–1855), the normal distribution is defined by a probability density function that follows a distinctive bell shaped curve. Continuous random variables following a normal distribution are more likely to have values near the peak of the curve than near the ends. In many situations, it is the logarithms of values, not the values themselves that follow a normal distribution. In this case the distribution is said to be lognormal. Another example of a widely used theoretical probability distribution is the uniform distribution, which is defined by minimum and maximum values. Each value in a uniform distribution has an equal probability of occurrence. The binomial distribution applies to discrete random variables.

Although the normal (and lognormal), uniform, and binomial distributions are among the most common probability distributions, there are many specialized distributions that are particularly well-suited for specific problems. The Pareto distribution, for example, is named after the Italian economist Vilfredo Pareto (1848–1923) and is used in many statistical problems that consist of many small values and relatively few large values. It has found applications in studies of the distribution of wealth, the distribution of wind speeds, and the distribution of broken rock sizes encountered in construction and mining.

The great value of theoretical probability distributions, especially the normal distribution, is that they facilitate the use of rigorous mathematical tests that scientists can use to evaluate hypotheses and understand uncertainties in experimental data. For example, how likely is it that two samples were drawn from the same population? How certain are regulators that water quality meets government standards? How precisely must a product be manufactured to ensure that there is less than 1 defect in 1,000,000? How reliable are the results of a public opinion survey? The answers to these kinds of questions are more precise if the sample distribution follows a theoretical distribution and parametric statistical tests can be used. Therefore, one of the first steps in the statistical analysis of data is to determine whether the data are normally (or lognormally) distributed.

Statistics or statistical tests that are tied to a theoretical probability distribution are known as parametric. Those that are independent of any theoretical distribution are known as non-parametric.

MINIMUM, MAXIMUM, AND RANGE

The most fundamental statistics that can be calculated from a set of observations are its minimum value, maximum value, and range, which is the difference between maximum and minimum values. If the set of observations comprises the entire population, then the minimum and maximum will represent the true values. If the observations are only a sample of a larger population, however, the true or population minimum and maximum will be smaller and larger, respectively, than the sample minimum and maximum.

Consider the following list of values as an example: 8.95, 6.93, 11.07, 10.21, and 10.31. In order to calculate the range, first identify the minimum and maximum values in the list. In this case, as in most real life applications, the minimum and maximum values are not the first and last values. The minimum and maximum values in this example are 6.93 and 11.07, so the range is 11.07 − 6.93 = 4.14.

AVERAGE VALUES

An average is defined as a number that typifies or characterizes the general magnitude or size of a set of numbers. In statistics, there are several different types of averages known as the mean, median, and mode. The word average itself, however, does not have a formal statistical definition and is generally not used in statistical work.

The most common kind of average is the arithmetic mean, which is found by adding together all of the numbers in a lists and then dividing by the length of the list. Using the same list of numbers as in the previous section, the arithmetic mean is $(8.95 + 6.93 + 11.07 + 10.21 + 10.31)/5 = 9.49$. Another kind of mean, the geometric mean, is calculated using the logarithms of the values. The geometric mean is calculated as follows: First, find the logarithm of each number in the sample or population. For the example list of five values used above, the natural (base $e = 2.7183$) logarithms are: 2.19, 1.94, 2.40, 2.32, and 2.33. Second, calculate the mean of the logarithms, which is $(2.19 + 1.94 + 2.40 + 2.32 + 2.33)/5 = 2.24$. Finally, raise e to that power, or $e^{2.24} = 9.37$. Any base can be used to calculate the logarithms as long it is used consistently throughout the calculation. Statisticians sometimes refer to the arithmetic mean of a population as its expected value.

Another kind of average, the median, is the number that divides the sample or population into two subsets of equal size. If the list of numbers for which a median is to be calculated is of odd length, then the median is found by ordering or sorting the values from smallest to largest and selecting the middle value. If the list is of even length, the median is the arithmetic average of the two middle values of the sorted list. The sorted version of the example list from the previous paragraph is 6.93, 8.95, 10.21, 10.31, and 11.07. The length of the list is odd and the middle value is in position $(5 + 1)/2 = 3$, so the median is 10.21.

Although sorting is a trivial computation for a short list of numbers, sorting large lists can be time consuming and the development of fast sorting algorithms has been an important contribution to applied mathematics and computer science. To illustrate how a simple sorting algorithm works, compare the first two values of the sample data set from the previous paragraph, 8.95 and 6.93. The second value, 6.93, is smaller than the first value, 8.95, so the positions of the two values are switched. Next, the third value, 11.93, is compared to the first two. Because 11.93 is greater than both of the first two values, none of their positions in the list are switched. The fourth value, 10.21, is then compared. It is greater than the first two values, 9.93 and 8.95, but smaller than the third value, 11.93. Therefore, the positions of 10.21 and 11.93 are switched. The same procedure is repeated until each value in the list is compared and, if necessary, put into the correct position.

If a population follows a normal distribution or uniform distribution, its mean will be equal to its median. Another way of saying this is that the ratio of arithmetic mean to median is 1. If a population follows a lognormal distribution, however, the mean will be larger than the median. Scientists analyzing data often calculate the ratio of arithmetic mean to median as a simple preliminary method of determining whether the data are likely to follow a lognormal distribution. This is not a rigorous statistical method, though, and the preliminary result is often followed by more sophisticated calculations.

Astute readers will have noticed that the mean and median values calculated as examples in this section are not equal, but almost certainly will not know that the five numbers used in the calculations were selected at random from a normal distribution with an arithmetic mean of 10. If the five numbers represent a normal distribution, why are the mean and median different and why does neither of them equal 10? The answer is a consequence of the law of large numbers, which states that the difference between expected and calculated values decreases towards zero as the number of trials (in this case the number of randomly selected numbers) grows large. In other words, small sample sizes are likely to yield sample statistics that differ from the true population statistics. If the example calculations had been carried out using a list of 1,000 or 10,000 numbers, the sample arithmetic mean would have both been very close to 10. The corollary of this is that the reliability of sample statistics is generally proportional to the sample size. The larger the sample, the more likely it is that the sample statistics are accurate reflections of the underlying population statistics. In most practical applications, however, sample sizes are limited by the amount of money available to pay for the study (especially in cases where expensive laboratory tests must be conducted). The job of the practical statistician in many cases is to strike a balance between the desired accuracy of statistical results and the amount of money available to pay for them.

The third kind of average, the mode, is the most frequently occurring value in a sample or population. If no value occurs more than once, then the sample or population has no mode. If one value occurs more than any other, the data are said to be unimodal. Data can also be multimodal if more than one mode exists. For example, the list of values 3, 3, 4, 5, 6, 7, 7 has modes of 3 and 7.

MEASURES OF DISPERSION

Statistical measures of dispersion quantify the degree to which the values in a sample or population are clustered or dispersed around the mean. To illustrate the need for measures of dispersion, consider two samples. The first is 2, 3, 4, 5, 5, 6, 7, and 8. The second is 2, 3, 5, 5, 5, 5, 7, and 8. Both samples have identical minima, maxima, ranges, means, and medians, but the numbers comprising the second are more tightly grouped around the mean value of 5 than those in the first sample.

The most common measure of dispersion is the variance, which is based on the sum of squares of differences between the sample values and their mean. For the first set of example values in the previous paragraph, the mean is 5 and the sum of squared differences is $(2 - 5)^2 + (3 - 5)^2 + (4 - 5)^2 + (5 - 5)^2 + (5 - 5)^2 + (6 - 5)^2 + (7 - 5)^2 + (8 - 5)^2 = 28$. If the list of numbers represents an entire population, then the sum of squared differences is divided by the length of the list (in this case $n = 8$) to find the population variance of $28 / 8 = 3.5$. If the list of numbers represents a sample of a population, however, the sum is divided by one less than the number of values ($n - 1 = 7$) to find the sample variance of $28 / (8 - 1) = 4.0$. Repeating the calculation for the second sample, the result is $(2 - 5)^2 + (3 - 5)^2 + (5 - 5)^2 + (5 - 5)^2 + (5 - 5)^2 + (5 - 5)^2 + (7 - 5)^2 + (8 - 5)^2 = 26$. Depending on whether the result is for a population or sample, the variance is either $26/8 = 3.25$ or $26/(8 - 1) = 3.71$. Therefore, the variance of the second sample is smaller than that of the first even though the two samples have the same mean, minimum, and maximum values.

Because the variance is calculated from squared terms, the units of the values being calculated must also be squared. If the units of measurement are length (meters, for example), then the variance would be expressed in terms of length squared. The use of squared terms also means that variances will always be positive values.

The denominator used to calculate the sample variance is slightly larger than that used to calculate the population variance in order to account for the uncertainty or bias inherent any time that a sample is used to make inferences about a population. If the data set for which a variance is being calculated is the entire population, then the mean value used in the calculation is the population mean and the calculated variance is therefore unbiased. If the data set is a sample or subset of the population, though, the mean value is only an estimate of the population mean. Therefore, any subsequent calculations must take into account the fact that the use of the sample mean adds some bias to the results. This is accomplished by using a slightly smaller number ($n - 1$ rather than n) in the denominator to produce an unbiased estimate of the variance. The effect of dividing by $n - 1$ rather than n will decrease as the sample size becomes large, which reflects the fact that a variance calculated from a very large sample is a more accurate representation of the population variance than one calculated from a small sample.

Another commonly used measure of dispersion is the standard deviation, which is simply the square root of the variance. As such, standard deviations have units of plus or minus (±) the original units of measure. A variance of 4.0 meters^2 is therefore equivalent to a standard deviation of ±2 meters. If the data being analyzed follow a normal distribution, then 68% of the values will fall within plus or minus one standard deviation of the mean, 95% will fall within two standard deviations of the mean, and 99.7% will fall within three standard deviations of the mean. If the data for which statistics are being calculated are measurements of error, for example the difference between the designed length and the actual length of an automobile part, then the standard deviation is often referred to as the root mean square or RMS error.

There are some situations in which the variance, and therefore the standard deviation, of a population is infinite. In such cases, attempts to calculate a variance will not converge on a single value as the sample size increases, and variances calculated using different samples of the same population will produce different results. It may still be possible, however, to calculate a statistic that is known as the average deviation, mean deviation, or mean absolute deviation. It is calculated in a manner similar to the variance, but the absolute values of each difference are used instead of their squares. The sum of absolute deviations of the sample 2, 3, 4, 5, 5, 6, 7, and 8 is thus Abs(2 − 5) + Abs(3 − 5) + Abs(4 − 5) + Abs(5 − 5) + Abs(5 − 5) + Abs(5 − 5) + Abs(7 − 5) + Abs(28 − 5) = 12, where Abs means "the absolute value of," and the average deviation is thus 12/8 = 1.5.

Statisticians have largely avoided the average deviation for two reasons. First, it is difficult to work with absolute values when performing mathematical derivations. Second, the trick of dividing through by $n - 1$ rather than n to produce an unbiased estimate does not work nearly as well as with the variance. Therefore, statistics books do not contain alternative population and sample formulations for the average deviation. For the large data sets commonly encountered by many scientists and engineers, however, the difference between dividing by n and $n - 1$ is small enough to be inconsequential. Therefore, the average deviation is a statistic that has theoretical limitations but can be a useful practical tool for large data sets, and particularly those for which the variance is infinite.

CUMULATIVE FREQUENCIES AND QUANTILES

Cumulative frequency is closely related to relative frequency probability and has many applications in real life statistics. It is defined as the number of occurrences in a sample that are less than or equal to a specified value. If the cumulative frequency is divided by the number of data in a sample, it is, following from the relative frequency definition of probability, known as the relative cumulative frequency, cumulative probability, or plotting position. For a sample consisting of n data sorted from smallest to largest, the relative cumulative frequency of data point m is often calculated as $m/(n + 1)$. Consider this sample of five values: 19, 7, 20, 10, and 17. To calculate the relative cumulative frequency, first sort the list from smallest to largest to obtain 7, 10, 17, 19, 20. The relative cumulative frequency of 7, the first value in the list, is thus $1/(5 + 1) = 0.17$, or 17%. The relative cumulative frequency of 10, the second value in the list, is $2/(5 + 1) = 0.33$, or 33%. This procedure is repeated for each element in the list until a relative cumulative frequency of $5/(5 + 1) = 0.83$, or 83%, is obtained for the largest value. Thus, 17% of the values in the sample are less than or equal to 7 and 83% are less than or equal to 20. If the sample is representative of the population from which it was drawn, the same relative cumulative frequencies apply to the population. This approach also assumes that relative cumulative frequency is being calculated for a sample, not a population, because the formulation allows for the proportion 1/n of the values to fall below the smallest value in the list and 1/n of the values to fall above the largest value in the list. It is attributed to the Swedish engineer Waloddi Weibull (1887–1979), whose statistical formulations are often applied to analyze the sizes of events in sequences (for example, the sizes of yearly floods along a river).

Quantiles, sometimes known as n-tiles, are the values that correspond to particular relative cumulative frequency values. Using the data from the previous paragraph, the 0.17th is 7 and the 0.83rd quantile is 20. If the sample size is small, some quantiles will be undefined. For example, there is no 0.10^{th} in the list of five values used in the previous paragraph because none of the values has a relative cumulative frequency of 0.10. If it can be shown that the sample was drawn from a known theoretical distribution, such as a normal distribution, then statisticians can calculate the value that theoretically corresponds to a given quantile. The 0.25, 0.50, and 0.75 quantiles are often referred to as the first, second, and third quartiles, whereas the 0.01, 0.02, 0.03, 0.99 quantiles are often referred to as percentiles.

The Weibull formula, $m/(n + 1)$, is only one of several different ways to calculate the cumulative probability. In fact, the Weibull formula is somewhat arbitrary. The 1 was added to the denominator because data were at one time plotted on special graph paper, known as probability paper, which did not allow values of 0 or 1. This is because, strictly speaking, it is impossible for the probability of an event occurring to take on either of those values. Probabilities can come very close to 0 or 1, but never reach them. Another approach, known as Hazen's method, uses the formula $(m - \frac{1}{2})/n$ and is widely used in hydrologic studies. If it can be inferred that a sample follows a normal distribution, the quantiles can be calculated using a formula specifically designed for normal distributions. For most practical statistical problems there is usually very little difference between the values calculated using different methods.

CORRELATION AND CURVE FITTING

Correlation describes the degree to which two or more sets of measurements are related. For example, there is a general correlation between the height and weight of people (especially if they are of the same age, sex, and location). Correlation does not require a perfect relationship, but rather a degree of relationship or correspondence. It is possible that any given tall person weighs less than any given short person, but on average tall people will weigh more than short people.

Statisticians calculate correlation coefficients to express the degree to which two variables are correlated. The most common form of correlation coefficient is called the Pearson correlation coefficient, and is calculated using sums of mean deviations for each variable. It is almost always represented by r or R. Correlation coefficients can range from -1 to $+1$. A correlation coefficient of $r = 0$ represents a complete lack of correlation between two variables, and points plotted on a graph to represent the two variables will appear to be randomly located. Variables with correlation coefficients of $r = -1$ or $r = +1$ plot along a perfectly straight line, with the sign of the correlation coefficient indicating whether the slope of the line is negative or positive. In real life, most correlations fall somewhere in between these two extremes.

If two variables are correlated, it is often useful to express the correlation in terms of the equation for a straight line or curve representing the relationship. The simplest relationship is one in which the two variables are related by a straight line of the form $y = b + mx$. Because it is rare for variables to be perfectly correlated, the challenge is to find the equation for the line that fits data the best. There are several ways to do this, and all of them incorporate some way of minimizing the differences between the line and the data points. Regression is a parametric, or distribution-dependent, procedure because it assumes that the differences to be minimized follow normal distributions. The general practice of finding the equation of the line that best represents the relationship between two correlated variables is known as regression or, more informally, curve fitting.

STATISTICAL HYPOTHESIS TESTING

In a previous example it was shown that the arithmetic mean of the numbers 8.95, 6.93, 11.07, 10.21, and 10.31 is 9.49. Could the numbers have been drawn at random from a normal distribution with a mean of 9 or less, even though the calculated sample mean is greater than 9? Possibilities such as this can be evaluated using statistical hypothesis tests, which are formulated in terms of a null hypothesis (commonly denoted as H_0) that can be rejected with a specified level of certainty. Statistical hypothesis tests can never prove that a hypothesis is true. They can only allow statisticians to reject null hypotheses with a specified level of confidence.

One common hypothesis test, the t-test, is used to compare mean values. It assumes that the values being used were selected at random from a normal distribution and that the variances associated with the means being compared are equal. It also takes into account the number of samples used to calculate the mean, because sample means calculated from a large number of values are more reliable than those calculated from a small number of values. The sample size is taken into account by using a probability distribution known as the t-distribution, which changes shape according to the number of samples. If the sample number is large, generally above 25 or 30, the t-distribution is virtually identical to the normal distribution.

To determine if the numbers 8.95, 6.93, 11.07, 10.21, and 10.31 are likely to have been drawn from a population with an arithmetic mean of 9 or less, first define a null hypothesis. In this case, the null hypothesis is that the arithmetic mean of the population from which the sample is drawn is less than or equal to 9. The result of the t-test, which can be performed by many computer programs, is a probability (p-value) of 0.27. This means that a person would be incorrect 27 out of 100 times if the population were repeatedly sampled and the null hypothesis rejected each time. Scientists often use a threshold (also known as a level of significance) of 0.05, so in this case the null hypothesis cannot be rejected because it is greater than either of those commonly used values. It can be tempting to interpret the failure to reject a null hypothesis at an 0.05 level of significance as a 0.95, or 95%, probability that the null hypothesis is true. But, this interpretation is inconsistent with the relative frequency definition of probability and should be avoided.

Similar tests can be conducted to compare the means of two samples (using a slightly different kind of t-test) or to compare the variances of two distributions (using an F-ratio test). In all cases, the tests are carefully structured so that the result is given as the probability of being incorrect if the null hypothesis is rejected.

CONFIDENCE INTERVALS

Another way to characterize the uncertainty associated with sample statistics is to calculate confidence intervals for the sample mean and variance. For the example of 8.95, 6.93, 11.07, 10.21, and 10.31, the confidence interval for the arithmetic mean at the 0.05 level of significance is 7.48 to 11.51. Calculation of the mean confidence interval relies on the t-distribution, so increased sample sizes will result in smaller confidence intervals. In other words, the larger the sample the more precisely the population mean can be estimated.

As above, the relative frequency definition of probability requires that this result be interpreted to mean that that true mean would be contained with the confidence interval 95 out of 100 times if samples of five were repeatedly drawn from the population. This is, strictly speaking, different than stating that there is a 95% probability that the population mean is between 7.48 and 11.51. The normal distribution from which the example values were drawn had a population mean of 10, so in this case the population mean did fall within the confidence interval. An analogous test can be performed to calculate confidence intervals for the F-ratio test.

If the variance of a population is known or can be estimated, the number of samples required to obtain a confidence interval of specified size can be calculated. Knowledge of the variance can come from other studies involving similar data or a small preliminary study.

ANALYSIS OF VARIANCE

Analysis of variance, which is often shortened to the acronym ANOVA, is a method used to compare several data sets. This is accomplished by comparing the degree of variability of measurements within individual sample sets to those among different sample sets to determine if their means are significantly different. The null hypothesis being tested is that all of the sample means are equal.

In biology and medicine, the different sample sets often represent different treatments (for example, does treatment with drug A produce better results than treatment with drug B or a placebo?). In geology, the samples

A mother with her triplets. The statistical chance of a woman having triplets without fertility treatments is about one in 9,000 births. SANDY FELSENTHAL/CORBIS.

might represent the sizes of fossils from different locations or the amount of gold in samples from several different rock outcrops. In political science, the samples might contain the ages of voters with different political tendencies (for example, are the average ages of liberal, moderate, and conservative voters significantly different?).

ANOVA assumes that the samples being compared are normally distributed (thus, like regression, it is a parametric procedure), that their variances are approximately equal, and that their samples are approximately the same size. Variances are calculated for each sample or treatment, and all of the samples are grouped together to calculate a total variance. ANOVA assumes that the total variance consists of two components: one resulting from random variance within each sample and the other resulting from variance among the different samples. The two variances are compared using an F-ratio test to determine whether the null hypothesis can be rejected at a specified level of significance. In the hypothetical case that all of the samples are identical, the variance among samples (and therefore the F-ratio) is zero. Thus, the null hypothesis would not be rejected. If the F-ratio is large, and depending on the sample sizes and desired level of significance, the null hypothesis may be rejected. As with all statistical tests, the F-ratio tests in ANOVA do not prove anything. They can only be used to reject or fail to reject the null hypothesis at a specified level of significance.

USING STATISTICS TO DECEIVE

The aphorism that there are "lies, damned lies, and statistics" is attributed to British statesman Benjamin Disraeli (1804–1881) and reflects the unfortunate fact that statistics can be accidentally or deliberately used to deceive just as easily as they can be used to illuminate and inform. Understanding how statistics can be accidentally or deliberately used to misrepresent data can help people to see through deceptive uses of statistics in real life.

Consider a group of four friends who graduated from the same college. Three of them earn $40,000 per year working as managers in a local factory, while the fourth earns $500,000 per year from his family's shrewd

Correlation or Causation?

Some of the most common examples of real life statistics are news stories describing the results of recently published medical or economic research. A newspaper article might give details of a study showing that men and women with college degrees tend to have higher incomes than those who have never attended college. A report on the evening news might explain that researchers have found a correlation between low test scores and excessive soft drink consumption among high school students. In both cases, variables are correlated but the studies do not necessarily prove that one causes the other to occur. In other words, correlation does not necessarily imply causation.

It is easy to think of reasons why people who obtain college degrees tend to make more money than those who do not. College degrees are required for many high paying jobs in science, engineering, law, medicine, and business. College graduates also know other college graduates who can help them to get good jobs and can take advantage of on-campus interviews. People who do not attend college, in contrast, are excluded from many high paying careers and may not have the same advantages as college students. This is not to say that there are no exceptions, because someone with a college degree may choose to take a low paying job for its intrinsic satisfaction. Social workers, teachers, or artists, for example, may have college degrees but earn less money than factory workers without degrees. Likewise, some multi-millionaires and even billionaires never completed college. What about the converse? Is it possible that high earnings cause people to become college graduates? In one sense, the answer is no. People usually attend college early in life, before they begin full-time careers, so it is unlikely that high earnings cause college attendance. It also seems unlikely that someone will make a sizable amount of money and, because of that, decide to attend college. It seems safe to conclude that, all other things being equal, college degrees are likely to cause higher earnings.

The other result, showing a correlation between soft drink consumption and low test scores, may be more difficult to explain. It is difficult to imagine that soft drink consumption alone causes a chemical or biological reaction that reduces intelligence and lowers test scores. But, there may be other factors to consider. It may be that students who like soft drinks place a higher priority on instant gratification than discipline, a quality that might also cause them to spend less time studying than students who consume few soft drinks. If that is the case, then both excessive soft drink consumption and low test scores are caused by another factor such as their parents' attitudes towards delayed gratification. If so, correlation would not reveal causation in this case.

investments in the stock market. What statistic best represents the income level of the four friends? The arithmetic mean is (\$40,000 + \$40,000 + \$40,000 + \$500,000)/4 = \$155,000, but in this case the arithmetic mean is not an accurate reflection of the underlying bimodal population. If anything, the median income of \$40,000 is more representative of most of the group even though it does not accurately reflect the highest salary. It is likewise strictly correct to state that the incomes of the four friends range from a minimum of \$40,000 to a maximum \$500,000, but that simple statistic does not convey the fact that most of the friends earn the minimum amount. It would therefore be true but misleading for a university recruiter to tell prospective students that a group of its graduates earns an average of \$155,000 per year or that graduates of the university earn as much as \$500,000 per year. A less deceptive statement that that the group earns between \$40,000 and \$500,000, and that three of them earn the minimum amount (or that the mode is \$40,000). But, this still does not paint an accurate picture. An even less deceptive statement would also explain that while the highest earner is indeed a graduate of that college, his income is tied to his family's investments and not necessarily related to his college education.

There are several kinds of clues that can help determine if statistics are deceptive. The first is use of only maximum or minimum values to characterize a sample or population, to the exclusion of any other statistics. Parties involved in a dispute may emphasize that reported values are as high as or as low as a certain figure without giving the range, mean, median, or mode. Or, someone hoping to use statistics to prove a point may cite a mean without mentioning the median, mode, or range. Another potential source of deception is the use of biased or misrepresentative samples, which may produce sample statistics that are not at all representative of the underlying population. Reputable statisticians will always explain how their samples were chosen.

A Brief History of Discovery and Development

The history of statistics dates back to the first systematic collection of large amounts of data about human populations in the sixteenth century. This included weekly data about deaths in London and data about baptisms, marriages, and deaths in France. The first book about statistics, titled *Natural and Political Observations Upon the Bills of Mortality*, was written by the English mathematician John Graunt (1620–1674) in 1662. His motivation was practical: London had suffered from several outbreaks of plague, and Graunt analyzed weekly death statistics (bills of mortality) to look for early signs of new outbreaks. He also estimated the population of London. British astronomer Edmond Halley (1656–1742), best known for the comet that bears his name, wrote about birth and death rates for the German city of Breslaw (sometimes spelled Breslau, and now Wroclaw, Poland). His results were used by the English government to set the prices of annuities, which provided regular payments similar to a retirement fund, according to the age and sex of the person. The government had previously lost a considerable amount of money when it sold annuities to young people using rates based on average life expectancy during times of plague and war, and the annuity holders failed to die quickly enough. The French mathematician Abraham de Moivre (1667–1754) worked in London and was also interested in the statistics of death and annuities, publishing the book *The Doctrine of Chances* in 1714. He is known as the first person to write about the important properties of the normal distribution, and also for predicting the date of his death.

The dawn of the eighteenth century was marked by an explosion of inquiry about statistics in probability, including important books by Karl Friedrich Gauss (1777–1855) and Pierre Simon Laplace (1749–1827). The normal distribution is often known as the Gaussian distribution in deference to his work. The Statistical Society was established in London in 1834, and five years later the American Statistical Association was established in Boston. Much of the theory that stands behind modern statistics, though, was not discovered until the early twentieth century by notables such as Karl Pearson (1857–1936), A.N. Kolmogorov (1903–1987), R.A. Fisher (1890–1962), and Harold Hotelling (1895–1973), for whom numerous statistical methods and tests are named. One of the most unusual statisticians of the early twentieth century was William S. Gosset (1876–1937), who wrote under the pseudonym Student. He is best known for the t-test and t-distribution, which is commonly referred to as Student's t.

Real-life Applications

GEOSTATISTICS

Geostatistics is a specialized application of statistics to variables that are correlated in space, and is based on a concept known as the theory of regionalized variables. It has important applications in fields such as mining, petroleum exploration, hydrogeology, environmental remediation, ecology, geography, and epidemiology.

Traditional statistics is concerned with issues such as sample size and representativeness, but does not explicitly address the observation that many variables are spatially correlated. Spatial correlation means that samples taken in close proximity to each other are more likely to have similar values than those taken great distances apart. The variable being sampled might be the distribution of insect types or numbers across a landscape, the physical properties that characterize a good petroleum reservoir or aquifer, the occurrence of valuable minerals (such as gold or silver) in different parts of a mine, or even real estate prices in different parts of a city. Whatever their discipline, people who use geostatistics measure some variable at a limited number of points (for example, places where oil wells have already been drilled or the locations of homes that have been sold in the past few months) but need to calculate values at locations where they have no measurements. This process is known as interpolation, and geostatistics provides a set of tools that interpolate values based on the distribution of known values at different locations.

Central to the theory and application of statistics is the variogram, which is a graphical representation of spatial correlation. It depicts the variance among samples located different distances from each other, as opposed to the variance of an entire group of samples without regard to their locations. To calculate a variogram, samples are generally grouped or binned. For example, samples located between 0 and 100 meters from each other are put into one group, samples located between 101 and 200 meters from each other are put into a second group, and so forth. The distance between samples is known as the separation distance or lag. A variance is calculated for each group of samples, and the results are then plotted on a graph as a function of the separation distance. This is traditionally done using the semi-variance, which is one-half of the variance, rather than the variance itself.

If a variable is spatially correlated, the semi-variances will increase with separation distance and eventually reach a constant value known as a sill. The separation distance at which the sill is reached is known as the variogram range. The semi-variance will, in theory, decrease to zero when the separation distance is zero. This is because if one repeatedly measured a value at the same location, the result should always be the same.

In real life applications, however, the result may differ if several samples are taken at the same location. If the values are chemical concentrations, for example, the differences may arise as a result of analytical errors or the inability to collect more than one sample (such as a scoop of soil) from exactly the same position. A non-zero semivariance at zero separation distance is known as a nugget or the nugget effect. This term dates back to the origin of geostatistics as a practical tool for mining engineers who needed to calculate the grade, or richness, of ore in order to determine the most efficient and economical way to run their mines. An unusually rich nugget or pod of ore might yield a very high grade, whereas rock or soil a very short distance away might have a much lower grade.

Once a variogram is developed, values can be interpolated using a process known as kriging, named after the South African mining engineer who invented the technique. Variograms can also be used as the basis for geostatistical simulation, which uses information about spatial variability to generate alternative realizations that are equally probable and poses the same statistical properties as the samples from which they are derived. A petroleum geologist might, for example, use geostatistical simulation to generate alternative realizations of an underground oil reservoir for which she has definite information from only a handful of wells. The exact nature of the oil reservoir between the existing wells is unknown, and geostatistical simulation provides a series of possibilities that can be used as input for computer models that determine how to most efficiently remove the oil.

QUALITY ASSURANCE

Statistics play a critical role in industrial quality assurance, and are often used to monitor the quality of products and determine whether problems are random occurrences or the result of systematic flaws that need to be corrected. All manufactured products will have some degree of variability. Components may be slightly shorter or longer than designed, not exactly the correct color, or prone to premature failure. Statistical process control can be used to monitor the variability of product quality by sampling components or finished products. If the results fall within pre-established limits (for example, as defined by a specified mean and variance), the process is said to be in control. If results fall outside of acceptable limits, the process is said to be out of control. Statistical quality analysts can also examine trends. If there is a gradually increasing number of unacceptable products, the underlying cause may be a piece of machinery that is gradually going out of adjustment or about to fail. Trends that fluctuate with time and appear to be correlated with factor shift changes may indicate human errors.

Six Sigma is an extension of statistical quality control that has evolved into a popular business philosophy. As it is used by many people, the term Six Sigma is nothing more than another way of saying that a process or procedure is nearly perfect or, among those who are slightly more mathematically inclined, that it produces no more than 3.4 failures per million opportunities. In the traditional manufacturing sense, each item produced on an assembly line is an opportunity to fail or succeed. In service-oriented fields such as retailing and health care, the opportunities might represent customer visits to a store or patient visits to a hospital.

The sigma in Six Sigma refers to the standard deviation of a normally distributed population, which is often represented in equations by the Greek letter sigma. The six has to do with the number of standard deviations required to achieve the desired standard of less than 3.4 failures per million opportunities.

Imagine that a bolt that is part of an airplane is designed to be exactly 10 centimeters long, but will still work if it is as short as 9.9 centimeters. Anything shorter than that will not fit and must be discarded. The owner of a machine shop hoping to supply bolts to the aircraft company collects samples of his product, carefully measures each bolt, and learns that the sample has a mean of 10 centimeters and a standard deviation of ±0.1 centimeter. If the owner collected a representative sample and bolt length that follows a normal distribution, then he can expect that 16% of the bolts will be too short. This is because 16% of a normal distribution is less than or equal to the mean minus one standard deviation, regardless of the size of the mean or the standard deviation. He can still provide bolts to the aircraft company, but would be forced to throw out 16% of his production to meet the standards. This amount of waste is inefficient and costs money, so the owner decides to adopt a Six Sigma policy.

To achieve Six Sigma, he must refine his bolt manufacturing process so that the standard deviation is small enough that only 3.4 out of each million bolts produced (or 0.00034%) are less than 9.9 centimeters. For a normal distribution, 0.00034% of the population is less than the mean minus 4.5 standard deviations, or 4.5 sigma. The average length of bolts produced in the machine shop, though, varies over time. This might be the result of seasonal temperature fluctuations (metal expands and contracts as its temperature changes), small variations in the composition of the metal used to make the bolts, or a host of other factors. Pioneering studies of electronics manufacturing

Cellular Telephones and Political Polls

Political pollsters have long relied on telephone surveys to sample public opinion on matters ranging from presidential elections to advertising effectiveness. As long as virtually everyone has a telephone, the population of a city, region, or nation can be sampled by randomly selecting telephone numbers and calling those people. Even people with unlisted telephone numbers are fair game because pollsters can use computers to generate and dial telephone numbers. Although there are some people without any telephone service, they generally represent less than 5% of the population.

The explosive growth of cellular telephone use, and particularly the increasing number of people who use only cellular telephones and do not have land line telephones, became an issue in the 2004 United States presidential election. During the months leading up to the election, some experts believed that a disproportionate number of people who used only cell phones were young voters. This presented a problem because political pollsters do not call cellular telephones. Federal law makes it illegal to use automated dialing machines to reach cellular telephones, and some state laws prohibit unsolicited calls to numbers at which the recipient will have to pay for the call (which includes most cellular telephones). If each voter is equally likely to have only a cellular telephone, then survey results will not be affected. If certain segments of the population, however, are more likely than others to be inaccessible to pollsters then the reliability of their polls decreases because their samples will be biased. The influence, if any, of young cellular-only voters on pre-election polls for the 2004 presidential election was never conclusively determined. The potential for poll bias as growing numbers of people abandon their traditional land line telephones for cellular phones, however, promises to be an important consideration in future elections.

processes showed that the mean value must be 6, not 4.5, standard deviations away from the acceptable limit in order to ensure no more than 3.4 defects per million products. In others words, an additional increment of 1.5 standard deviations is added to account for the fluctuations. Hence the association of the name Six Sigma with a defect rate of 3.4 pieces per million. In terms of the bolt manufacturer, this means that he must improve his manufacturing process to the point where the standard deviation of bolt lengths is $(10.0 - 9.9)/6 = 0.017$ centimeters.

PUBLIC OPINION POLLS

Public opinion polls, particularly political polls during major election years, are another real life application of statistics in which samples consisting of a few hundred people are used to predict the behavior or sentiments of millions. Careful selection of a representative sample allows pollsters to reliably forecast outcomes ranging from consumer product demand to election outcomes.

Modern public opinion polling starts with carefully selected questions designed to elicit specific opinions. For example, asking a voter whether she likes Candidate A may elicit a different response than asking the same voter if she dislikes Candidate B, even if Candidate A and Candidate B are the only choices. Interviewers are trained to ask questions in a neutral, rather than suggestive or leading, manner. The selection of people to be interviewed, known as sampling, begins with the generation of random telephone numbers. Known business telephone numbers and cellular telephone numbers are removed from the list, and random number generation ensures that every residential telephone number has an equal probability of being called even if it is not listed in the telephone directory. In a national poll, the list of telephone numbers is then sorted by state and county and the number of telephone numbers called for each state or county is proportional to its population. Because there may be more than one eligible respondent in each residence, interviewers may ask to speak to the person who has had the most recent birthday. Women are more likely than men to provide complete and usable responses, so interviewers ask to speak to male household members more often than female household members to account for that bias.

The number of people interviewed is estimated using a standard formula based on the normal distribution. The formula predicts that the uncertainty of results (often referred to as the margin of error) for a random sample of 500 people, which is a common size for a nationwide political poll in the United States, is ±4.4%. The uncertainty is inversely proportional to the square root of the sample size, so increasing the sample size to 5000 (a factor of 10) decreases the margin of error to ±1.3% (a factor of 3.4). Decreasing the sample size to 50 would increase the margin of error to ±14%. Thus, the often used sample size of 500 represents a compromise that provides relatively reliable results for a reasonable expenditure of time and money.

Once the required number of responses have been obtained, the results are broken down into groups according to the age, race, sex, and education of the respondent. The results for each group are weighted according to census results in order to arrive at a final result that is representative of the population as a whole. For example, if 30- to 40-year-old Asian males who graduated from college comprise 2% of the population but represent 3% of the poll respondents, then the results are adjusted downward so that they do not unduly influence the outcome.

Perhaps the most difficult political polling problem is the identification of so-called likely voters. Pollsters will ask respondents if they are likely to vote in an upcoming election, but there is no guarantee that the respondent will follow through. Unexpected bad weather, in particular, can reduce the number of voters and skew results if different parts of the country are affected. Good weather in states with many conservative voters may compound bad weather in states with many liberal voters, or vice versa. Unexpected mobilization of large blocs of voters with vested political interests, for example religious or labor groups, may also invalidate pre-election polls. Thus, the political pollster is faced with the problem of trying to sample a population that will not exist until election day.

Potential Applications

The potential applications of statistics in real life will increase as society continues to rely on technological solutions to social, environmental, and medical problems. Optimization methods based on statistics are becoming increasingly more important as airlines strive to become more competitive. Advance knowledge of the likely weight of passengers and their luggage, or the number of passengers who are likely to miss their flights, can help an airline to utilize its resources in the most effective manner possible. High tech manufacturing calls for rigorous quality assurance procedures to ensure that expensive and complicated electronic components don't fail, especially those used in situations where failure may have life-threatening consequences. The explosive growth of the Internet during the 1990s led to the creation of a new field known as data mining, which involves the statistical analysis of extremely large data sets containing many millions of records, that will no doubt continue to grow as the prevalence of electronic commerce increases.

Where to Learn More

Books

Best, Joel. *Damned Lies and Statistics: Untangling Numbers from the Media, Politicians, and Activists.* Berkely: University of California Press, 2001.

Graham, Alan. *Teach Yourself Statistics (2nd edition)* Chicago: McGraw-Hill, 2003.

Huff, Darrell. *How to Lie With Statistics (reissue).* New York: W.W. Norton, 1993.

Periodicals

Langer, Gary. "About Response Rates," *Public Perspective.* vol. 14 no. 3 (2003): 16–18.

Web sites

ABC News. "ABC News' Polling Methodology and Standards." 2005. <http://abcnews.go.com/US/PollVault/story?id=145373&page=1> (April 9, 2005).

UCLA Department of Statistics. "History of Statistics." August 16, 2002. <http://www.stat.ucla.edu/history/> (April 9, 2005).

Subtraction

Overview

Subtraction is the inverse operation of addition. It provides a method for determining the difference between two numbers; put another way, it is the process of taking one number from another to determine the amount that remains. While the basics of this fundamental process are taught at the preschool level, subtraction provides a foundation for many aspects of higher mathematics, as well as a conceptual basis for some cutting-edge methods of developing new technology. In addition, subtraction provides answers to a wide array of practical daily questions in areas ranging from personal finance to athletics to making sure one gets enough sleep to remain healthy.

Fundamental Mathematical Concepts and Terms

A subtraction equation consists of three parts. The solution or answer to a subtraction equation is called the difference. While this term is commonly known, the other two elements of a subtraction equation also have labels, albeit far less well-known ones. The starting value in a subtraction equation is called the minuend, while the second term is called the subtrahend. Thus, a subtraction equation is formally labeled: minuend − subtrahend = difference. Simple two-place subtraction problems can be solved by subtracting each column individually, beginning at the right and working progressively left. The equation 49 − 21 is solved by evaluating 9 − 1 for the right value and 4 − 2 for the left value to produce a final answer of 28.

Complications in this simple process arise when borrowing and carrying become necessary, as in the equation 41 − 28. Because 8 cannot be directly subtracted from 1, it becomes necessary to borrow ten from the next place, in this case the value 4. This operation is made possible by applying the distributive property of mathematics, that describes how values can be distributed in multiple ways and that in this example insures that the value 41 is equivalent to the expression 30 + 11. Following this operation, solving this equation is simply a matter of subtracting 8 from 11 and 2 from 3 using the same column by column approach demonstrated in the initial example. Subtraction equations using large values may require multiple instances of borrowing in order to produce a solution, though the method used to solve these equations is identical to that used for simpler equations.

A second complication arises when subtraction involves negative numbers. While the physical world does not contain negative quantities of any physical object, some measurement systems include negative values, the

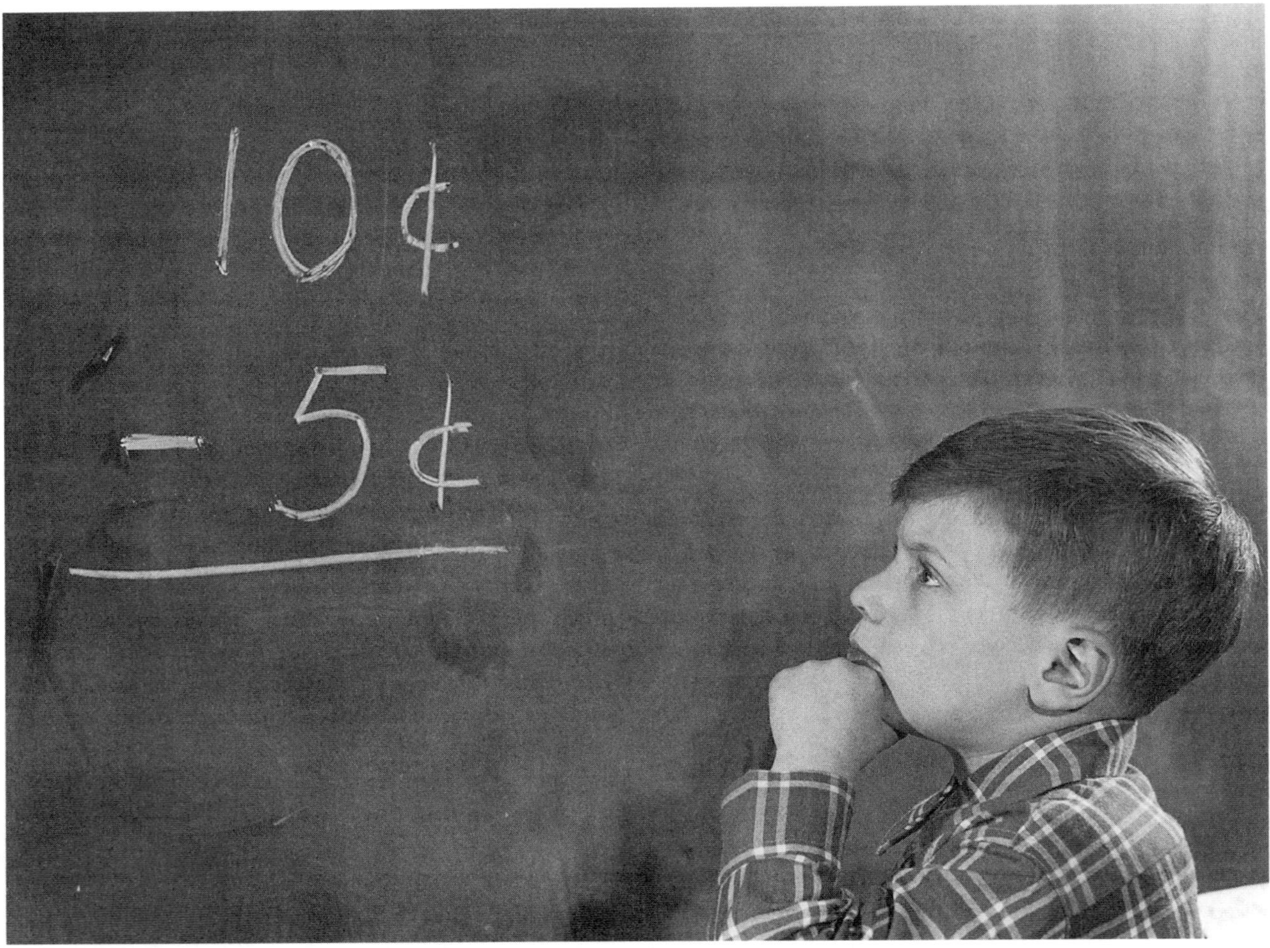

While the basics of this fundamental process are taught beginning at the preschool level on up, subtraction provides a foundation for many aspects of higher mathematics, as well as a conceptual basis for some cutting-edge methods of developing new technology. WILLIAM GOTTLIEB/CORBIS.

most common of these being the modern system of temperature measurement. Whether dealing with Fahrenheit or Celsius, both systems measure temperature with values gradually falling to a value of 0 long before temperatures stop decreasing; in both systems, the temperatures reach zero and simply begin again, this time with the number values labeled negative and decreasing as the temperature cools, such that −10 degrees is colder than 10 degrees.

Now suppose that we wish to find the difference between a day's high and low temperatures, or the temperature range for that day (also called the diurnal temperature). If the high and low temperatures are both positive, this is accomplished by simply subtracting the low temperature from the high temperature to find the difference. However, if the low value happens to be negative, this process must be handled differently. In order to subtract a negative number, we simply add the absolute value of that number; if we wish to subtract −14, we accomplish this by adding 14. Applying this convention to a day where the high is 40 and the low is −9, we solve this equation: $40 - (-9)$, that we convert to $40 + 9 = 49$, the difference in the two measured temperatures and the temperature range for the day. This same process can be used for any temperature system that does not have an absolute 0 point, as well as in any other type of measure that uses both positive and negative values. Among modern temperature scales, the only one that does not require this type of adjustment is the Kelvin temperature scale, in which 0 represents the coldest any object can ever become, and the point at which molecules have a minimum of molecular motion (many texts incorrectly state that at absolute zero motion ceases. However, this is incorrect because there is still vibratory motion). For comparison, 32 degrees Fahrenheit (32°F) equals 0 degrees Celsius (0°C), and equals 273 Kelvin (273 K).

Because carrying is frequently required to resolve subtraction equations, most people find subtraction harder to perform than addition. For this reason, a different type of borrowing and carrying is sometimes employed to simplify mental subtraction. In an equation such as 41 − 29, the first step requires borrowing ten and adding it to the 1, the step at which most mistakes are made, and where a simple shortcut can help avoid errors. This shortcut is based on the fact that the simplest number to subtract from any value is 0, and this shortcut takes advantage of this fact. To apply this shortcut to the equation 41 − 29, we simply change the 29 to 30 by adding one. Then, we can easily evaluate the new equation, 41 − 30, to get 11, to which we add back the one extra that we subtracted to reach the correct total of 12. This process can be quickly learned, and with practice becomes routine, helping improve the accuracy of mental arithmetic.

A Brief History of Discovery and Development

Subtraction has been used for millennia, initially being calculated with counting sticks, stones, or other items, and later using early tools such as the counting table and the abacus. However, the written notation for subtraction, the modern minus symbol, came into use much more recently. In England during the 1400s, the dash as a minus symbol was first used to mark barrels that were under-filled, signifying that the marked barrels had missing or inadequate contents. By the 1500s, this notation had migrated from barrels into mathematical notation as the accepted symbol for subtraction, and has remained in use ever since.

The modern method of solving subtraction problems can be traced as far back as the 1200s, when this method was originally called decomposition; not until the 1600s did the term "borrowing" come into use. Two other subtraction methods were also taught well into the twentieth century, though these are largely forgotten today. One fairly intense debate arose during the early 1900s, dealing with the proper notation for subtraction. While students today are taught to cross out values and write in new ones above them as part of the borrowing process, this practice did not appear widely in American textbooks prior to the middle of the twentieth century. Before this adoption, an ongoing debate raged over the use of these hash marks, or crutches as they were originally called. Critics argued that subtraction should be accomplished without the use of this pejoratively labeled aid; one 1934 math text went so far as to give examples of equations performed both with and without "crutches," labeling the version without crutches the preferred method and noting that teachers should not allow students to use crutches when solving problems. Advocates of crutches, many of them school teachers, based their argument on simple utility, countering that the use of crutches aided students in calculating correct results with fewer errors. A 1930s study published by researcher William Brownell offered strong evidence that the teachers were right, and that using crutches or other notations to keep track of borrowing did reduce errors in subtraction. Almost immediately following this study, textbooks teaching the crutch notation method of subtraction became the norm, and this technique continues to be used today.

Real-life Applications

SUBTRACTION IN FINANCIAL CALCULATIONS

Profit is the amount earned from a business transaction, and can be found using subtraction. In the simplest form, profit is determined by subtracting cost from selling price; for the up-and-coming lemonade merchant who takes in \$6.75 from her customers after spending \$2.25 on lemonade mix, cups, and ice, a simple profit calculation of \$6.75 − \$2.25 reveals a positive outcome or profit of \$4.50. However, profit calculations are rarely this simple, and in many cases, unplanned costs can subtract significant amounts from the final profit earned.

Consider a beginning entrepreneur trying to make a start on E-bay. This young businessman purchases the latest Tony Hawk PlayStation game at a garage sale for \$14.00. Because he already owns a copy of this game, he is eager to sell it on E-bay for a quick profit. He lists it on the auction site with free shipping and a "Buy-it-now" price of \$19.95 that he calculates will give him a quick \$5.00 profit after paying his expenses. The game sells quickly, the seller ships it to the buyer, and then sits down to calculate his profits.

The beginning point of this calculation is the amount of income received, often called gross income, that in this case is \$19.95. From this starting value, the seller must subtract all his expenses to find his actual profit, sometimes referred to as net income. He begins with his cost for the game, that was \$14.00; 19.95 − 14.00 = 5.95. From this value, he then subtracts his other costs, such as postage of \$1.45; 5.95 − 1.45 = 4.50. The seller was surprised to find that the padded envelope he needed was more expensive than he expected, at 75 cents; 4.50 − .75 = 3.75. Other fees also must be subtracted, and while most of these are small, they begin to accumulate. E-bay fees including a listing fee, "Buy-it-now fee," additional photo fee, and final sale fee totaled 1.75; 3.75 − 1.75 = 2.00. The final surprise for the young businessman comes when he receives his electronic billing statement and learns that the service charged him 3% of the total sale price of $19.95, or 60 cents; 2.00 − .60 = 1.40. The final profit left after subtracting all expenses is $1.40, far less than he had hoped. What appeared to be a fairly profitable business transaction turned out to be a near-loss when all the relevant expenses were correctly subtracted.

TAX DEDUCTIONS

One of the more enjoyable uses of subtraction involves the use of tax deductions. Throughout history, most taxpayers around the world have complained that taxes are too high. In the American federal tax system, several items may be subtracted from total income before taxes are calculated, and in many cases, the net tax savings from these items can be thousands of dollars.

The standard U.S. Federal Income Tax form is called Form 1040. On the first page of this form, taxpayers enter the total amount of their earnings for the year. However, before paying taxes, numerous items are subtracted, reducing the taxable income as well as the actual income tax paid. For instance, taxpayers are allowed to take a personal exemption for each family member; for tax year 2004, this exemption is $3,100, meaning that a family of four can subtract $3,100 four times, for a total reduction in taxable income of $12,400. Contributions to an Individual Retirement Account are often deductible up to a maximum limit (e.g., $3,000 per person), and self-employed individuals (those who don't work for a company) can deduct their costs of health insurance from their taxable income. In many cases, students can deduct tuition and textbook costs up to the maximum allowed limit as well. Finally, expenses such as mortgage interest on a home loan can be deducted prior to calculating the actual tax bill.

Only after all these items and others are deducted, or subtracted from gross income, is a final value reached. This value, called taxable income, is the actual amount on which federal taxes are calculated. Because so many items can be subtracted before calculating taxes, a typical family of four might easily reduce its taxable income by $20,000 or more by following the tax instructions carefully. Because the tax system is designed with these subtractions as an expected part of the process, failing to claim these deductions is equivalent to voluntarily paying more income taxes than required, something very few taxpayers have any interest in doing. Modern tax software has made the previously tedious process of tax filing far simpler and more accurate.

Along with electronic tax filing, some tax services offer to give filers their tax refund immediately, in the form of a refund anticipation loan or RAL. RALs are offered to tax filers who don't want to wait for their tax refund to arrive. While RALs may be a useful tool for situations in which money is needed immediately, an RAL can significantly reduce the amount of the final refund. For example, a consumer expecting a tax refund who requests an RAL would typically have to subtract several fees, including an application fee that averages about $30, and a loan fee that can range from $30 to more than $100. For 2005, a refund of $2,050 incurred an average fee of $100, which reduces the total refund to $1,950. While this reduction seems small, it represents a 5% fee for borrowing this money until the actual refund arrives from the IRS. Because the average refund is now deposited in less than two weeks, this loan equates to an annual percentage rate of roughly 187%. In 2003, 11% of taxpayers took RALs, costing themselves more than $1 billion in fees for these short-term loans that many consumer advocates criticize as an unreasonable effort to charge taxpayers interest on their own money.

Rebates are a popular method of selling an item for less than its original price in order to attract buyers. Rebates come in several forms. Most new cars today are sold with a manufacturer's rebate, meaning that the sticker price on the window of the car is automatically reduced by subtracting the rebate amount. This rebate is in addition to the normal amount subtracted from the sticker price by most car dealers. Automobile rebates are paid automatically to any buyer, and are given at the time of purchase. Information on actual dealer costs and available rebates can be found at numerous online car buying sites.

Another popular form of rebate is the mail-in rebate. These rebates are frequently offered on electronic equipment and other high-priced items, particularly in the case of older merchandise that manufacturers wish to clear out of inventory. A mail-in rebate is not paid at the time of purchase; instead, the purchaser is required to complete one or more rebate forms and mail these forms, along with specific pieces of documentation, to a processing center. If the documents are submitted correctly and prior to the offer's deadline, a check is normally mailed to the buyer within a period of four to six weeks.

Why are mail-in rebates so popular with manufacturers, and why do companies use rebates instead of simply reducing the price of the products? Consumers behave in predictable ways, and most rebate programs save manufacturers money due to a phenomena researchers call slippage, in that many customers never redeem their rebates. Estimates vary on just how high slippage rates are, and the rate is influenced by factors such as the size of the rebate, the length of time allotted to redeem it, and the difficulty of complying with the program rules. However, on average, rebate redemption rates for small items can be as low as 2%, while for larger rebates in the $50 to $100 range, redemption levels typically hover around 50%. The benefit of rebates to the manufacturer are obvious: they can advertise a much lower price, knowing that half or fewer of the buyers will get this lower price, while the rest will pay the full, unrebated amount. Rebates can be a wonderful bargain for those who follow through on them. However, for many buyers, the promised reduction in price is never realized due their own unwillingness to follow through on the process.

While most highways can be driven free of charge, toll roads require a driver to pay for the privilege. While using a toll road has traditionally meant stopping to throw a handful of coins into a basket or waiting for an attendant to make change, many toll roads now provide the option to pay electronically without stopping. These systems, with names such as Pike Pass in Oklahoma and FasTrak in California, allow a user to purchase a small electronic unit to mount in her vehicle; this unit can then be filled by paying in advance and then used like a debit card while driving. To use the automated systems, drivers typically change into a specific lane that is equipped with sensors to read data from the user's transmitter. Using this identification data, the system automatically subtracts the proper toll amount from the user's account; in many cases, the system automatically sends a reminder e-mail or letter when the balance drops below a set limit. Drivers using these systems not only avoid the hassle of carrying correct change with them and waiting in line to pay, some states also give them a reduced toll rate for using the automatic system. In addition to saving 5–10% on their tolls, drivers in Oklahoma also enjoy the pleasure of paying the toll while never dropping below the 75 mile per hour posted speed limit on the state's tollways.

A countdown clock on the Eiffel Tower in Paris marking the last 100 days before the year 2000. Countdown clocks use simple subtraction to countdown to zero. AP/WIDE WORLD PHOTOS. REPRODUCED BY PERMISSION.

SUBTRACTION IN ENTERTAINMENT AND RECREATION

One of the more entertaining uses of subtraction is a process known as a countdown, in that a large starting value is gradually reduced by one until it finally reaches zero. Countdowns are used in a variety of settings in that people need to know in advance when a particular event will happen. Countdowns are perhaps best known for their use in space exploration, where an enormous clock traditionally ticks off the final seconds until liftoff. While this process provides dramatic footage for television news, the use of countdowns, which typically start several days before launch, is actually a method of insuring that the complex series of events required for a successful launch are completed on time and in the proper sequence. Space launch countdowns normally include several planned holds, during which the countdown clock stops for a set period of time while various checks are made.

Countdowns are also used for recreational purposes. Each year, millions of people across the globe eagerly count down the final seconds until the arrival of a new year, celebrating its arrival with cheers, hugs, and toasts. Hockey players, banished to the penalty box for rule violations, sit and impatiently wait for their penalty time to count down to zero so they can re-enter the game. Top ten lists, including television host David Letterman's long-running version, are often used to poke fun or entertain by leading the audience gradually down from ten to one, and weekly top 20 countdowns guide music fans gradually to number one, the week's top song.

Golf Handicaps While most sports force players to compete head-to-head without any adjustment to the score, a few events attempt to level the playing-field by adjusting player totals. Golf is one of the more popular sports in which subtraction is used to allow players of differing skill levels to compete on an equal basis. Using a system known as handicapping, a golfer's handicap index is assigned based on a series of ten recent rounds he has played. Using these game scores, a difficulty rating for the courses on which they were played, and a complex formula, an authorized golf club can issue an official handicap index to a player. Using this index, each player can then calculate his handicap for a particular course, meaning he is given strokes and can subtract a specific number of shots from his score. Using this system, a golfer who normally scores 76 and a golfer who normally scores 94 can compete fairly on the same course. By subtracting the proper number from each score, each golfer is able to arrive at an adjusted score and compare how well or how poorly he played that particular course that day.

Track and Field One measure of an athlete's performance is his vertical jump. Vertical jump is not a measure of how high an athlete can leap in absolute terms, because this result is strongly influenced by an athlete's height and arm-length; rather, vertical jump is a measure of how high an athlete can propel himself from a standing start, relative to his standing height; for this reason, it provides a better measure of absolute jumping ability than a simple measure of how high a leaping athlete can reach.

Vertical jump is calculated using subtraction. First, an athlete's standing reach is measured by having him stand flat-footed and reach as high as possible with one arm. Then, the athlete's jump reach is measured by having him stand and jump straight up without taking a step. True vertical jump is calculated using the following equation: Jump Reach − Standing Jump = Vertical Jump. For reference, professional basketball players typically have a standing vertical jump of 28–34 inches, meaning their final reach height is almost three feet higher than their standing reach. Jumping, like most other athletic skills, can be improved with training. Because of the explosive nature of jumping, performance is often improved using both strength-building and power-enhancing forms of exercise.

Pop Culture

Each December, millions of people around the world plan for a new year by making one or more New Year's resolutions. While many of these resolutions focus on addition, such as making more money, spending more time with family, or playing more golf, the two most popular resolutions for 2005 both involved subtraction. The second most popular resolution in 2005 was to lower payments by reducing personal debt. The most popular resolution has stood atop the list for some time, and will probably remain there: more people chose subtracting pounds, or losing weight, than any other New Year's resolution for 2005.

Weight Loss and Dieting

Because losing weight is such a popular goal, one might assume that many people are reaching this goal and losing weight. In truth, the popularity of the goal is probably tied to the increasing incidence of obesity; as of 2000, approximately two-thirds of United States adults were defined as overweight or obese, and predictions suggest that this number will continue to rise. Most of the hundreds of methods of subtracting pounds involve subtracting from what is being eaten. Some diets reduce intake of fats while others restrict intake of carbohydrates. While debate continues to rage on which plans work best (and that do not work at all), one piece of advice seems to make sense: reducing the amount of food on one's plate helps many people eat less. This simple subtraction can provide a solid starting point for any weight-loss plan, and has been shown to lead to weight loss even without any other behavioral changes.

Sleep Management

Before the invention of the electric light bulb, Americans slept an average of nine hours per night; today, the average is one to two hours less. While doctors and sleep experts recommend that teenagers get 8.5–9 hours of sleep each night, the average teenager in America gets far less. Sleep experts say that each person has a set need for sleep each night, and that each hour of missed sleep adds up to create a sleep deficit. This deficit describes how far in debt a person is in terms of sleep and represents needed sleep hours that have been subtracted and applied to other activities. While being a few hours overdrawn on sleep is not an immediate danger and can usually be made up over a weekend of sleeping late, the long-term impact of inadequate sleep can be serious. As the sleep deficit grows, a variety of negative physiological outcomes become more likely, including obesity, high blood pressure, reduced productivity at work, poor mood, and increased an likelihood of accidents at home, at work, and while driving. While sleep time can be subtracted over the short-term without major impact, the sleep account must eventually be rebalanced by adding additional hours of sleep to the account.

Subtraction in Politics and Industry

DOOMSDAY CLOCK

One famous countdown clock has been ticking for more than half a century, though this clock has actually moved only a few minutes during that time, and has occasionally run backwards. In June of 1947, the Bulletin of the Atomic Scientists, an academic journal dealing with atomic power and physics, placed on its cover a clock, with the hands showing seven minutes until midnight. In a lengthy editorial inside, the journal described this so-called Doomsday Clock, in which midnight signaled the destruction of mankind by atomic weapons. The Doomsday Clock stirred a great deal of discussion with its appearance during the earliest years of the atomic age.

In the years since 1947, the Clock has made many appearances on the journal's cover, with the minute hand moving either forward or backward depending on the state of world events. In 1949, after the Soviet Union detonated its first atomic weapon, the clock advanced four minutes, displaying three minutes before midnight. Four years later, following the test detonations of thermonuclear devices in both the Eastern and Western hemispheres, the hands advanced again, reaching two minutes until midnight. During the following years, events including new arms treaties and the rekindling of old conflicts nudged the minute hand repeatedly backward and forward. The signing of the Strategic Arms Reduction Treaty (START) in 1991 moved the clock to seventeen minutes till midnight, its earliest point ever. At its last appearance in 2002, the clock stood once again at seven minutes till midnight.

Engineering Design

As popular as weight loss goals are for individuals, subtracting pounds or ounces can also become a major goal in industry. During the design phase of the Apollo moon missions, NASA became concerned that the Lunar Module, the ship that would carry two astronauts on the final leg of the trip to the moon's surface, was significantly overweight. Major redesigns began, and, by reducing the size of the observation window, cutting the thickness of the craft's skin, and making other changes, the craft's weight was significantly reduced. However, in order to reach the specified weight target, Grumman, the craft's builder, resorted to extraordinary measures, at one point actually paying company engineers a bonus for each ounce they were able to shave off the craft's weight. The efforts of these professionals were successful, and the lunar module performed as designed.

Weight reduction is also a priority in the automobile industry. In order to meet fuel economy goals, most automobile manufacturers have made significant changes to their designs in order to subtract from the vehicle's total weight. In many cases, steel has been replaced with aluminum, which is more expensive, but far lighter; in other cases, plastics or lightweight carbon composites have been introduced in order to reduce weight. One extreme example of this type of engineering weight loss involves a revolutionary car, General Motors' EV1, the first totally electric production car. Introduced in 1996, the EV1 was also faced with extraordinarily tight weight limits in order to reach its target mass of under 3,000 pounds (1,360 kg). Toward this end, GM engineers adopted a variety of changes to subtract weight from the vehicle. Among the solutions was the decision to use aluminum for the frame and wheels, shaving more than 300 pounds (136 kg) off the weight of traditional steel parts, and the choice of a non-traditional material, magnesium, for the steering wheel and seat-backs. While the EV1 was not a commercial success, GM's experience in cutting weight during its development has led to applications in other vehicles. According to one calculation, an automaker can subtract $4.00 from a car's cost for each pound of weight it manages to remove from the design.

Potential Applications

While the basic process of subtraction itself offers few potential breakthroughs, the concept of removing items from a collection in order to reach an objective remains useful, and one early application of this principle is already producing impressive breakthroughs. Evolutionary design is a technique that allows computers to consider millions or billions of possible solutions to a complex problem to arrive at an optimal solution. In many ways, this process is similar to the concept of natural selection, in which the stronger predator survives to reproduce and pass his genes on to succeeding generations while the weaker predator is eliminated from the gene pool.

Antenna Design

The field of antenna design is unfamiliar to most people. However, the ability to design lightweight, efficient antennas is critical to the space program and other industries. One challenge in this endeavor has been that antenna design requires a deep understanding of the field, limiting this work to a relative handful of experts. A second limitation is that even these experts are not always certain how to improve the design of a specific antenna. Evolutionary design accepts that the present understanding of how to improve antennas is limited; this process instead simply creates and evaluates so many different choices that it is likely to produce a useful one.

The evolutionary design process begins with a researcher creating a group of antennas with different combinations of shapes and sizes, that are then mathematically described for the software. Next, the software applies random mutations to these beginning antennas, such as lengthening some and giving others more or fewer arms. After that, the resulting antennas are tested for performance. Using the results of this testing, the more effective models are kept, while the poorest performers are replaced with new samples similar to the good performers. Then, the process of mutating the designs, testing the resulting models, and retaining the best versions is repeated. After this process of evolutionary improvement has occurred for thousands of generations, a single model eventually emerges that offers the best possible combination of performance traits.

In the case of this small, one-inch square antenna designed for satellite use, more than ten hours of supercomputer time was required to assess millions of possible configurations; by comparison, an expert antenna designer would have needed twelve years working full-time to process the first 100,000 designs. Further, given the strange appearance of the antenna, which resembles little more than a collection of strangely bent paper clips, it seems doubtful that a human designer would ever have proposed such a configuration. The secret to this unique design process lies in a radically advanced form of subtraction that allows removal of the every design except the very best ones, allowing those designs to be further enhanced. Future uses of this technique are anticipated in producing such developments as computer chips that can heal themselves in the case of malfunction, and improved components for implantable medical devices.

Where to Learn More

Books

Brownell, W.A. *Learning as Reorganization: An Experimental Study in Third-grade Arithmetic.* Durham, NC: Duke University Press, 1939.

Periodicals

Ross, Susan, and Mary Pratt-Cotter. "From the Archives: An Historical Perspective" *The Mathematics Educator.* (2000): 10 (2).

Shaw, Mary, Richard Mitchell, and Danny Dorling. "Time for a smoke? One cigarette reduces your life by 11 minutes." *British Medical Journal.* (2000): 320 (53).

Web sites

About Golf. Golf handicaps, an overview. <http://golf.about.com/cs/handicapping/a/handicapsummary.htm> (March 19, 2005).

Bulletin of the Atomic Scientists. Doomsday Clock. <http://www.thebulletin.org/doomsday_clock/timeline.htm> (March 17, 2005).

Centers for Disease Control. Obesity Trends Among U.S. Adults Between 1985 and 2003. <http://www.cdc.gov/nccdphp/dnpa/obesity/trend/maps/obesity_trends_2003.pdf> (March 19, 2005).

Federation of American Scientists. Strategic Arms Reduction Treaty. <http://www.fas.org/nuke/control/start1> (March 19, 2005).

Internal Revenue Service. Form 1040. <http://www.irs.gov/pub/irs-pdf/f1040.pdf> (March 17,2005).

The Math Lab. Subtraction in your head! An algebraic method for eliminating borrowing. <http://www.themathlab.com/Pre-Algebra/basics/subtract.htm> (March 19, 2005).

National Air and Space Museum Oral History Project. Interviewee: James E. Webb, November 4, 1985. <http://www.nasm.si.edu/research/dsh/TRANSCPT/WEBB9.HTM> (March 19, 2005).

National Sleep Foundation. Myths and Facts About Sleep. <http://www.sleepfoundation.org/NSAW/pk_myths.cfm> (March 19, 2005).

Spaceref.com. Press Release: NASA Evolutionary Software Automatically Designs Antenna. <http://www.spaceref.com/news/viewpr.html?pid=14394> (March 19, 2005).

U.S. Department of Energy; EV America. General Motors EV1 Specifications. <http://avt.inel.gov/pdf/fsev/eva/genmot.pdf> (March 18, 2005).

Symmetry

Overview

Objects that have parts that correspond on opposite sides of a dividing line are said to have symmetry.

Fundamental Mathematical Concepts and Terms

If a spatial operation can be applied to a shape that leaves the shape unchanged, the object has a symmetry. There are three fundamental symmetries: translational symmetry, rotational symmetry, and reflection symmetry.

An example of translational symmetry can be seen in lengths of rope or in the patterns on animals. If the rope is closely inspected, a braided pattern can be seen. By moving along the rope a bit further, the same pattern is seen again; thus the rope has translational symmetry. This pattern is very important for climbers, if the braided pattern is distorted in any way the force will no longer be evenly distributed along its length and it can break at this point under load. For this reason, climbing ropes will often have brightly colored patterns in their braiding to help the climber spot any deviations from this symmetry.

Imagine a sunflower that is the object of an operation, and the operation can be applied to its rotation around the center of the flower. If it is rotated so that the petals line up again so that it will look the same as before, the sunflower pattern is said to be "symmetric under rotation." Symmetries are probably the easiest patterns in nature for us to see and also the most common. The reason that nature has used symmetry in such abundance is that it allows complex objects to be constructed from simpler shapes, greatly reducing the amount of information that needs to be stored and processed to build the object.

Your whole body has reflection symmetry along the center. This symmetry can be seen if you stand by a reflective shop window, or large mirror, so that one half of your body is hidden from view and the other half is reflected. To an observer it looks as if you are whole because humans have a biological symmetry (often distorted or fused in the case of internal organs such as the heart) that roughly corresponds to an imaginary plane through the sagittal suture of the skull that divides the body onto left right planes.

Other symmetries can be built by repeated application of these basic symmetries, for example, the teeth of a zipper have a symmetry made by reflection and translation. This symmetry is called glide-reflection.

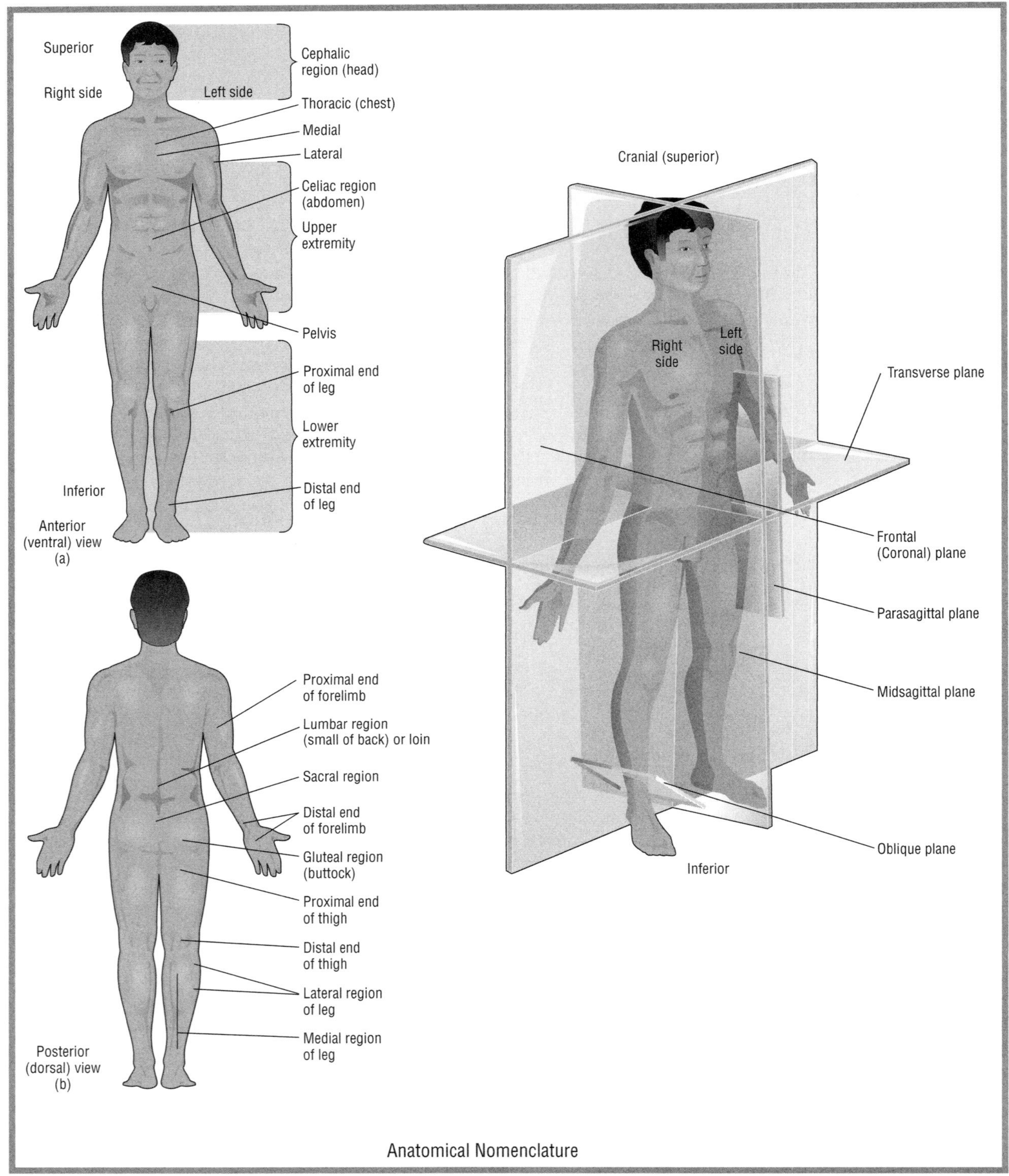

A plane through the sagittal suture establishes a plane of left and right symmetry for the human body. ILLUSTRATION BY ARGOSY. THE GALE GROUP.

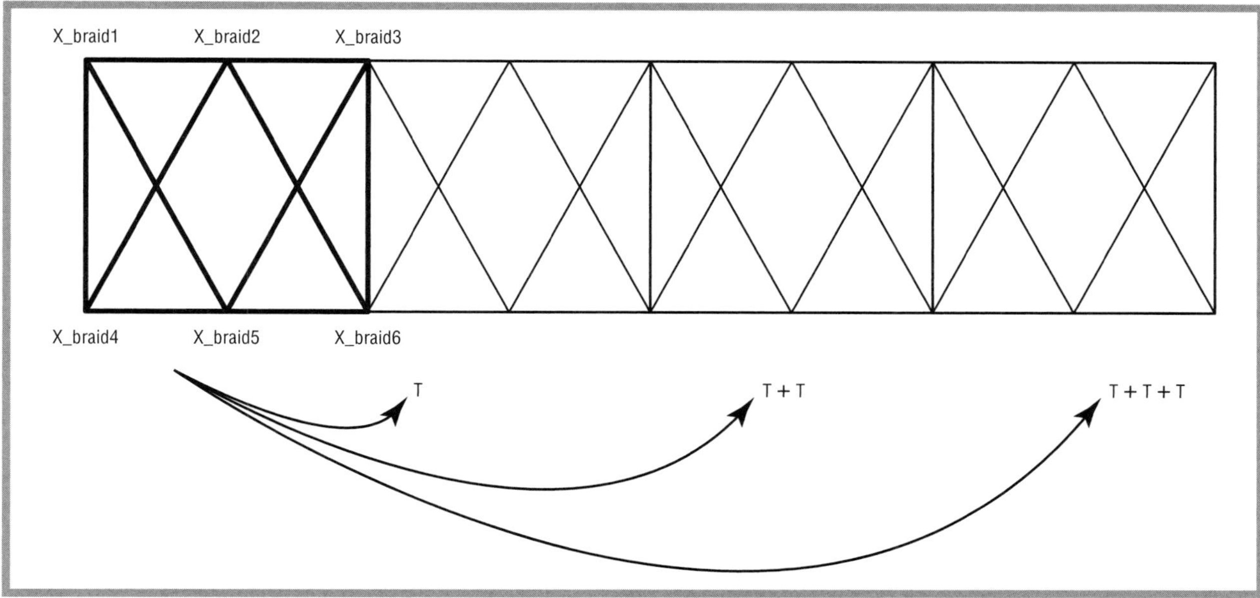

Figure 1.

EXPLORING SYMMETRIES

To understand the nature of translation, rotation, and reflection symmetry, one must first define how these operations act on an object. If an object is defined by a set of points, an operation can be defined by its action on these points.

Let us start with translation, the basic braided pattern of a rope can be recorded by a number of points which can be grouped together into a set called X_braid. As a simple braiding, imagine the rope has a repeating pattern made from two crosses inside by a box. This pattern can be represented by points as the set of points X_braid. The act of translation will be to copy and shift each of the sets by a fixed distance T. If the translated points, X_new = X_braid + T match the current braiding on the rope at that point X_new = X_current, then the translation, T, was symmetric. In our example this means that the translated "two cross and box" pattern matches the current pattern at that point on the rope. This translation can be applied as many times as we like, if our rope is long enough, and our new pattern will always match the braiding at that point. (See Figure 1.)

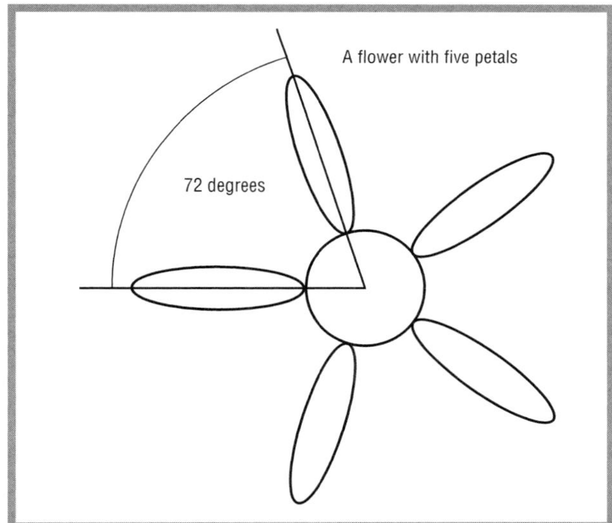

Figure 2.

For rotational symmetry, using our flower pattern we can find the relation between the angle the flower is rotated and the number of petals on the flower. Start by marking one of the petals with a cross so the rotation can be seen. If there are n petals and each rotation takes us to the next petal, it will take n rotations for the all the petals to be marked, a 360-degree rotation. The angle of one rotation that moves the cross from one petal to the next is therefore 360/n.

As an example, think of a flower pattern with 5 evenly spaced petals. The smallest rotation that will leave the flower pattern unchanged is 360 / 5 petals = 72 degrees. So, if we wanted to draw a flower with five petals that has a rotational symmetry, each petal must be spaced exactly 72 degrees from the next. (See Figure 2.)

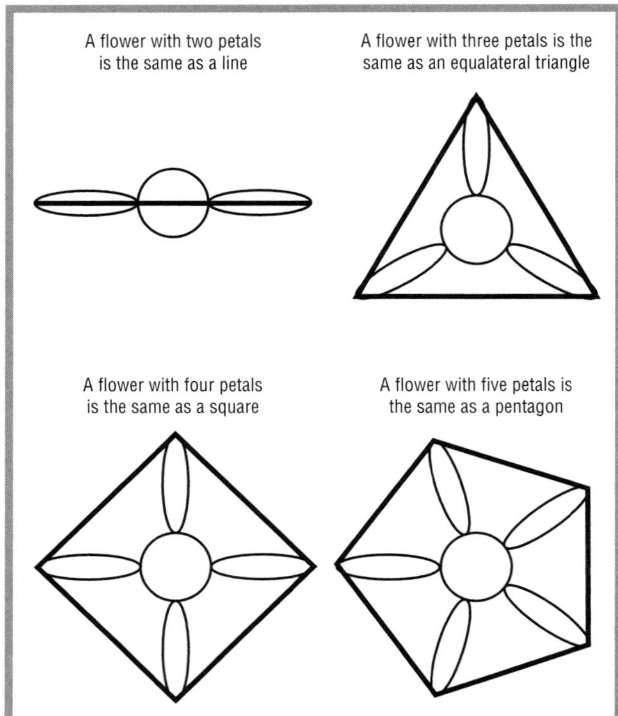

Figure 3.

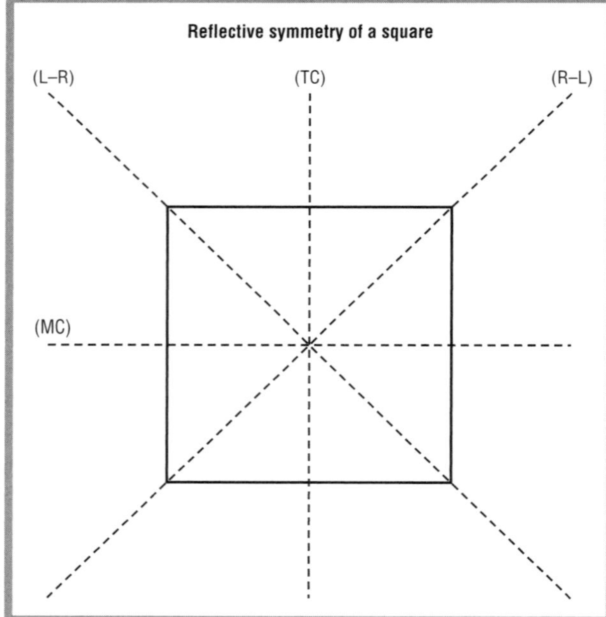

Figure 4.

Consider a flower pattern that is rotationally symmetric with four petals; this means that each petal will be spaced at 90 degrees from the next for the formula. Another shape that has four points that are each separated by 90 degrees around the center is a square. All of our flower patterns with rotational symmetry can be represented by geometrical shapes such as this. For example, a flower with two petals is identical to a line, with three it is identical to an equilateral triangle (a triangle where each side has an equal length), with four a square, with five a pentagon and so on. (See Figure 3.)

The operation for a reflection is defined by drawing a line that acts as a mirror. This is easier to see if it is done in stages for every point in our object. The first stage is to draw a line from a point through the mirror line; this is called the line of reflection and must cross the mirror line at exactly 90 degrees. The next step is to measure the distance along the line of reflection from the point to the mirror line. The reflection of the point is made by drawing a point on the opposite side of the mirror line at an equal distance along the line of reflection.

If an object is placed in front of the mirror line and we generate a number of reflected points behind the mirror line and join them up we simply have made a reflection of the object but this is not a symmetry of the object as the reflected points are not matched up with the original object. However, if we can place the mirror line in the center of the object and all the points match up we have a reflective symmetry, for example a mirror line drawn down the center of a photograph of a face almost shows this symmetry.

An example of perfect reflection symmetry is a square. Using the reflection operation, there are four lines that can be drawn that will keep the shape of a square. The first is from the top left hand corner to the bottom right hand corner (L-R), from the center of the top edge to the center of the bottom edge (TC), the top right corner to the bottom left corner (R-L) and along the center of the middle left to right edge (MC). Now, mark the top left corner with a cross as was done with the flower pattern to see the effect of each reflection. Under the (L-R) reflection the cross will not move, under (TC) the cross is reflected in the top right corner, under (R-L) the cross is reflected to the bottom right corner and under (MC) it is reflected to the bottom left corner. This is identical to the effect of the cross under rotation; our square has to be rotated four times to bring the cross back to the starting position and this is also the number of lines of reflection symmetry. (See Figure 4.)

Geometric objects with rotation symmetry, lines, triangles, and squares etc. have an identical symmetry to reflection. The number of reflection planes for a geometric shape is given by the number of rotations needed to make the object turn one full circle, 360 degrees. This is simply equal to the number of corners, or petals if we are using a flower shape. The angle from the center between two opposing corners is given by the rotation formula, 360/n.

An equilateral triangle, or a flower with three petals, has 3 lines of reflection and rotational symmetry every 45 degrees from the formula. A square, or a flower with four petals, will have 4 lines of reflection and rotational symmetry every 90 degrees and so on.

Real-life Applications

ARCHITECTURE

Nature is not the only one to use this tool, mankind uses it extensively as well. Look at most architecture and you will see symmetries used in the construction. The architect uses symmetry to distribute the forces in the building in a manageable way and for artistic reasons to give the building an appealing elegance. Many psychological studies have shown the human concept of symmetry is closely related to perceptions of beauty when humans evaluate shape as either beautiful or ugly.

The sun sets under the Arc de Triomphe in Paris, providing a display of symmetry for tourists and drivers on the Champs-Elysees. AP/WIDE WORLD PHOTOS. REPRODUCED BY PERMISSION.

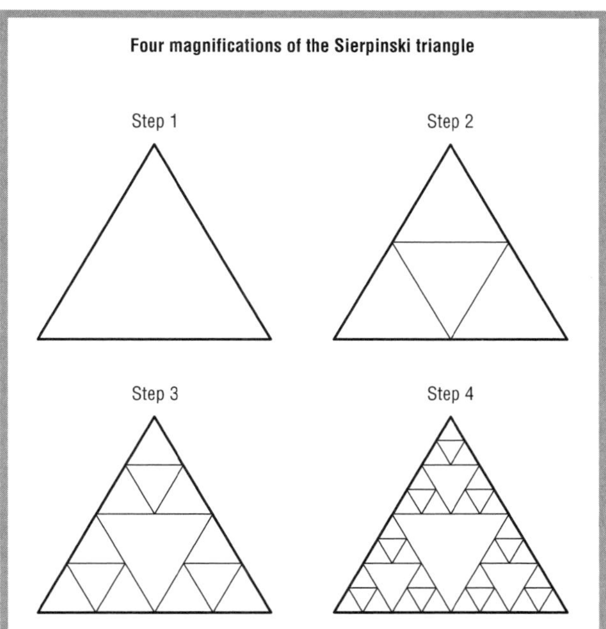

Figure 5: Sierpinski triangles.

Symmetry and Perceptions of Beauty

Symmetry has artistic importance as it is deeply embedded in our experience of the world and how it should look. It gives us strong feelings of how beautiful or ugly something is. Non-symmetric and symmetric shapes can affect us quite deeply, and these effects are exploited by modern artists. For an example Pablo Picasso often challenged perceptions of symmetry with a stunning effect.

FRACTAL SYMMETRIES

There is another special form of symmetry that is common in the natural world that is called fractal symmetry, or scaling symmetry. If a fractal is scaled up or down by a certain magnification it will look exactly the same as before. In nature this type of symmetry can be seen in many plants. A tree, for example, has a thick truck that divides into a number of branches; each branch then divides into a smaller branches and so on. It is possible to imagine that if one of these branches were scaled up it would look like a tree itself. This scaling symmetry is finding uses in many areas of science, such as weather patterns, the stock market and earthquake prediction, that are too complex to predict with normal means.

An example of a simple fractal shape is the Sierpinski triangle. To draw a Sierpinski triangle, start by drawing an equilateral triangle. Along the middle of each edge make a point. Each of these points is then joined by a line.

This divides the triangle into three smaller triangles. Repeat the operation along the middle edges of these triangles, only for the triangles that point upwards. This continues forever. (See Figure 5.) No matter how closely we magnify the triangle, we will always find that it is made from equilateral triangles exactly like the one we started with.

IMPERFECT SYMMETRIES

The symmetries seen in nature are rarely perfect. Consider the rotational symmetry of a flower; on closer inspection each petal will have a unique pattern and marks that define it from the rest. The translation symmetry of the marks on animals and the reflection symmetry of the human body is never quite perfect. A human face made with a perfect reflection symmetry by a computer will look odd. These imperfections probably arise due to perturbations in the natural replication process. It seems that these imperfections are important for giving natural objects unique identities.

SYMMETRIES IN NATURE

There are other forms of symmetry that can help us understand the way our world works. These symmetries are often highly abstract and mathematical but offer deep insight into science and nature. Our day-to-day experience of electricity and magnetism show that they exhibit very different characteristics, but under certain operations they can be shown to be the same, hence they are symmetric. This unified force is called electromagnetism. The study of this abstract symmetry has, amongst other things, allowed the development of radio technology used to carry mobile phone calls and television signals. The study of symmetries and their properties in mathematics is called group theory.

Key Terms

Reflection: The operation of moving all the points to an equal distance, on the opposite side of a line of reflection.

Rotation: The operation of moving all the points of an object through a fixed angle around a fixed point.

Translation: The operation of moving each point a fixed distance in the same direction.

Where to Learn More

Web sites

Picasso, Pablo. "Girl in Boat," 1938 <www.abcgallery.com/P/picasso/picasso205.html> (January 30, 2005).

Youngzones.org "Symmetry." <http://math.youngzones.org/symmetry.html> (January 21, 2005).

Tables

Overview

A table consists of a series of rows and columns that are used to organize various types of data.

Tables serve many uses in daily life, providing a teaching tool for basic math problems or easy access to the solutions to more complex equations, and displaying information in a logical format that enables anyone access to data they might otherwise be unable to determine themselves.

Fundamental Mathematical Concepts and Terms

Table headings, both along the top of columns and the start of rows, set the parameters for what sort of information the table will provide, and answers to each problem or question can be found by tracing each column and row to their point of intersection. The actual mathematical equation or work that has been done to provide each answer is done behind the scenes, and the table itself only displays the starting points and the solutions. In the case of highly complicated mathematics, the table provides the solutions to readers unable to perform the steps to find the answers on their own. Even if the equations are less complex, a table can save the time necessary to work out the solution by displaying the work that has been done earlier.

The amount of information provided by the table depends on the subject matter. Tables used for basic mathematical equations, such as addition or multiplication, generally cover basic numbers from 1 to 10, or 12, or else series of numbers, such as 10s or 100s. Complicated mathematical equations, such as logarithms, may be illustrated over many tables, each one covering a small number of problems. Other tables, where the mathematical application is less obvious, include the information relevant for that specific function. For instance, time tables for train departures would be based on the length of the day and the frequency of train trips.

A Brief History of Discovery and Development

The earliest records of the use of tables date back to Mesopotamia in approximately 3000 B.C., and the Sumerians, who used symbols to denote different goods and kept track of quantities of each item in table form on tablets. Between 2000 and 1600 B.C., the Old Babylonian period,

Types of tables

Tables fall into various categories, including mathematical, scientific, and astronomical. They can be used to communicate schedules, to translate weights and measures from one standard to another, or to convert currencies between countries. When designing a Web site, tables can help provide underlying structure as part of the coding of a Web page. Tables are available that convey various statistical information, including salary ranges for different jobs or for different regions; religious or political demographics by town, state, or country; types of disabilities and percentage of the population affected; health trends and life expectancies. Financial information is frequently presented in table format, from the transactions in your checking account, to interest rates, to income tax brackets. Tables provide ready-made data that enables you to get the information you need without extensive calculations or research.

Addition Table

+	**0**	**1**	**2**	**3**	**4**	**5**	**6**	**7**	**8**	**9**	**10**
0	0	1	2	3	4	5	6	7	8	9	10
1	1	2	3	4	5	6	7	8	9	10	11
2	2	3	4	5	6	7	8	9	10	11	12
3	3	4	5	6	7	8	9	10	11	12	13
4	4	5	6	7	8	9	10	11	12	13	14
5	5	6	7	8	9	10	11	12	13	14	15
6	6	7	8	9	10	11	12	13	14	15	16
7	7	8	9	10	11	12	13	14	15	16	17
8	8	9	10	11	12	13	14	15	16	17	18
9	9	10	11	12	13	14	15	16	17	18	19
10	10	11	12	13	14	15	16	17	18	19	20

Figure 1: Addition table.

two types of multiplication tables came into use. The first kind simply listed the multiples for a single number, while the second combined a series of these smaller tables on one tablet, illustrating a number of multiplication equations.

Single tables listed a principal number, denoted as p, then indicated the solutions for multiplying p by numbers 1 through 20. Mesopotamians operated on a base 60 system of numbers, and therefore could have calculated through 59p, but instead their tables went in increments of 10 following 20p, and to determine a number in between, one simply combined equations, so that 26p was the result of adding 20p with 6p. Combined tables included a variety of single tables on one tablet, with the individual tables generally written in descending order according to the primary number illustrated.

Examples of later tables were discovered in early Egyptian texts, listing the values of letters used as place holders in equations that contained fractions. Early Arabic astronomers utilized tables to keep track of variables used in calculating planetary positions necessary for astrology. Then in the eleventh century, Hebrew astronomers corrected errors in observations based on inconsistent calculations, and began recording constants in tables to ensure that all future observations were based on the same set of calculations.

As mathematical equations grew more complex, tables were used to record trigonometric functions, such as values for sine and cosine, and eventually logarithms and differential equations. In 1627, Johannes Kepler printed the first modern astronomical tables, the Rudolphine Tables, which provided accurate, complex planetary calculations based on longitude and latitude of the stars, previously not available.

Real-life Applications

MATH SKILLS

Tables are commonly used to teach basic math skills, such as addition and multiplication—equations that are eventually committed to memory. For addition, a series of numbers run across the top of the columns and down the start of the rows as headings. Then each number across the top is added to each number down the side and the result entered at the intersection of the column and row. The table is used to drill students in the basic addition skills. (See Figure 1.)

In multiplication, the numbers along the top of the columns are instead multiplied by those down the side of the rows, with the solution again placed at the intersection. This format allows certain patterns to become clear, often providing tricks that help students memorize the answers to the equations. For instance, when multiplying

Multiplication Table

×	0	1	2	3	4	5	6	7	8	9	10	11	12
0	0	0	0	0	0	0	0	0	0	0	0	0	0
1	0	1	2	3	4	5	6	7	8	9	10	11	12
2	0	2	4	6	8	10	12	14	16	18	20	22	24
3	0	3	6	9	12	15	18	21	24	27	30	33	36
4	0	4	8	12	16	20	24	28	32	36	40	44	48
5	0	5	10	15	20	25	30	35	40	45	50	55	60
6	0	6	12	18	24	30	36	42	48	54	60	66	72
7	0	7	14	21	28	35	42	49	56	63	70	77	84
8	0	8	16	24	32	40	48	56	64	72	80	88	96
9	0	9	18	27	36	45	54	63	72	81	90	99	108
10	0	10	20	30	40	50	60	70	80	90	100	110	120
11	0	11	22	33	44	55	66	77	88	99	110	121	132
12	0	12	24	36	48	60	72	84	96	108	120	132	144

Figure 2: Multiplication table.

numbers by a factor of 9, as the first digit of the solution increases by one, the second digit decreases by one, so that $4 \times 9 = 36$, $5 \times 9 = 45$, $6 \times 9 = 54$, and so on. (See Figure 2.)

Similar tables can be used for both subtraction and division.

For more complicated math equations, tables list solutions that might normally require complicated calculations or even the use of a computer, or values for variables used in complex equations. Examples include trigonometry, logarithms, and differential equations.

OTHER EDUCATIONAL TABLES

Another table studied in school is the periodic table. Used in chemistry, this table displays the various elements according to abbreviation, their placement providing information such as atomic weight. Similar substances are grouped together, such as gases, liquids, and those elements that are synthetically crafted.

Probability and statistics findings can also be displayed using a table. Available data provides the material for headings, while the resulting odds for each combination fill in the body of the table. In genetics, this sort of information can be applied to the Punnet square. This small table illustrates the likelihood that various genetic traits will be inherited by a child, based on what genetic material each parent might donate, and whether that material is dominant or recessive. An example of this using two peapod plants, one tall carrying a recessive short gene (Tt) and one short (tt), where tall is the dominant trait, results in half of the second generation plants being tall and half of them short. (See Figure 4.)

CONVERTING MEASUREMENTS

Units of measurement vary from country to country, with some nations using metric measurements and others utilizing a standard of feet, yards, miles and so on. There are calculations that allow one to translate inches into centimeters or yards into meters, but tables that illustrate these transitions eliminate the need to recall the relationship between the two forms and provide a shortcut to performing the math. (See Figure 5.)

Tables can be used to illustrate not only equivalent measures of distance, but weight, liquid and solid capacity, or temperature, by converting pounds to kilos, quarts to liters, and degrees in Fahrenheit to degrees in Celsius. Fractions can be converted into decimals, with the table also illustrating the equivalent percentages. Cooking measurements, such as cups or tablespoons, may be listed as weights, in ounces, pounds, grams, and kilos, enabling a cook to translate recipes printed using American measurements into their own more familiar European measurements, or vice versa.

I	II	IIIb	IVb	Vb	VIb	VIIb	VIIIb			Ib	IIb	III	IV	V	VI	VII	0
1	2	3	4	5	6	7	8	9	10	11	12	13	14	15	16	17	18
H																	He
Li	Be											B	C	N	O	F	Ne
Na	Mg											Al	Si	P	S	Cl	Ar
K	Ca	Sc	Ti	V	Cr	Mn	Fe	Co	Ni	Cu	Zn	Ga	Ge	As	Se	Br	Kr
Rb	Sr	Y	Zr	Nb	Mo	Tc	Ru	Rh	Pd	Ag	Cd	In	Sn	Sb	Te	I	Xe
Cs	Ba	La*	Hf	Ta	W	Re	Os	Ir	Pt	Au	Hg	Tl	Pb	Bi	Po	At	Rn
Fr	Ra	Ac**	Rf	Db	Sg	Bh	Hs	Mt	Uun	Uuu	Uub		Uuq		Uuh		Uuo
Lanthanides *			Ce	Pr	Nd	Pm	Sm	Eu	Gd	Tb	Dy	Ho	Er	Tm	Yb	Lu	
Actinides **			Th	Pa	U	Np	Pu	Am	Cm	Bk	Cf	Es	Fm	Md	No	Lr	

Figure 3: The Periodic Table of the Elements.

Parent Pea Plants ("P" Generation)		Offspring ("F1" Generation)	
Genotypes:	Phenotypes:	Genotypes:	Phenotypes:
Tt x tt	tall x short	50% (2/4) Tt	50% tall
		50% (2/4) tt	50% short

Figure 4: Punnet square. This small table illustrates the likelihood that various genetic traits will be inherited by a child, based on what genetic material each parent might donate, and whether that material is dominant or recessive.

Linear measure

1 mil	0.001 inch	0.0254 millimeter
1 inch	1,000 mils	2.54 centimeters
12 inches	1 foot	0.3048 meter
3 feet	1 yard	0.9144 meters
5.5 yards or 16.5 feet	1 rod (or pole or perch)	5.029 meters
1 mile	5,280 feet	1.6094 kilometers
40 rods	1 furlong	201.168 meters
8 furlongs	1 mile	1.6094 kilometers
3 miles	1 league	4.83 kilometers
	1 millimeter	0.03937 inches
10 milimeters	1 centimeter	0.3937 inch
10 centimeters	1 decimeter	3.937 inches
10 decimeters	1 meter	39.37 inches or 3.2808 feet
10 meters	1 decameter	393.7 inches or 32.8083 feet
10 decameters	1 hectometer	323.083 feet
10 hectometers	1 kilometer	0.621 mile or 3,280 feet
10 kilometers	1 myriameter	6.21 miles

Figure 5.

FINANCE

The financial industry makes use of tables as a way of conveying information for nearly every type of transaction. Checkbook registers are structured as very basic tables, with each row making up a separate transaction in your checking account, and the columns indicating what type of transaction has taken place—the writing of a check, a deposit of funds, the addition of interest, amounts, and whether the transaction was of a tax-deductible nature, with the final column keeping a running tally of the account balance. Statements for bank accounts are also structured as tables, with dates and transactions indicated in each row, and the columns separating deposits from withdrawals, interest, and an updated balance.

Banks list interest rates in tables, both for certificates of deposit (CDs) and other investment accounts, and for their advertised loan rates for cars or mortgages. In the

Local and Universal Time

Due to the shape and movement of Earth, the planet is divided into time zones based upon the amount of time it takes to make one complete rotation of the sun—approximately twenty-four hours. What time zone you are in determines the time of day at any given point, with your own time zone considered "local time." Tables list the various different time zones and enable you to determine what the corresponding time is in another part of world at a glance. An international timetable makes it possible to see the time of day anywhere in the world, simply by comparing the time zones. In addition, a detailed timetable will include information that accounts for the use of daylight savings time in the summer, as some parts of the world do not observe this manual change of the clocks, and those who do sometimes begin and end this period on different dates.

Coordinated Universal Time (UTC), formerly known as Greenwich Mean Time, sets the time standard for the entire world. It is essentially solar mean time, and moves at a variable rate to account for the fact that the planet does not rotate around the sun in precisely twenty-four hours, but rather is off by a few seconds that eventually add up. Because of this, UTC will occasionally leap ahead by several seconds in order to even out the time with the actual rotation of the planet. All other clocks in all other time zones are set to correspond to UTC. (See Figure A.)

Standard time zone conversions

Conversions from UTC to some US time zones: * = previous day

UTC (GMT)	Pacific standard	Mountain standard	Central standard	Eastern standard
00	4 pm *	5 pm *	6 pm *	7 pm *
01	5 pm *	6 pm *	7 pm *	8 pm *
02	6 pm *	7 pm *	8 pm *	9 pm *
03	7 pm *	8 pm *	9 pm *	10 pm *
04	8 pm *	9 pm *	10 pm *	11 pm *
05	9 pm *	10 pm *	11 pm *	12 mid
06	10 pm *	11 pm *	12 mid	1 am
07	11 pm *	12 mid	1 am	2 am
08	12 mid	1 am	2 am	3 am
09	1 am	2 am	3 am	4 am
10	2 am	3 am	4 am	5 am
11	3 am	4 am	5 am	6 am

Figure A.

case of CDs, the table will list a row for each locked-in time period, as interest rates differ based on the length of time the account is held. Corresponding interest rates are listed in the next column, followed by a column for APY, or the actual period yield that reports the amount of interest you would earn when accounting for compounding. For instance, a CD that is deposited for a preset period of six months might earn a 2.52% interest rate, which would result in an actual earned rate of 2.55%. In the case of money market accounts, while there is no set time period, some banks offer greater interest rates for larger deposits to encourage customers to keep more money in the institution. They display these advantages in table form, listing each deposit increment, followed by the interest rate given on that amount. (See Figure 6.)

Loan information is also listed in tables. Car loans vary both in time period and interest rate amount, so a table might list the average interest rate offered for a loan that is spread over 36 months, 48 months, and 60 months. Mortgage rates offer even more choices, including the length of the loan—anywhere from 15 to 30 years—and whether a loan rate is fixed or variable. Sometimes the

	Current	1 Month Prior	3 Month Prior	6 Month Prior	1 Year Prior
1-Month	1.25	1.25	1.13	1.06	0.89
6-Month	2.55	2.37	2.01	1.81	1.34
1-Year	3.28	3.01	2.63	2.33	1.71
2-Year	3.66	3.33	3.11	2.95	2.23
5-Year	4.25	4.09	3.97	4.03	3.49
3-Month Jumbo	2.15	2.06	1.82	1.63	1.19
6-Month Jumbo	2.82	2.62	2.26	1.98	1.46
1-Year Jumbo	3.49	3.18	2.81	2.5	1.85
2-Year Jumbo	3.9	3.52	3.27	3.12	2.34

Source: CDs provided by Bankrate.com

Figure 6: Interest and earning table.

Ideal Weight for Men

Height (in shoes)	Small Frame	Medium Frame	Large Frame
6'4"	162 to 176 lb	171 to 187 lb	181 to 207 lb
6'3"	158 to 172 lb	167 to 182 lb	176 to 202 lb
6'2"	155 to 168 lb	164 to 178 lb	172 to 197 lb
6'1"	152 to 164 lb	160 to 174 lb	168 to 192 lb
6'	149 to 160 lb	157 to 170 lb	164 to 188 lb
5'11"	146 to 157 lb	154 to 166 lb	161 to 184 lb
5'10"	144 to 154 lb	151 to 163 lb	158 to 180 lb
5'9"	142 to 151 lb	148 to 160 lb	155 to 176 lb
5'8"	140 to 148 lb	145 to 157 lb	152 to 172 lb
5'7"	138 to 145 lb	142 to 154 lb	149 to 168 lb
5'6"	136 to 142 lb	139 to 151 lb	146 to 164 lb
5'5"	134 to 140 lb	137 to 148 lb	144 to 160 lb
5'4"	132 to 138 lb	135 to 145 lb	142 to 156 lb
5'3"	130 to 136 lb	133 to 143 lb	140 to 153 lb
5'2"	128 to 134 lb	131 to 141 lb	138 to 150 lb

Figure 7: Generalized "healthy" weight ranges for men.

tables are used to illustrate trends in loan rates as well, listing a column for current rates and another for rates that were offered the previous week. If rates are rising, the table is an easy way to encourage customers to make a decision before prices go even higher by showing how much they have changed in a short period of time.

Other investments display necessary information in table form. The business section of most large newspapers includes the most recent closing prices for the various stock markets. These enormous tables go on for pages and list row after row of company names, followed by column after column of information regarding how each stock is trading, including the most recent price, the price the day before, percent that price has changed, high and low prices over the last year, and other pertinent information for investors. Anyone following the trends of a particular company has only to know what exchange it trades on and they can locate the stock information within the table.

The United States Federal Government provides their own set of financial tables to the public each year—the most recent tax tables. These tables are designed to help you determine how much income tax you owe on your previous year's salary. The tables are divided into sections based on your adjusted gross income (AGI), with each row indicating a salary range, such as between $18,000 and $18,050, followed by columns for the amount of tax owed based on whether you are filing as an individual, a married person filing with their spouse, a married person filing alone, or the head of your household. The tax tables provide tax payment information for salaries ranging from a single dollar to just under $100,000. For earned income over $100,000, there are additional tables that explain how to calculate taxes owed based on different, broader ranges of income.

The United States government also provides information about average government salaries, displaying their findings in tables. These figures are available for executive positions and mid-level jobs, or according to location, or for very specific posts, such as administrative law judges or law enforcement officials. Each row lists the pay grade category of the position, while the columns indicate the various raises available at that job level. General job statistics, not limited to government employees, are also available through census findings.

HEALTH

Certain basic health information often appears in tables. Healthy weight ranges for both men and women of varying heights are typically displayed in table format. Columns are labeled according to the size of the person's frame—small, medium, or large—and then each row lists a height. Weight ranges are listed for each body type. (See Figure 7 and Figure 8.)

Ideal Weight for Women

Height (in shoes)	Small Frame	Medium Frame	Large Frame
6'	138 to 151 lb	148 to 162 lb	158 to 179 lb
5'11"	135 to 148 lb	145 to 159 lb	155 to 176 lb
5'10"	132 to 145 lb	142 to 156 lb	152 to 173 lb
5'9"	129 to 142 lb	139 to 153 lb	149 to 170 lb
5'8"	126 to 139 lb	136 to 150 lb	146 to 167 lb
5'7"	123 to 136 lb	133 to 147 lb	143 to 163 lb
5'6"	120 to 133 lb	130 to 144 lb	140 to 159 lb
5'5"	117 to 130 lb	127 to 141 lb	137 to 155 lb
5'4"	114 to 127 lb	124 to 138 lb	134 to 151 lb
5'3"	111 to 124 lb	121 to 135 lb	131 to 147 lb
5'2"	108 to 121 lb	118 to 132 lb	128 to 143 lb
5'1"	106 to 118 lb	115 to 129 lb	125 to 140 lb
5'0"	104 to 115 lb	113 to 126 lb	122 to 137 lb
4'11"	103 to 113 lb	111 to 123 lb	120 to 134 lb
4'10"	102 to 111 lb	109 to 121 lb	118 to 131 lb

Figure 8: Generalized "healthy" weight ranges for women.

Expected life spans are also listed in tables. The numbers of years a person is expected to live is based on many factors, such as family history, eating and exercise regimens, whether or not they smoke, year they were born, and so on. However, by distilling all of this information, it is possible to come up with an average life expectancy for a person of a given age, and these are what are published in the tables. The information is useful to individuals planning for retirement, as it helps them determine how many years they will need to support themselves based on their investments and savings. Life insurance costs are also based on a person's age and how long they are expected to live.

TRAVEL

There are numerous uses for tables when traveling. Schedules are frequently illustrated in table format. In order to determine what train or bus to take, one must consult the timetable that lists the various train or bus stations and then shows the progression of each vehicle from its starting point to the destination. The schedules are determined based on distance between stations and the amount of time it takes to travel between points, but travelers can simply consult the schedule rather than working out the distances mathematically.

Travelers going to the beach may consult a table to determine when high or low tide will occur. The tables take into account time of day and day of the month, seasons, the cycle of the moon—everything that affects the transition of the tide. The equations are solved in advance and the results published in a table so that beach-goers have no need to understand the math necessary to determine the tide's movements.

Currency varies from country to country, and travelers need to determine how much things cost regardless of their location. Banks issue tables that estimate the conversion rate between currencies, providing a translation from one monetary denomination to another for basic round numbers, such as a hundred dollars and its equivalent in various other currencies. This enables travelers to determine quickly how much they are spending without performing complicated math equations.

Potential Applications

DAILY USE

Tables can be used to assist with many day-to-day tasks. Computer spreadsheet programs provide a template that can apply to various uses, whether or not they use mathematical equations.

If you are interested in keeping track of your finances—both how much money you earn from various

Single Life Expectancy
(For Use by Beneficiaries)

Age	Life Expectancy	Age	Life Expectancy
56	28.7	84	8.1
57	27.9	85	7.6
58	27.0	86	7.1
59	26.1	87	6.7
60	25.2	88	6.3
61	24.4	89	5.9
62	23.5	90	5.5
63	22.7	91	5.2
64	21.8	92	4.9
65	21.0	93	4.6
66	20.2	94	4.3
67	19.4	95	4.1
68	18.6	96	3.8
69	17.8	97	3.6
70	17.0	98	3.4
71	16.3	99	3.1
72	15.5	100	2.9
73	14.8	101	2.7
74	14.1	102	2.5
75	13.4	103	2.3
76	12.7	104	2.1
77	12.1	105	1.9
78	11.4	106	1.7
79	10.8	107	1.5
80	10.2	108	1.4
81	9.7	109	1.2
82	9.1	110	1.1
83	8.6	111 and over	1.0

Figure 9: Average life expectancies.

sources, and how much you spend on both necessities and pleasure—a simple table can provide the format. Each row can be numbered with a day of the month; the column headings might include such labels as job income, allowance, interest, and gifts for incoming funds, and car payment, gas, insurance, housing, food, utilities, clothing, and entertainment for expenditures. If you wish to become even more detailed, you can break out certain categories further. For example, entertainment might become movies, concert tickets, eating out, sports activities, and travel. At the end of the month, the spreadsheet program will enable you to calculate for different factors, such as money earned over the month, total money spent, money spent on necessities, and money spent on luxuries. This can be very useful if you are looking to cut back and save for a big-ticket item, such as a new stereo or a car. The table allows you to see at a glance where you have to spend your money, and where you might eliminate a few costs.

Tables do not always have to deal with numbers. Although they originally were used for mathematical purposes, the structure applies to many other things. Schedules are a prime example of this. While some schedules, such as those for transportation or attending classes, deal with times and dates, others might simply distribute tasks. A table could assign chores to different household members on a rotating basis, with each column labeled with a task and each row indicating the week. Each person determines when they are scheduled to do the dishes, vacuum, or take out the trash by finding their name at the intersection of the chore and the week it has been assigned to them.

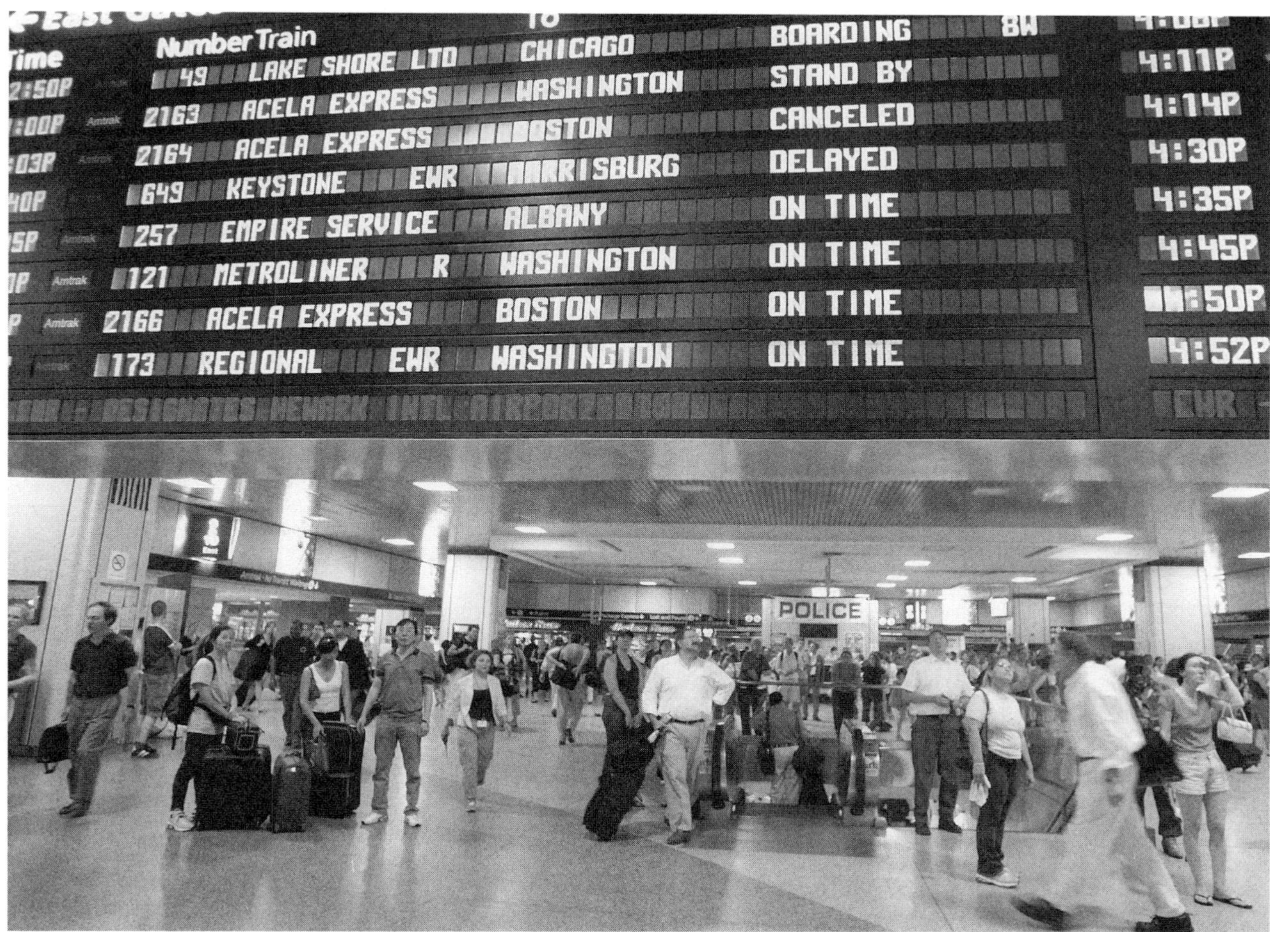

Changeable table showing New York City's train schedule. REUTERS/CORBIS.

Where to Learn More

Books

Auth, Joanne Buhl. *Deskbook of Math Formulas and Tables: A Handy Reference to Math Formulas, Metric Tables, Terminology and Everyday Problem Solving.* New York, NY: Van Nostrand Reinhold Company, 1985.

Fitzpatrick, Gary L. *International Time Table.* Metuchen, NJ: The Scarecrow Press, Inc., 1990.

Grattan-Guinness, Ivor (editor). *The Norton History of the Mathematical Sciences: The Rainbow of Mathematics.* New York, NY: W.W. Norton and Co., 1998.

Web sites

Baby Steps through the Punnet Squares. <http://www .borg.com/~lubehawk/psquare.htm> (April 5, 2005).

Bloomberg.com. <http://www.bloomberg.com/markets/rates/> (April 3, 2005).

IRS Web site. <http://www.irs.gov/> (April 1, 2005).

Math.2.org. <http://www.math2.org.> (April 5, 2005).

Mesopotamian Mathematics. <http://it.stlawu.edu/~dmelvill/mesomath/> (April 2, 2005).

Periodic Table Styles. <http://chemlab.pc.maricopa.edu/periodic/styles.html> (April 5, 2005).

S.O.S. Mathematics Tables and Formulas. <http://www.sosmath.com/tables/tables/html> (April 5, 2005).

United States Office of Personnel Management. <http://www.opm.gov/oca/05tables/index.asp> (April 5, 2005).

Universal Time: UTC, UT1. <http://www.hartrao.ac.za/nccsdoc/slalib/sun67.hxt/node219.html> (April 5, 2005).

University of Michigan Health System. <http://www.med.umich.edu/1libr/primry/life15.htm> (April 5, 2005).

Key Terms

Array: A rectangular arrangement of numerical data in rows and columns, as in a matrix.

Logarithm: The power to which a base number, usually 10, has to be raised to in order to produce a specific number.

Trigonometry: A branch of applied mathematics concerned with the relationship between angles and their sides and the calculations based on them. First developed as a branch of geometry focusing on triangles during the third century B.C, trigonometry was used extensively for astronomical measurements. The major trigonometric functions, including sine, cosine, and tangent, were first defined as ratios of sides in a right triangle.

Topology

Overview

Topology is a branch of mathematics that studies the shapes of objects. More specifically, topology is concerned with how portions of an object do not change when a change in overall shape occurs. Two objects are considered to be the same if they can be changed to the other form without being cut or torn. An example is a bowl and a plate. At least in the imagination, it is possible to change a bowl into a flat plate by pressing down on the curved surface of the bowl (of course, you would not want to try this in your kitchen with your parents' best china).

Another way to visualize topology is to look at a map of the freeway system of a typical large city. Dozens of freeways intersect and fan out in different directions. Looking at the map in a topological manner would involve determining how the lines would connect to get a driver to a given destination. This has nothing to do with how far the journey would be, which would be more in the realm of geometry.

Looking at the freeway map from the viewpoint of topology, it would be much more appropriate to ask if the beginning and end of the journey were connected by the same road, or whether several roads had to be taken to reach the destination.

A topological freeway journey has to do with shape. That is the heart of topology; the shape of an object and how the object can be changed in shape without changing the *properties* of its shape. For example, if you hold a beach ball in your hands and pull with both hands to stretch the ball, the formerly spherical ball is now a football shape. This new shape is called an ellipsoid. Even though the sphere and the ellipsoid are different in shape, they have the same topology.

Other questions that can be asked about an object's topology include: are there holes in the object? Is the object hollow, and does the object have a limit, like a balloon, or does it reach to infinity.

Thinking about topology can be a bit confusing and mind-bending. Topology has been described as being "rubber-sheet geometry." That means that it is fine to shrink an object, or twist it or stretch it, because topology is not concerned with how close one molecule is to another. Changes like cutting, pasting, or puncturing, however, which take molecules from one part of the object and move them to another part of the object, are not part of studying topology.

An object's topology can be different depending on how it is viewed. Let us return to the dinner plate example to see an illustration of this concept. If a plate is lying on a dinner table and you look at it before you sit down,

A commuter passes a map of the London Underground (subway) system showing how the lines and stations connect (topological information), rather than distances, directions, or other geometric information. TOBY MELVILLE/REUTERS/CORBIS.

it appears like a circle. But, if you walk ten feet away and then look at the same plate, its shape may well be more like the elliptical football; the plate will look much wider than long. Two very different views are produced by an object whose shape has not changed.

Topological variations like the dinner plate example are one reason why learning how to draw can be a difficult process. Even though the shape of a plate from across the room is elliptical, the mind can still interpret a plate as being a circle. So, when drawing a table, the artist can actually fight against his or her brain telling him to draw a circle.

A dinner plate is a solid object, whose shape cannot be changed except by breaking it. But consider a piece of bread dough. The dough can be squeezed, rolled, flattened and stretched. A circle of dough can be changed to a rectangle, triangle, sphere, or other shapes. From a topological point of view, all the shapes of the piece of dough are the same. This is because to change from one form (a circle, for example) to another form (a triangle, for example) does not involve a rearrangement of the molecules that make up the object. A dough circle is the same as a shape like a dough triangle, as portions of the circle can be tugged outward to form the triangle. Likewise, a dough triangle is the same topology as a dough sphere, as the triangle can be rolled around the form a sphere.

Topology also involves the shape that can be created in space by the movement of an object. A good example of this is the hands of a clock. As the hour, minute, or second hands move from the 12 o'clock position around the dial and back to 12 o'clock, they create a circle.

Real-life Applications

VISUAL ANALYSIS

Anyone who has looked at a graph of scientific or other information has an appreciation for topology, even if they have not realized it. The relationships between two or more factors (for example, retail sales, day of the week, and age of shopper) can be detailed without the use of a graph. But, plotting the information in graph form, with the resulting hills and valleys, makes it much easier for people to interpret and to understand the information.

VISUAL REPRESENTATION

Topology is used to measure and evaluate magnetic and electrical fields. Topology can determine the shape of the fields. The changing shape of Earth's magnetic field varies the field's reactions to incoming solar particles and radiation. These changes often impact the performance of electronic navigation and communication devices.

Topology is also used in physics to describe string theory or to construct models of the shape of the universe that conform to observed data.

COMPUTER NETWORKING

Topology is critically important in designing computer network systems. The layout of workstations, hubs, switches, and servers constitutes the physical topology of the network and greatly impacts the capability and speed of transmitting data.

I.Q. TESTS

A number of psychology, medical, and I.Q. type tests utilize topological puzzles to assess visual and coordination skills as they relate to thinking or manipulative skills.

For example, the handcuff puzzle has been around for over 250 years. Two people, some lengths of rope, and some space to move around in are required. Each person uses a length of rope as handcuffs. The rope has loops at either end to fit over each hand. As well, each length of rope should be long enough to allow each person to move around without tripping.

As each person puts the rope handcuffs on, the lengths of rope are themselves looped together. The challenge is then to get themselves apart from each other without removing the handcuffs, or damaging the ropes.

But how can the ropes be separated from each other when they are looped together and when the ropes cannot be cut?

Here is the solution to the handcuff puzzle. Let us pretend that you are one of the handcuffed pair of people. You take the other person's rope and move it along yours until the rope is lying on one of your arms (make it the right arm). The other person's rope should not be wrapped around your rope. It should just be lying along your right arm. Now, take your left hand and reach through the handcuff around the right wrist. Grab the other person's rope, which is still lying along your right arm. Pull the rope through the handcuff over the right hand. Now let the rope go back through the handcuff. You should now be separated from the other person.

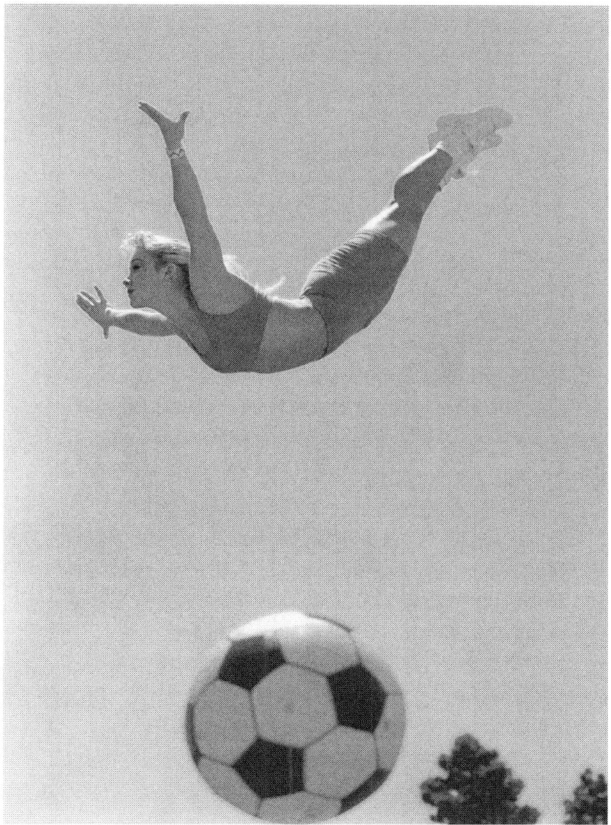

Gymnast competes for airspace with a soccer ball-shaped balloon. Topology helps mathematicians to characterize diverse shapes.

The basis of the trick is topology. As a bowl can be made to form a plate without altering the surface, so can the arrangement of the ropes be altered without the need for cutting or other damage.

MÖBIUS STRIP

At first it seems absurd; a strip of paper that has only one side. But that is the magic of the Möbius strip. Once again, at the heart of this "magic" is topology.

A Möbius strip is easy to make. A strip of paper is closed into a loop. Before the ends of the looped strip are taped together, however, one end is given a twist to produce a looped piece of paper with a half-twist in it.

Now comes the fun part. When a pen or pencil mark is made down the middle of the strip all the way around, the mark will be on both sides of the paper. It is the twist that does it, as it makes the mark change from one side of the paper strip to the other side. In other words, a Möbius strip only has one side.

Likewise, the strip only has one edge. And, when the strip is cut down the middle, the result is one long strip instead of two separate strips.

The Möbius strip is not just a novelty. This form of topology used to be part of car engines, in the form of the fan belt. By having a twist in the belt, any strain imposed on the belt would be directed evenly over the whole surface of the belt.

Where to Learn More

Books

Gamelin, T.W., and R.E. Greene. *Introduction to Topology.* Mineola: Dover Publications, 1999.

Periodicals

Collins, G.P. "The Shapes of Space," *Scientific American* (July, 2004) 291: 94–103.

Web sites

Weisstein, E.W. "Topology." *MathWorld* <http://mathworld.wolfram.com/Topology.html> (September 5, 2004).

Trigonometry

Overview

Trignometry is the study of relationships among the sides and angles of triangles, and derives its name from the Greek word for triangle, *trignon.* Real life uses of trigonometry include navigation, land surveying, global positioning system (GPS) applications, robotics, and the design of structures such as buildings and bridges. The height of Mt. Everest, the world's tallest peak, was calculated using trigonometry long before it was scaled by mountaineers.

Plane trigonometry involves trigonometric relationships that occur on a flat plane such as a piece of paper, whereas spherical trigonometry involves trigonometric relationships that occur on spheres such as planets. If the area being studied is small compared to the size of the sphere, it is often possible to obtain acceptably accurate results by using plane trigonometry for spherical problems. The grid systems that land surveyors use when laying out construction sites or locating property boundaries, for example, are formulated by assuming that Earth's curved surface can be represented by a series of flat planes. Although it is relatively easy to perform trigonometric calculations using points that lie on any one of the flat planes, it is very difficult to perform calculations using points that lie on more than one of the planes.

Fundamental Mathematical Concepts and Terms

MEASURING ANGLES

Plane angles are measured using wedge-shaped increments representing a fraction of a circle. The most common unit of angular measurement is the degree, which is denoted by the symbol °. A circle consists of 360°. Thus, an angle of 1° is 1/360 of a circle. For very accurate measurements of angles, degrees can be expressed in decimal form (for example, 10.5°) or in terms of minutes and seconds of arc. Each degree is divided into 60 minutes and each minute is divided into 60 seconds. In most branches of mathematics, including geometry and trigonometry, positive angles are measured counter-clockwise from the positive x-axis. Angles measured clockwise from the x-axis are negative.

The use of 60 rather than some other number, for example 10 or 100, to subdivide angles into minutes and seconds dates back to the Babylonian civilization, which arose around 1800 B.C. The Babylonians used a base 60 number system for their financial and scientific calculations, so it was natural for them to divide each degree into 60 seconds and each second into 60 degrees. The reason

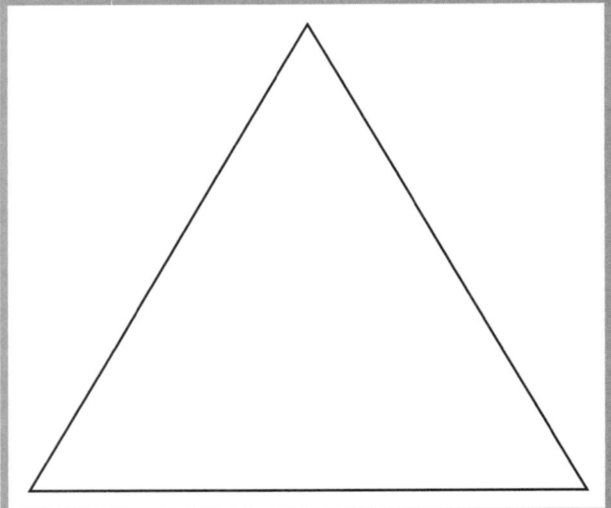

Figure 1: Equilateral triangle.

behind the division of circles into 360°, however, is less obvious. One explanation is that the Babylonians recognized that the sun follows a nearly circular path through the sky each year and that each Babylonian year consisted of 360 days. Thus, each degree in a geometric circle corresponded to one day in the Babylonian calendar. Another possible reason has been inferred from a Babylonian clay tablet unearthed in 1936, which describes geometric relationships within a circle circumscribed around a hexagon. The Babylonians knew that the perimeter of a hexagon is exactly six times the radius of a circumscribed circle, which led them to divide each of the six sides of the hexagon into 60 units, giving a total of 360° in a circle. The Babylonians also used relationship between hexagons and circumscribed circles to make a remarkably accurate estimate of the value of π. Regardless of the reason why circles were divided into 360°, it is a convenient integer because is divisible by 2, 3, 4, 5, 6, 8, 9, 10, 12, 15, 18, 20, 24, and 30. As such, fractions such as 1/2 or 1/5 of a circle can be represented by an integer number of degrees.

Degrees are not the only units that can be used to measure angles. In calculus and computer programs, angles are often measured using units called radians. There are 2π (or approximately 6.28) radians in a circle because the circumference of a circle with a radius of 1 is 2π. Thus, 1 radian represents one increment of length along the circumference of the circle, and is equal to approximately 57.3°. The need to quickly perform trigonometric calculations in battle led the German army to use the mil, which is short for milliradian, as a unit of angular measurement during World War II. It was subsequently adopted by other armies and is now the standard for angular measurement in military applications. A mil is 1/1,000 radian and is equal to the angle formed by a triangle 1,000 meters long and 1 m wide. This relationship applies only to small angles, so 1 radian cannot be defined in terms of a triangle that is 1 m long and 1 m wide. There are 6,283 or 2,000 π mils in a circle, so $1° = 17$ mils. The German army and modern NATO armies use an approximate value of 6,400 mils in a circle, while the Soviet Union adopted an approximation of 6,000 mils in a circle. Binoculars, gunsights, and other instruments were calibrated and marked with angular measurements in mils in order to estimate distances. A truck that is 10 m long occupies 10 mils in the field of vision, and is approximately $10/10 \times 1{,}000 = 1{,}000$ m away. If the same truck occupies 20 mils, it is $10/20 \times 1{,}000 = 500$ m away.

A third form of angular measurement is the grad, which is a metric unit of angular measurement rarely used in the United States. One grad is defined as 1/400 of a circle, so it is slightly smaller than a degree. Instead of being divided into minutes and seconds, grads are divided into centigrads and milligrads.

Angles less than 90° are referred to as acute angles and those greater than 90° are referred to as obtuse angles. Angles that are exactly equal to 90° are known as right angles. The right in right angle is an outgrowth of the Latin word rectus, an adjective meaning correct or proper, that has found its way into English words such as direct, correct, erect, and rectify. A likely explanation is that a 90° angle is called a right angle because it is upright or erect, as in a wall that forms a right angle with the floor and ceiling in a house. This explains why there are no left angles. Right angles are sometimes described as orthogonal, which is derived from the Greek words for right (ortho) and angle (gonia).

TYPES OF TRIANGLES

Plane triangles can be classified according to the relative lengths of their three sides or the angles between the sides. In either case, three kinds of triangles are recognized. If classified according to the lengths of their sides, triangles are equilateral, isosceles, or scalene. If classified according to their angles, triangles are acute, right, or obtuse. Regardless of the triangle type, the sum of the three angles in a plane triangle must always add up to 180° or π radians. If two triangles are identical, they are said to be congruent. If they are of the same shape but different sizes, so that their angles are identical but the lengths of their sides are different, they are said to be similar.

Equilateral triangles have three sides of equal length and, as a consequence, three angles of equal size. (See Figure 1.) Although the lengths can be of any size, the three angles in an equilateral triangle are all 60°. Isosceles triangles have two sides of equal length and a third side that is shorter than the other two. Two of the angles in an isosceles triangle are equal to each other and the third, located opposite the shortest side, is always smaller than the other two angles. Scalene triangles have three unequal sides and three unequal angles, with the largest angle opposite the longest side and the smallest angle opposite the shortest side. (See Figure 2.)

Acute triangles are defined by three acute angles, with no restrictions regarding the lengths of the sides. Thus, equilateral and isosceles triangles are acute triangles. Obtuse triangles contain one obtuse angle, and scalene triangles are necessarily also obtuse triangles. Because a triangle must consist of three angles that sum to 180°, a triangle cannot contain more than one obtuse angle. Right triangles are defined by the presence of one 90°, or right, angle. (See Figure 3.) The side opposite the right angle is known as the hypotenuse and, because of the right angle, the sum of the two remaining angles must always be 90°. Just as an obtuse triangle cannot contain more than one obtuse angle, a right triangle cannot contain more than one right angle.

Spherical triangles are formed by the intersection of three curved lines, or arcs, on the surface of a sphere.

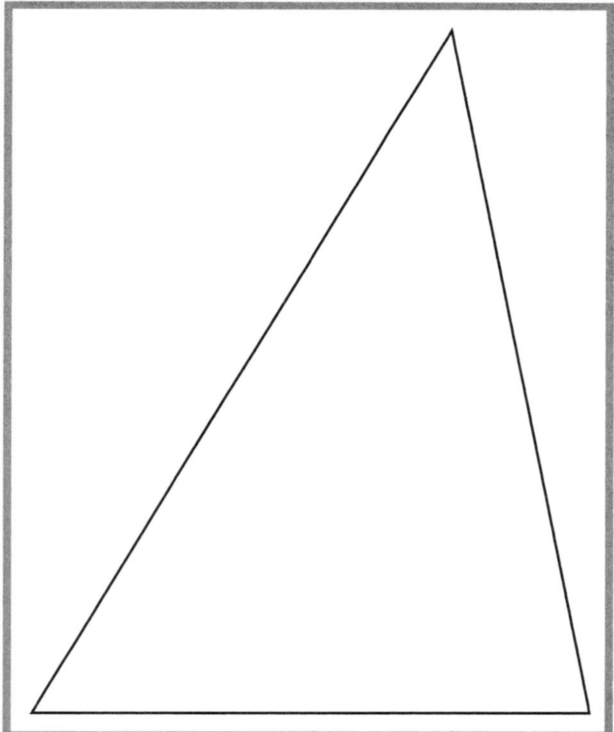

Figure 2: Scalene triangle.

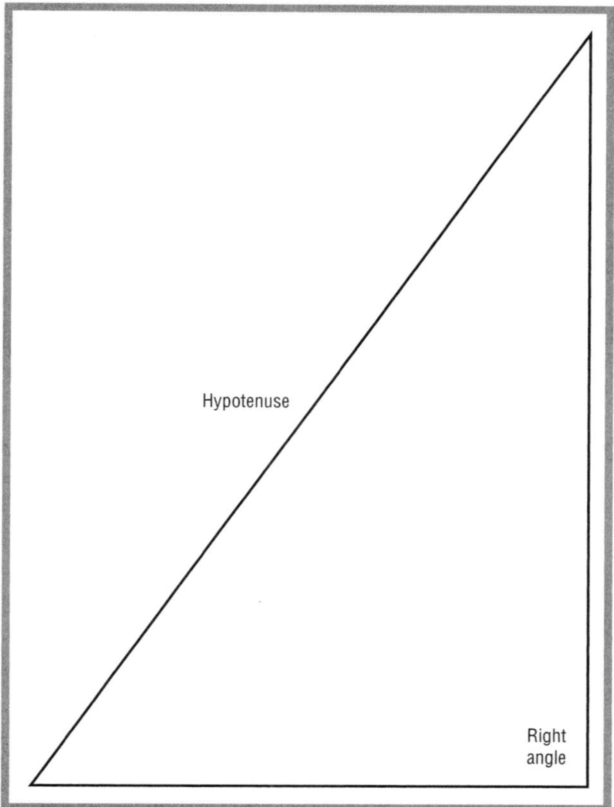

Figure 3: Right triangle.

Unlike the angles comprising a plane triangle, the angles inside of a spherical triangle do not always add up to a fixed value of 180°. Instead, they add up to a value between 180° and 540° (or π to 3π radians), and the difference between the sum of the angles and 180° is known as the spherical excess. As the surface area of the sphere becomes large relative to the size of the triangle, the spherical excess decreases towards zero and the spherical triangle becomes much like a plane triangle. This is why small areas of Earth's curved surface can be mapped as if they were planes. The second difference between spherical and plane triangles is that because a sphere wraps around on itself, the three intersecting arcs form one interior spherical triangle and one exterior spherical triangle. The sum of the angles of the outer spherical triangle always falls between 540° and 900° ($3\ \pi$ and $5\ \pi$ radians). The third difference is that the lengths of the sides of spherical triangles can be measured in degrees as well as units of length.

PYTHAGOREAN THEOREM

Some of the most widely used real life applications of trigonometry are based on the Pythagorean theorem, which relates the lengths of the three sides of a right

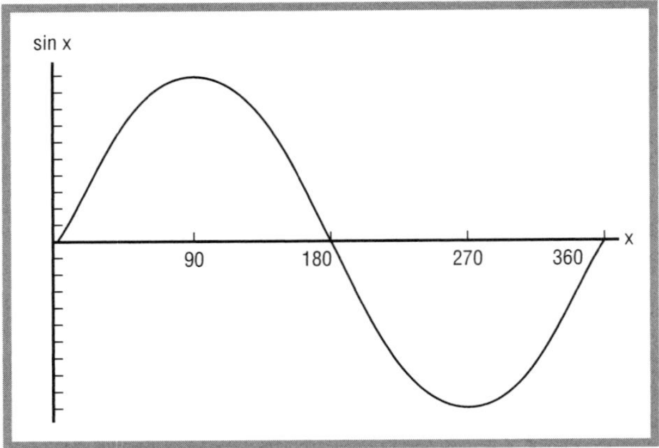

Figure 4: Sine curve.

triangle. Pythagoras (560–480 B.C.) was a Greek mathematician and philosopher who founded a school of religion and philosophy in the city of Croton. Pythagoras and his followers believed that geometric properties could always be expressed as ratios or products of whole numbers. They were troubled that the length of the hypotenuse of a right triangle is not a simple multiple of the lengths of the sides. Instead, the square of the hypotenuse is the sum of the squares of the two sides. If the lengths of any two sides of a right triangle are known, the Pythagorean theorem allows the length of the third side to be calculated.

If the length of the hypotenuse of a right triangle is C, and the lengths of the two sides are A and B, then the Pythagorean theorem is $A^2 + B^2 = C^2$. In the case of a triangle with sides of $A = 3$ and $B = 4$, $C = 5$ and the Pythagorean theorem is $3^2 + 4^2 = 5^2$. In this case the three lengths are related using only whole numbers. In the special case in which both $A = 1$ and $B = 1$, however, the theorem is $1 + 1 = 2$. Because $C^2 = 2$, $C = \sqrt{2}$, which is often referred to as Pythagoras's constant. The problem that faced Pythagoras and his followers was that $\sqrt{2}$ is an irrational number that cannot be expressed as a ratio of whole numbers such as ½ or ⅓, which violated their belief that geometric relationships must be expressible in terms of whole numbers. It is said that Hippasus, a follower of Pythagoras, was thrown overboard by other Pythagoreans when he proved during an ocean voyage that $\sqrt{2}$ is irrational. However disappointing it was to the Pythagoreans, the discovery of irrational numbers was an important step in the progress of mathematics that led to the acceptance of concepts such as π.

TRIGONOMETRIC FUNCTIONS

Real life applications of trigonometry almost always involve the use of three trigonometric functions that relate the sides and angles of right triangles. Those three functions are the sine, cosine, and tangent of an angle. To define the trigonometric functions, use the letter C to represent the length of the hypotenuse of a right triangle and the letters A and B to represent the lengths of the other two sides. It does not matter which of the two shorter sides is A and which is B, as long as C is the hypotenuse. Next, use the lower-case letters a, b, and c to represent the three angles. Assign the letters so that angle a is opposite side A, angle b is opposite side B, and angle c is opposite side C (which is the hypotenuse).

The sine of an angle, which is almost always abbreviated as sin, is the ratio of the side opposite the angle and the hypotenuse. In other words, $\sin a = A / C$ and $\sin b = B / C$. The actual numerical value of the sine function will depend on the size of the angle. For example the sine of 30°, which is abbreviated as sin 30°, is 1/2. One of the reasons that the sine function is an important tool for scientists, engineers, and mathematicians is that it is periodic, meaning that it repeats itself at regular intervals. It begins with $\sin 0° = 0$, increases until it reaches a peak at $\sin 90° = 1$, decreases to $\sin 180° = 0$ and then $\sin 270° = -1$, and then increases to $\sin 360° = 0$. This pattern repeats itself indefinitely every 360°. A plot of the sine function for angles ranging from 0° to 360° produces a wave-like line that is known as a sine curve. (See Figure 4.)

Another important trigonometric function is the cosine, which is defined as the ratio of the side adjacent to an angle and the hypotenuse. Using the same definitions as in the previous paragraph, the cosine of a, which is usually written as cos a, is B / C. Likewise, $\cos b = A / C$. The cosine function follows a curve that is identical in shape to a sine curve that has been shifted 90° to the left or right along the horizontal axis, starting with $\cos 0° = 1$ and decreasing to $\cos 90° = 0$ and $\cos 180° = -1$, then increasing smoothly to $\cos 270° = 0$ and $\cos 360° = 1$. Like the sine curve, the cosine curve repeats itself indefinitely.

The identical shapes of the sine and cosine curves can be expressed by the relationship $\sin \tau = \cos(\tau - 90°)$ or, equivalently, $\cos \tau = \sin(\tau + 90°)$, in which τ is any angle measured in degrees. The same relationship works for angles measured in radians if the 90° is changed to $\pi/2$ radians. Another way of writing the relationship between the sine and cosine of an angle is to use a variation of the Pythagorean theorem, which is that $\sin^2 \tau + \cos^2 \tau = 1$. Thus, if a person knows the sine of an angle then he or she can easily calculate the cosine, or vice versa.

Sine and cosine curves are important in many scientific and engineering problems involving waves in space or time. For example, a seismologist analyzing the vibrations recorded during a large earthquake considers his or her seismogram to consist of many different sine and cosine waves added together to produce the complicated shaking. An oceanographer can analyze the waves traveling across a body of water using sine and cosine functions. Even Earth's topography can be simulated as a collection of sine and cosine waves added together. Although values of sines or cosines can be calculated by drawing a triangle to scale, measuring the lengths of the appropriate sides, and dividing, this is not an efficient approach for practical problems. An engineer who needs to draw a graph of a sine curve, for example, would have to draw many triangles in order to calculate the value of sin τ at enough values of τ to produce a smooth curve. Until computers and scientific calculators became widespread in the last half of the twentieth century, people who needed to find the values of trignometric functions looked them up in mathematical handbooks that contained tables of values for each function. Since then, however, so-called trig tables have virtually disappeared and values are almost always calculated using a calculator or a computer. Whereas most scientific calculators will allow their users to choose whether angles are specified in degrees, radians, or grads, computer languages generally require angular measurements to be specified in radians.

The third basic trignometric function is the tangent, which is the ratio of the sides opposite and adjacent to an angle. Again using the same variables as above, the tangent of a, or tan a, is A / B and tan b = B / A. As the size of the angle decreases, so does its tangent. The tangent of any angle that is an even multiple of 90° (for example, 0°, 180°, and 360°) is 0. As the size of the angle increases, the tangent increases until it reaches ∞ for angles that are odd multiples of 90° (for example, 90°, 270°, and 450°). The tangent of an angle can also be defined in terms of its sine and cosine, tan a = sin a / sin b. One of the reasons why angles measured in radians are useful in science and engineering is that if the angle is very small, the angle will very nearly equal to its tangent. For example, the tangent of an angle measuring 1/100 radian is 1/100. Because of this, some complicated equations involving the tangents of angles can be made much simpler. It is also the basis for the use of mils as angular measurements that can be used to easily calculate distances.

In addition to the three basic functions (sine, cosine, and tangent), there exist three reciprocal functions: cosecant, secant, and cotangent. The cosecant of angle a is csc a = 1/sin a, the secant of a is sec a = 1/cos a, and the cotangent of a is cot a = 1/tan a. Notice that the cosecant is the reciprocal of the sine and the secant is the reciprocal of the cosine, which can be confusing. The cosecant, secant, and cotangent can help to simplify complicated equations involving trigonometric terms. Two other trigonometric functions, the versed sine (versin a = 1 − cos a) and the exsecant (exsec a = sec a − 1), have fallen out of use.

Over the years many mnemonics have been proposed to help students remember which sides of a triangle are associated with each of the functions. To use the mnemonics, first take the first letter of each word in the following equations: Sine = Opposite / Hypotenuse, Cosine = Adjacent / Hypotenuse, and Tangent = Opposite/ Adjacent. This will form the combination SOHCAHTOA, which can be used to invent sentences consisting of words beginning with those letters, for example Some Old Horses Chase And Hunt 'Til Old Age or Silly Old Harry Caught a Herring Trawling Off America. Another approach is to use only the first letters of the names of the sides, which form the combination OHAHOA. Two common mnemonics involving those letters are Old Houses Always Have Old Attics and Oscar Has A Heap Of Acorns. Regardless of how clever they are, it may take more effort to memorize the mnemonics and their meanings than to simply memorize the basic definitions.

LAW OF SINES

The law of sines relates the sides and angles in any triangle, regardless of whether or not it is a right triangle. Again using upper case letters to represent the sides of a triangle and lower case letters to represent the angles opposite those sides, the law of sines is A / sin a = B / sin b = C / sin c = 2R, where R is the radius of a circle circumscribing the triangle. Any two of the terms separated by equal signs can be combined to perform trigonometric calculations. For example, if A = 1 cm and a = 30°, then 1 / sin 30° =2 R. Because sin 30° = 1/2, the law of sines requires that 1 / 1/2 = 2 R or, simplifying the equation, R = 1. Calculations can also be formed using only sides and angles. For example if A = 1 cm and a = 30°, as before, and b =50°, then the law of sines can be written as 1 cm / sin 30° = B / sin 50°. Knowing that sin 30° = 0.5000 and sin 50° = 0.7660, it follows that B = 1.53 cm.

A Brief History of Discovery and Development

Although the Babylonians developed the unit of angular measurement known today as a degree, knew many geometric techniques, and gave angular coordinates for stars, the Greek astronomer Hipparchus (180–125 B.C.) is generally known as the father of trigonometry. He constructed tables of chords, which are line segments joining two points along the circumference of a circle, and it is said that he wrote a 12–volume treatise on chords. Although that work has never been found, it has been described by other Greek writers, and the series of volumes appears to have been the first written about trigonometry. The analysis of chords was an important development in the history of trigonometry because they are related to the sine and cosine functions. To illustrate this, consider a straight line connecting any two points along the circumference of a circle. Draw a line from one of the points to the center of the circle, and then from the center to the other end of the chord. This forms a triangle, and the length of the chord is related to the angle formed within the circle.

The Greek astronomer Menelaus (70–130 A.D.), about whom little is known, wrote a treatise on spherical trigonometry and its applications to astronomy. Like Hipparchus, he worked with chords rather than the modern trigonometric functions. He was a contemporary of Ptolemy (85–165 A.D.), another astronomer and author of a book most commonly known as the Almagest, which included tables of chords computed in %½° increments. Ptolemy also described how to use his tables of chords to solve trigonometric problems. Despite his great contributions to science and mathematics, Ptolemy is best remembered for his geocentric theory stating that the Sun and planets revolve around Earth. At about the same time, astronomers in India were using the precursor to the modern sine function rather than chords in their calculations. Unlike the modern sine function, the sine function developed in India was based on the length of one leg of a right triangle and not the ratio of the leg to the hypotenuse.

Muslim astronomers took the lead during the Middle Ages, building upon the work of their Greek and Indian predecessors. In particular, they began using all six modern day trigonometric functions (sine, cosine, tangent, secant, cosecant, and cotangent). Muslims continued to use lengths rather than ratios in their trigonometric functions, but also appear to have started using circles with a radius of 1, rather than the Babylonian value of 60, in their derivations. This produced the same values for the trigonometric functions that are used today. In addition to being used for astronomical calculations, their results helped the faithful to determine the direction to Mecca for prayers five times each day.

Latin translations of Arabic books did not make their way to Europe until the twelfth century. During the thirteenth century, German astronomer Georges Joachim proposed that trigonometric functions should be expressed as ratios rather than the lengths of lines. This was an important contribution because it meant that the values of the functions would depend only on the angle and not the actual lengths of the triangle legs. Like other scholars of his time, Joachim adopted the Latin name Rheticus. Subsequent work by European scholars included the development of many relationships involving multiple angles and powers of trignometric terms, which laid the groundwork for much of European science in the following centuries. Trigonometry also played an important role in Isaac Newton's invention of the calculus, which included ways to write trigonometric functions of an angle as infinite series involving powers of the angle. With the invention of the calculus, trigonometry was absorbed into the larger field of mathematics known as analysis.

Real-life applications

NAVIGATION

Pilots, mariners, and mountaineers all use trigonometric concepts to find their way from one point to another. In navigation, positive angles are measured clockwise from North and are known as azimuths. Azimuths convey direction, so they can range from 0° to 360°, and the word azimuth is sometimes used synonymously with the word heading. The azimuth of a line running from south to north is 0° and the azimuth of a line running from north to south is 180°. This distinction is critical in navigation. In other applications, it may not be critical to distinguish the direction. For example, it does not matter whether the boundary of a country runs from north to south or south to north.

A related term, which was in common use before computers and calculators existed, is bearing. Like an azimuth, a bearing conveys the direction of a line. The difference is that a bearing is always an acute angle measured from north or south. The reason for this was that the trigonometric tables necessary for navigation calculations contained values ranging from 0° to 90°, so it was impossible to look up a value for, say, an azimuth of 140°. Instead of using an azimuth for the obtuse angle of 140° measured from north, a navigator would use its acute supplementary angle, which is $180° - 140° = 40°$. In order to avoid confusing this value with an azimuth of 40°, the bearing includes the axis from which the acute angle is measured (south in this case) and the direction in which it was measured (towards east in this case). Thus, the complete expression for the bearing would be south 40° east, or S40°E. One way to ensure that an azimuth is never confused with a bearing is to always write an azimuth using three digits and a bearing with two. Thus, an azimuth of 40° can be written 040° to avoid confusion with a bearing of S40°E or N40°E. Any angle greater than 90° cannot be a proper bearing.

In real-life problems involving travel over large distances, Earth's curvature becomes important and spherical, rather than planar, and trigonometry must be used. Coordinates for navigation over long distances are given in terms of latitude and longitude, which are angular measurements. Trigonometry is used to calculate the distance between the starting and ending points of a journey, taking into account that the path follows the surface of a sphere and not a straight line. The latitude and longitude of waypoints along a journey can also be calculated using trigonometry.

Navigation on Earth is complicated by the fact that the North Magnetic Pole, to which compass needles are attracted, does not coincide with the North Geographic Pole. The North Magnetic Pole is located in far northern Canada. For very approximate navigation, for example if a hiker wants to know if she is generally headed north or south, the fact that the geographic and magnetic poles are different does not make much difference. For any kind of precise navigation or mapmaking, however, the difference is important. The difference between true north, which is the direction to the North Geographic Pole, and magnetic north, which is the direction to the North Magnetic Pole, is known as magnetic declination. It is shown as an angle on topographic maps and navigational charts. Magnetic north is about 20° east of true north in the northwestern United States and about 20° west of true north in the northeastern United States. The line of zero declination runs through the Midwestern part of the country. In other areas of the world, the magnetic declination can be as great as 90° east or west in the far southern hemisphere. The North Magnetic Pole moves from year to year as a consequence of Earth's rotation, so the magnetic declination also changes over time. Government agencies responsible for providing navigation aids track the movement of the North Magnetic Pole, and maps are continually revised to reflect changing declination. Measurements by the Canadian government show that the North Magnetic Pole moved an average of 25 miles (40 km) per year between 2001 and 2005.

A simple trigonometric calculation illustrates the error that can occur if magnetic declination is not taken into account. The distance off course will be the distance traveled multiplied by the sine of the magnetic declination. In an area where the magnetic declination is 20°, therefore, a sailor following a course due north would find herself 34 kilometers (21 mi) off course at the end of a 100-kilometers (62 mi) trip. The longer the distance traveled, the farther off course the traveler will be. If the magnetic declination is only 10°, however, the error will be 100 km $\times$ sin 10° = 17 km (11 mi).

VECTORS, FORCES, AND VELOCITIES

Vectors are quantities that have both direction and magnitude, for example the velocity of an automobile, airplane, or ship. The direction is the azimuth in which the vehicle is traveling and the magnitude is its speed. Using trigonometry, vectors can also be broken down into perpendicular components that can be added or subtracted. Take the example of a ferry that carries cars and trucks across a large river. If there are ferry docks directly across from each other on opposite banks of the river, the captain must steer the ferry upstream into the current in order to arrive at the other dock. Otherwise, the river current would push the ferry downstream and it would miss the dock. If the velocities of the river current and the ferry are known, then the captain can calculate the direction in which he must steer to end up at the other dock. The velocity of the river current forms one leg of a right triangle and the velocity of the ferry forms the hypotenuse (because the captain must point the ferry diagonally across the river to account for the current). If the current is moving at 5 km/hr and the ferry can travel at 12 km/hr, the angle at which the ferry needs to travel is found by calculating its sine. In this case, the sine of the unknown angle is 5 / 12 = 0.4167. The angle can then be determined by looking in a table of trignometric functions to find the angle that most closely matches the calculated value of 0.4167, by using a calculator to calculate the sines of different angles and comparing the results, or by using the arc sine (asin) function. Each of the trigonometric functions has an inverse function that allows the angle to be calculated from the value of the function. In this case, the answer is asin 0.4167 = 25°. In other words, the captain must point his ferry 25° upstream in order to account for the current and arrive at the dock directly across the river.

Another real-life application of vectors and trigonometry involves weight and friction. Automobiles and trains rely on friction to move uphill or remain in place when parked, and friction is required in order to hold soil and rock in place on steep slopes. If there is not enough friction, cars will slide uncontrollably downhill and landslides will occur. Even if a car is traveling downhill, friction is required to steer. In the simplest case, the traction of a vehicle or the resistance of a soil layer to landsliding depends on three things: the weight of the object, the coefficient of friction, and the steepness of the slope. The weight of the object is self-explanatory. The coefficient of friction is an experimentally measured value that depends on the two surfaces in contact with each other and, in some cases, temperature or the rate of movement. The value used before movement begins, for example between the tires of a parked car and the pavement or a soil layer that is in place, is known as the static coefficient of friction. Once the object begins moving, the coefficient of friction decreases and is known as the dynamic coefficient of friction. Some typical examples of coefficients of friction are 0.7 for tires on dry asphalt, 0.4 for tires on frosty roads, and about 0.2 for tires on ice. The coefficients of friction for soils involved in landslides can range from about 0.3 to 1.0, with most values around 0.6.

Because weight is a force that acts vertically downward, trigonometry must be used to calculate the components of weight that are acting parallel to the sloping surface. The frictional force resisting movement parallel to the slope is $\mu \times w \times \cos \tau$, where w is the weight, μ is the coefficient of friction, and τ is the slope angle. The component of the weight acting downslope is $w \times \sin \tau$. Division of the frictional resisting force by the gravitational driving force gives the expression $\mu / \tan \tau$. If the result is equal or greater than 1, the car or soil layer will not slide downhill. If it is less than 1, then downhill sliding is inevitable. If the coefficient of friction for tires on dry asphalt is 0.7, then parked cars will slide downhill if the slope is greater than 35°. If the road is covered with ice, however, the coefficient of friction is only 0.2 and cars will slide downhill on slopes greater than 11°.

SURVEYING, GEODESY, AND MAPPING

Land surveyors and cartographers make extensive use of trigonometry in their work. Land surveyors are responsible for establishing official property boundaries and locations, for example the legal location of a piece of property being bought or sold. They also perform topographic surveys that depict the elevation of Earth's surface and important features such as stream channels, roads, utility lines, and buildings. Location maps showing the locations of oil, gas, and water wells also fall within the scope of land surveying. Because their work is used in legal transactions such as real estate sales and applications to drill wells, land surveyors are licensed by government agencies and must have a good knowledge of trigonometry. Geodetic scientists perform work that is similar to that done by land surveyors, but over much larger distances and in some cases with much greater demands for accuracy. They are responsible for establishing the networks of known points that land surveyors rely upon in their daily work. Cartographers use information provided by surveyors, geodetic scientists, and others to produce maps. One of the great challenges of cartography is the development of techniques to represent the three dimensional surfaces of Earth and other planets on two dimensional pieces of paper or computer monitors.

In order to determine the exact location of a property line or other feature, a surveyor begins at a point with a known location, known as a point of beginning. These are generally small metal discs or monuments established by government agencies such as the U.S. Geological Survey or the U.S. Coast and Geodetic Survey, and for which the location has been carefully determined in advance. The discs marking a point of known location are called benchmarks. Surveyors determine locations by using optical and electronic instruments to accurately and precisely measure angles and distances from a benchmark to the points for which locations must be determined. The angles and distances are plotted to create a map from which the locations of new points, for example the corners of a rectangular piece of property, can be calculated.

One of the techniques used by surveyors is triangulation, which allows them to determine the location of a point without actually occupying it. This is done by accurately measuring the length of a line between two known points, which is known as a baseline. The azimuth to the unknown point is measured from each of the known points, forming an imaginary triangle. The surveyor knows the length of one leg of the triangle and two of its angles, and can use trigonometry to calculate the lengths of the remaining two legs. This provides the location of the unknown point. The surveyors then move their instruments and use the newly calculated lengths as the sides of two more triangles, repeating the process to create a network of benchmarks. The locations of some or all of the points are calculated more than once, using different baselines each time, in order to improve the accuracy of the survey.

One of the greatest surveying projects of all time was the Great Trigonometric Survey, which was undertaken to map India when it was a British colony during the nineteenth century. Because of the great distances involved, the surveyors used specially manufactured theodolites, which are surveying instruments designed to measure angles. Surveyors look through a telescope at the center of the theodolite to align it with their target, and then read angles measured both horizontally and vertically. The theodolites manufactured for the Great Trigonometric Survey used circles 36 inches in diameter (the larger the circle, the more accurately angles can be measured) that were read through microscropes to achieve extremely high accuracy. Theodolites used for ordinary surveying, by way of comparison, have circles just a few inches in diameter. After the measurements were made, elaborate trigonometric calculations were made by hand in order to calculate the horizontal and vertical distances between points. In 1852, the Great Trigonometric Survey measured the height of a mountain known as Peak XV. Its

Surveyors use basic trigonometry while taking measurements at Kirinda, about 140 miles (225 km) southeast of Colombo, Sri Lanka. AP/WIDE WORLD PHOTOS. REPRODUCED BY PERMISSION.

height and location were calculated using a triangulation from six different locations, each of them at least 100 miles (160 km) from the peak. The results showed it to be the highest mountain on Earth, with a summit elevation of 8,850 m (29,035 ft). In 1856 Peak XV was renamed in honor of George Everest, the previous Surveyor General of India. It would be more than 100 years after its discovery before Sir Edmund Hillary and Tenzing Norgay reached the summit of Mt. Everest in 1953.

Even when global positioning system (GPS) receivers are used to determine the locations of unknown points, the locations of known points are used to increase accuracy. This is done by placing one GPS receiver over a known point such as a benchmark and using a second receiver at the point for which a location must be determined. In the United States, one of the known points might be a continuously operating reference station, or CORS, operated by the government and providing data to surveyors over the internet. Data from the two receivers are combined, either in real time or afterwards by post-processing, to obtain a more accurate solution that can be accurate to a millimeter or so. Although it may not be obvious because the calculations are performed by microprocessors within the GPS receivers and on computers, they require extensive use of trigonometric functions and principles.

Once the locations of points or features are determined, they must be plotted on a map in order to be visualized. If the area of concern is relatively small, the map can be constructed using an orthogonal grid system of perpendicular lines measuring the north-south and east-west distance from an arbitrary point. If the area to be mapped is large, however, then trigonometry must be used to project the nearly spherical surface of Earth onto a flat plane. Over the centuries, cartographers and mathematicians have developed many specialized projections involving trigonometric functions. Some are designed so that angles measured on the flat map are identical to those measured on a round globe; some are designed so that straight-line paths on the globe are preserved as straight lines on the planar map.

COMPUTER GRAPHICS

Both two- and three-dimensional computer graphics applications make heavy use of trigonometric relationships and formulae. Rotating an object in two dimensions, for example a spinning object in a video game or the text in an illustration, requires calculation of the sine and cosine of the angle through which the object is being rotated. Graphics objects are typically defined using points for which x and y coordinates are known. In some cases, the points may represent the ends of lines or the vertices of polygons. Many computer programs that allow users to rotate objects require the user to enter an angle of rotation or use a graphics tool that allows for freehand rotation in real time. Each time a new angle is entered or the mouse is moved to rotate an object on the screen, the new coordinates of each point must be quickly calculated behind the scenes.

Rotation of graphical objects in three dimensions is much more complicated than it is in two dimensions. This is because instead of one angle of rotation, three angles must be given. Although there are several different conventions that can be used to specify the three angles of rotation, the one that is most understandable to many people is based on roll, pitch, and yaw. These terms were originally used to describe the three kinds of rotation experienced in a ship as it moves across the sea, and were adopted to describe the motion of aircraft in the twentieth century. Roll refers to the side-to-side rotation of a ship or aircraft around horizontal axis. An aircraft is rolling if one of its wings goes up and the other goes down. Pitch refers to the upward or downward rotation of the bow of a ship or the nose of an aircraft. As the bow or nose goes up, the stern or tail goes down and vice versa. The final component of three-dimensional rotation is yaw, which refers to the side-to-side rotation of the nose or bow around a vertical axis. Just as in two-dimensional graphics, the simulation of three-dimensional rotation by a computer program requires that trigonometric functions be calculated for each of the three angles and applied to each point or polygon vertex. Three-dimensional graphics are also more complicated because the shape of each object being simulated must be projected onto a two dimensional computer monitor or other plane, just as Earth's spherical surface must be projected to make a map.

CHEMICAL ANALYSIS

Chemists rely on trigonometry to analyze unknown substances using methods such as Fourier transform spectroscopy. Spectroscopes are instruments that break down the electromagnetic radiation absorbed or emitted by a substance into a collection of component wavelengths known as a spectrum. Spectroscopes attached to telescopes, for example, allow astronomers to learn the chemical composition of distant stars by analyzing the color of their light. In the laboratory, chemists use a variety of spectroscopic techniques to determine the chemical composition of substances. Invisible forms of electromagnetic radiation such as infrared radiation can also be used for spectroscopy.

In conventional spectroscopy, the wavelength of electromagnetic radiation to which a sample is subjected is varied over a period of time. The result is a graph showing the response of the sample to different wavelengths of radiation. Fourier transform spectroscopy is different because the sample is subjected to many different wavelengths at once, which produces a complicated combination of responses to those many wavelengths of radiation. Its advantage is that Fourier transform spectroscopy is much quicker than conventional spectroscopy. Its disadvantage is that a series of complicated calculations known as a Fourier transform, which is based on sines and cosines, must be performed in order to make the results useful. Although the precursor of the Fourier spectroscope was invented in 1880, real life Fourier transform calculations are so time consuming that practical applications of Fourier transform spectroscopy were limited until digital computers became common in the 1950s. The subsequent discovery of an especially efficient computational method known as the fast Fourier transform, or FFT, in 1956 was a major advance that led to the practical development of Fourier transform spectroscopy. Fast Fourier transforms have since become an important computational tool for digital audio processing, digital image processing, and seismic data processing.

Potential applications

Trigonometric principles and calculations have applications in virtually every discipline of science and engineering, so their importance will continue to increase as technology continues to grow in importance. Global positioning system (GPS) receivers embedded in cellular telephones, vehicles, and emergency transmitters will allow lost travelers to be located and criminal suspects to be tracked. Fast Fourier transforms will help to advance any kind of computation involving waveforms, including voice recognition technologies. Trigonometric calculations related to navigation will also become even more important as global air travel increases and the responsibility for air traffic control is shifted from humans to computers and GPS technology.

Key Terms

Chord: A straight line connecting any two points on a curve.

Hypotenuse: The longest leg of a right triangle, located opposite the right angle.

Where to Learn More

Books

Downing, Douglas. *Trigonometry the Easy Way*, 3rd Ed. Hauppauge, New York: Barron's Educational Series, 2001.

Keay, John. *The Great Arc: The Dramatic Tale of How India was Mapped and Everest was Named.* New York: HarperCollins, 2000.

Maor, Eli. *Trigonometric Delights (Reprint edition).* Princeton, New Jersey: Princeton University Press, 2002.

Schneider, Philip, and David H. Eberly. *Geometric Tools for Computer Graphics.* San Francisco: Morgan Kaufman, 2002.

Web sites

The Institute of Navigation. "Navigation Education Materials." August 21, 2000. <http://www.ion.org/satdiv/education.cfm> (May 9, 2005).

Stern, David P. "(M-7) Trigonometry—What is it good for?" November 25, 2001. <http://www-istp.gsfc.nasa.gov/stargaze/Strig1.htm> (May 5, 2005).

University of Cambridge. "Our Own Galaxy: The Milky Way." Cambridge Cosmology. 2005. <http://www.damtp.cam.ac.uk/user/gr/public/gal_milky.htm> (October 19, 2002).

Vectors

Overview

A vector is a mathematical object that contains two or more numbers in an ordered set. For example, [5 6 8] is a vector. Vectors containing two or three numbers (that is, vectors in two or three "dimensions") can be drawn as arrows. Vectors describe things that have more than one measurable feature: an arrow, for instance, has a certain length and points in a certain direction. Vectors are used in physics, medicine, engineering, and animation to describe forces, positions, speeds, changes in speed, electric and magnetic fields, gravity, and many other physical quantities. Vectors are also used in the field of linear algebra, along with the arrays of numbers called matrices, to stand for more abstract quantities. The rules for doing math with vectors are called "vector algebra."

Fundamental Mathematical Concepts and Terms

TWO-DIMENSIONAL VECTORS

A vector containing two numbers is termed a "two-dimensional" vector. A two-dimensional vector can be drawn as an arrow. To see how a pair of numbers can stand for an arrow (or the other way around), picture an arrow 25 centimeters (cm) long that has been drawn near one corner of a piece of paper. (See Figure 1.)

With the arrow drawn as in Figure 1, its base and tip are 3 cm apart as measured along the bottom edge of the

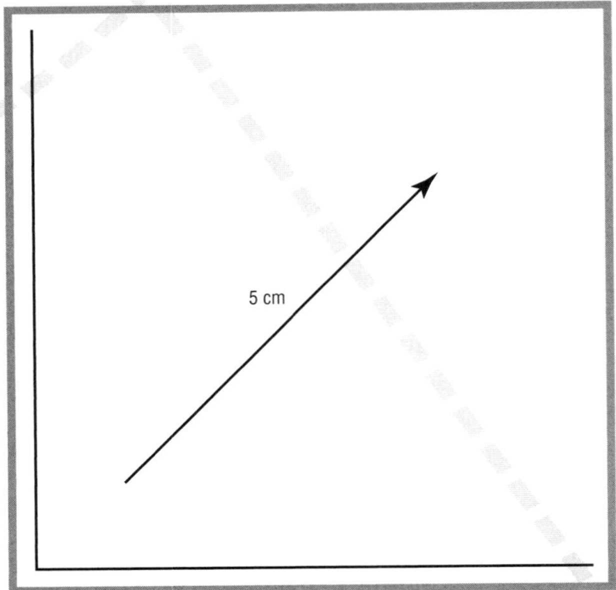

Figure 1. The arrow is not drawn to scale (that is, it is not exactly 5 cm long).

paper and 4 cm apart as measured along the side. (See Figure 2.)

The arrow can therefore be described by the vector [3 4]. Since this arrow is 5 cm long and is drawn at a 53° angle to the bottom edge of the paper, it could also be described by writing down its length and its angle: 5 cm, 53°. It cannot be described using fewer than two numbers. It is therefore called a "two-dimensional" vector.

Having more than one dimension—that is, consisting of more than one number—is what makes a vector different from a plain number (also called a "scalar"). A scalar can describe a measurement that consists of a single number, like the mass of a rock, but anything that pushes or points in a certain direction is best described by a vector. The force of a dancer's shoe against a floor, for example, is a vector because it has both a strength or "magnitude"—how hard the shoe is pushing against the boards—and a direction. This force can either be pictured as an arrow pushing against the sole of the shoe or written out in numbers. The push of a rocket motor, the speed and direction of a moving object, the strength and direction of an electric or magnetic field—all these things, and many more, are best treated as vectors.

THREE-DIMENSIONAL VECTORS

Many things in the real world cannot be described using two-dimensional vectors because they exist in three-dimensional space, namely, the ordinary space in which we live and move. Imagine, for example, an arrow placed inside a glass box. The arrow is too big to lie down in the box, but it can just fit with its base in one corner and its point in another. Drawing the edges of the box as dashed lines, we develop the image in Figure 3.

The length of this arrow and the direction it is pointing can be described by three numbers (also called "dimensions"). For example, if the glass box is 4 cm wide, 4 cm deep, and 8 cm tall, then writing down these numbers tells us exactly where the two ends of the arrow are. Writing down these three numbers gives us a numerical vector, [4 4 8]. In real life, many forces must be described using three-dimensional vectors like this one. For example, the force of a dancer's shoe against the floor is a three-dimensional vector because it might push on at the floor at any angle.

To write vectors in higher dimensions, all one has to do is write more numbers between the brackets, like [4 4 8 10 2]. Although it is not possible to draw pictures of vectors in higher dimensions, such vectors have many uses in physics, economics, and other fields.

THE MAGNITUDE OF A VECTOR

A vector's "magnitude" is basically its size. If a vector connects two points on a flat surface or in a three-dimensional space, then its magnitude gives the distance between those two points—the number we would measure if we stretched a measuring tape between them. If a vector

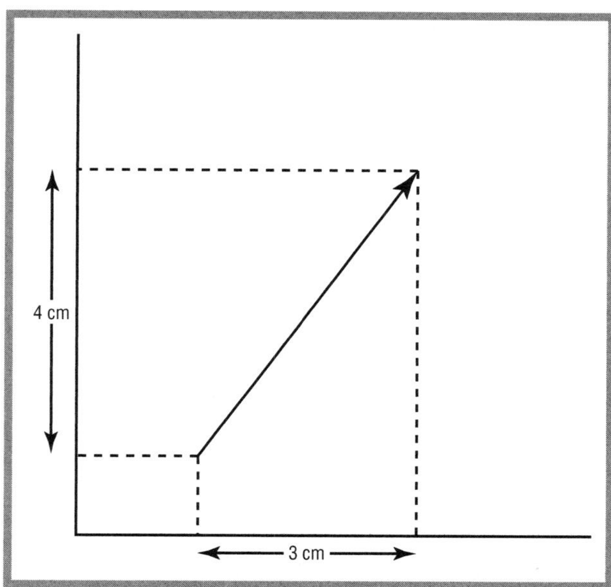

Figure 2.

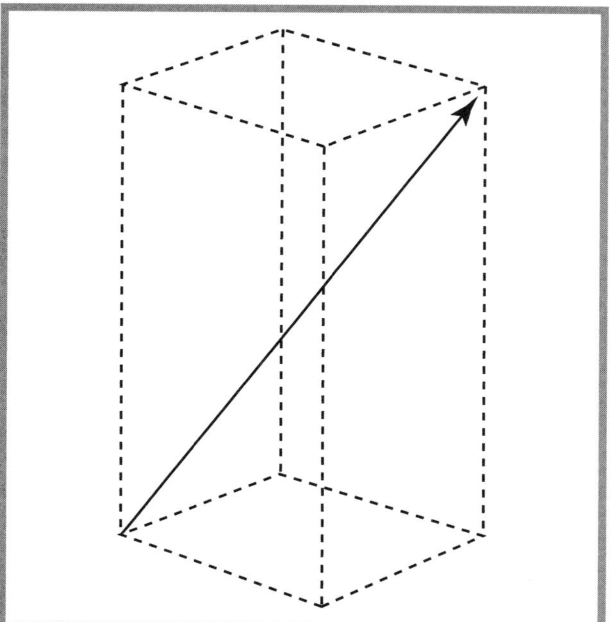

Figure 3.

Air traffic controllers use vectors (orders to fly a certain compass heading for a certain distance or time) to sequence aircraft during takeoff and landing. DAVID LAWRENCE/CORBIS.

stands for a force, then its magnitude gives the strength of that force. If a vector stands for an object's motion, then its magnitude gives how fast the object is moving.

The magnitude of a vector can be found by drawing it or building a model of it and measuring its length with a ruler, but it is easier to find the magnitude mathematically. To find the length of a vector that is known in numerical form, like the three-dimensional vector [4 4 8], first add the squares of all its parts or components. (The "square" of a number is that number multiplied by itself.) For example, to find the magnitude of [4 4 8], first calculate $(4 \times 4) + (4 \times 4) + (8 \times 8) = 96$. The second step is to find the square root of this sum. (The "square root" of a number is a second number that, when multiplied by itself, gives the first number. Square roots can be found using a calculator.) In this case, the sum is 96 and the square root of 96 is 9.798 because $9.798 \times 9.798 = 96$. Therefore, the magnitude of the vector [4 4 8] is 9.798. In the example where the arrow touching the two diagonally opposite corners of the box is described by the vector [4 4 8], the length of the arrow is 9.798 centimeters.

VECTOR ALGEBRA

Vectors can be added or multiplied according to the rules called "vector algebra." Addition and multiplication are different for vectors than for ordinary, "scalar" numbers.

Imagine that we want to add a second arrow or vector to the 5 cm vector shown earlier. This second vector is also 5 cm long, but points down rather than up. (See Figure 4.)

Let us call the first vector A and the second vector B. To perform vector addition of A and B, place them tip to tail and draw a third vector, C, from the base of the first arrow to the tip of the second, forming a triangle as depicted in Figure 5.

C is the vector sum of A and B. In vector algebra, the sum is written $A + B = C$. The vector C can also be found by adding the numerical versions of A and B. Here, A can be written [3 4] and B can be written [3 − 4]. C is found by adding the first dimension of A and B to give the first dimension of C, and adding the second dimension of A and B to give the second dimension of C: $C =$ *[60].*

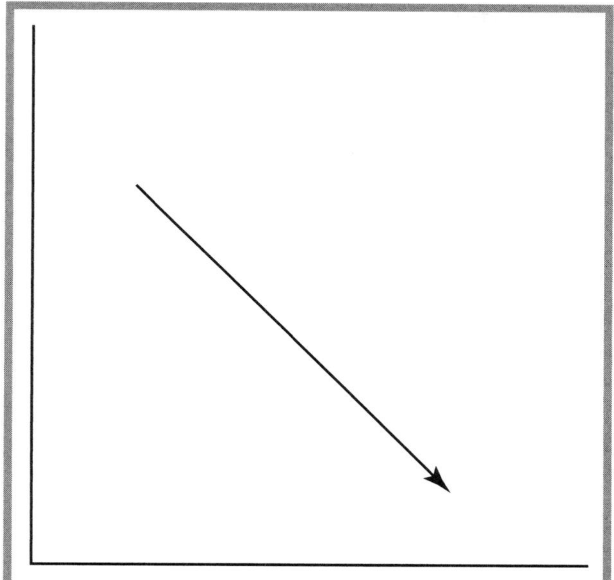

Figure 4.

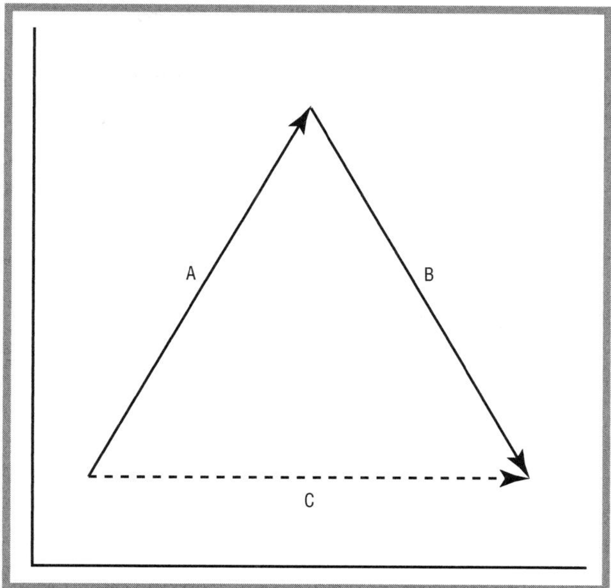

Figure 5.

Addition of vectors with more than two dimensions is done the same way.

Vector multiplication is more complicated. There is only one kind of multiplication for scalars—the $2 \times 2 = 4$ kind of multiplication—but there are two kinds of vector multiplication. The first, simpler kind is called the "inner product" and is written $A \cdot B = c$ (where A and B are vectors having the same number of dimensions and c is an ordinary number, a scalar). The inner product is found by adding up the products of the dimensions of A and B. For example, if $A = [3\ 4]$ and $B = [5\ 6]$, then $A \cdot B = (3 \times 5) + (4 \times 6) = 39$. The dot product of two vectors is a scalar, not a vector.

The second kind of vector multiplication is called the "vector product" or "cross product" of the two vectors. Some knowledge of trigonometry—the mathematical study of triangles—is needed to fully understand the vector product. The vector product of A and B is written $A \times B = C$, where A, B, and C are all vectors having the same number of dimensions. The magnitude and direction of C is calculated separately. The magnitude of C is given by multiplying the magnitude of vector A (written $|A|$), the magnitude of vector B (written $|B|$), and the sine of the angle between A and B (written τ): $A \times B = |A| \times |B| \times \sin\tau$. As for direction, C points at right angles to the plane containing the two vectors being multiplied. For example, if A and B are drawn on this piece of paper, then C sticks straight up out of the paper (or straight down into it, depending on the directions of A and B).

The elements of vectors may be functions rather than numbers. In the field known as "vector analysis," the methods of calculus are applied to such vectors.

VECTORS IN LINEAR ALGEBRA

We have already seen that it is easy to write a vector having more than two or three dimensions, such as [4 4 8 10 2]. This particular vector happens to have five dimensions. The rules for doing math with five-dimensional vectors are the same as for two- or three-dimensional vectors, but when vectors have more than three dimensions nobody tries to think of them as "arrows." In linear algebra, vectors with hundreds of dimensions may be used.

A Brief History of Discovery and Development

The idea of using vectors (arrows) to stand for velocities and the idea of adding two vectors by placing the arrows tip to tail to form a triangle were known to Greek thinkers over two thousand years ago. (This method of adding velocity vectors is also known as the "parallelogram of velocities" method, since the same answer can be found either by making the triangle or by making a parallelogram that has the two vectors as two of its edges.)

However, it was not until the 1600s that scientists began to handle many vector quantities, such as velocity, force, momentum, and acceleration. In the 1700s and 1800s, science also began to deal more with other vectors in optics and electricity. This gradually forced the invention of better ways to handle vectors. In the late 1700s and 1800s, mathematicians struggling to deal with the question of "complex" numbers created two-dimensional vector algebras. The first important three-dimensional vector algebra was invented in the 1840s, at about the same time that matrix algebra was being created to handle matrices and vectors of higher dimensions. Since that time, the term "vector algebra" has come to refer mostly to the rules and symbols for handling vectors only (especially vectors of two, three, or four dimensions), while "matrix algebra" has come to refer to the rules and symbols for handling both vectors and matrices (rectangular arrays of numbers). Today, vectors are used through business, mathematics, and science.

Real-life Applications

3D COMPUTER GRAPHICS

3D ("three dimensional") computer graphics is the art or science of creating an imaginary world of spaces and objects inside a computer using numbers, then producing flat images from that imaginary world that can be shown on a screen or printed on paper. Popular movies such as *Shrek* and *The Incredibles* are produced using 3D computer graphics. The animators who make these movies rely on vectors at every stage. Vectors are used in 3D computer graphics to stand for the locations of points in the imaginary space, the edges of objects, the way objects are moving (velocities), and the way motions are changing (accelerations). Vector algebra is also used to create two-dimensional (flat) images of the computer's imaginary 3D world as seen from any angle desired. This involves the use of the dot product ($A \cdot B = C$, where A, B, and C are vectors) to find the "projections" of vectors defining object edges in the 3D world—that is, what all the vectors defining the edges of an object in the 3D world look like when seen from a certain angle. Some computer graphics programs also allow for realistic physics, that is, for "objects" in the imaginary world to fall or respond to pushes as if they were physically real. This, too, requires the use of vector algebra, since the motions of objects in response to forces are calculated using vectors.

DRAG RACING

To design a race car, engineers must take into account that a car is not a rigid block: it has a suspension system that allows its wheels to move. A car therefore changes overall shape temporarily whenever a force is applied to it, like acceleration, braking, or turning. All these forces must be treated as vectors.

When a car accelerates, several things happen at once. First, assuming that the car is rear-wheel drive, the rear tires push against the pavement. If they push too hard, they start to slip or spin, which is fine if the driver just wants to lay down rubber, but bad if he wants to win a race. The second thing that happens is that the force of the car's weight (a vector that points straight down, toward Earth's center) and the force accelerating the car (a vector pointing forward along the road) add to a total force vector that does not point straight down. Nor does it point straight through the car's center of mass. This effect is called "weight shift." Weight shift causes the car to "lift" in the front and "squat" in the rear, as if weights had been piled on the rear of the car. Because of weight shift, the front tires press on the road with less force and the rear tires press on the road with more force. This is good, up to a point, because more pressure on the back tires means that the car can accelerate faster without starting to slip. It is bad, however, if so much force is taken off the front tires that they rear right up off the road, which makes it impossible for the driver to steer. A drag racing car, therefore, must be designed using vector analysis so that when it is accelerating as hard as it can, the front wheels remain on the road—barely.

LAND MINE DETECTION

According to the United Nations, there are 60 to 100 million land mines buried around the world. These kill about 10,000 people a year—most of them civilians, not soldiers fighting wars—and maim about twice as many. Detecting landmines is therefore an urgent problem.

One method is to use radar in small, handheld units that are swept back and forth over the ground. The radar unit sends burst of radio waves down into the ground. Since landmines are made of metal and are usually round in shape, they reflect the radar differently than rocks, dirt, or differently-shaped metal objects such as wires and pipes. Still, it is not easy, even using a computer to analyze the radar reflections, to tell whether the radar is seeing a landmine or just "clutter," by which mine detection experts mean ground containing anything but land mines.

Vector analysis is being used to help make the detection of land mines more reliable. First, each signal bounced back from the ground is turned into a vector (an ordered series of numbers). This is done by finding the radar signal's "spectrum," a set of numbers that says how much of the signal's energy consists of vibrations at different rates of speed: a signal with many vast vibrations will have energy at the high end of its spectrum, a signal with slow vibrations will have energy at the low end of its spectrum, and so on. Each radio echo from the ground therefore gives a vector.

The goal is to decide whether each vector consists of an echo from a land mine, or not. If the vectors were two-dimensional, then each could be plotted as an arrow on a piece of paper. Or, more simply, the tip of the arrow alone could be plotted, as a dot. The vectors from landmines are not always the same, so if many are plotted the dots will not be located all at one place, but in a fuzzy, rounded area called an "ellipse" (even though its outline it may not have the exact shape of a true geometric ellipse). The computer's job is to figure out whether any given vector is inside the ellipse, which would show a landmine was present. In practice, the vectors used in landmine detection may have from 40 to 128 dimensions, rather than two, but researchers still refer to the "ellipse" or "ellipsoid" (egg-shaped volume) in which the land mines are to be found.

SPORTS INJURIES

A part of the body often injured in sports is the ACL (anterior cruciate ligament). This is a tough, ribbon-like structure that helps keep the top part of the knee joint from slipping forwards and back (other ligaments keep the knee from slipping from side to side). ACL injuries are more common in sports that involve "cutting"—changing direction suddenly while running.

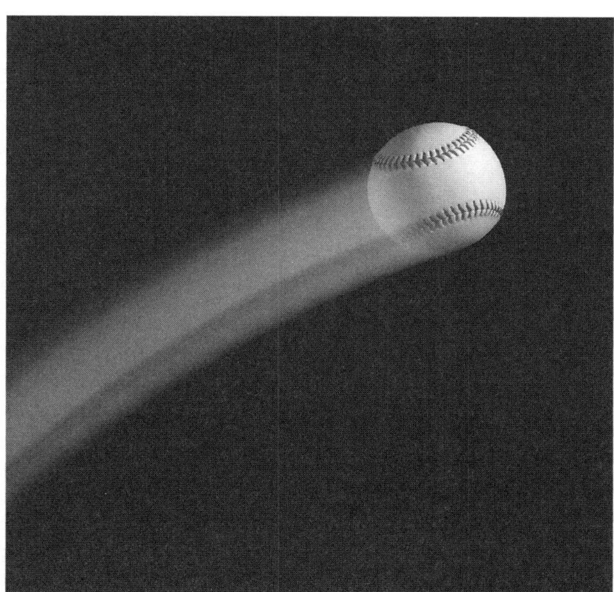

A baseball in motion. Its velocity at any instant is a three-dimensional vector. W. CODY/CORBIS.

Key Terms

Scalar: A directionless quantity with magnitude (e.g., speed as opposed to velocity which is a vector with both magnitude and direction).

Vector: A quantity consisting of magnitude and direction, usually represented by an arrow whose length represents the magnitude and whose orientation in space represents the direction.

Movement technique refers to how an athlete moves her or his body when making sports maneuvers. To study movement techniques, researchers begin by making measurements of male and female athletes in motion. They do so by attaching bright dots to various points on the athletes' bodies and then taking three-dimensional movies of them cutting, pivoting, jumping, and making other motions. They also place "force platforms" under the athletes' feet, flat plates that measure the strength and direction of the force vectors exerted by the athletes' feet against the ground.

These methods only produce raw data—lots of numbers, but no explanations. To understand what the raw data really mean, researchers must use many mathematical tools. One is vector analysis. Vectors are used to calculate the forces acting on the skeleton. They are also used to calculate the angle at which the knee is bent when the athlete's foot hits the ground (which is important to understanding how the knee is stressed). First, three-dimensional vectors standing for the length and position of the upper and lower parts of the leg are calculated from the video. Then the dot product of these two vectors is calculated by the method described earlier in this article. Since the dot product of two vectors depends on the angle between them, finding the dot product (plus one more mathematical step using trigonometry) gives us the angle.

Studies using these methods have shown that the knee is more vulnerable to ACL injury when it is less bent.

Where to Learn More

Books

Barnett, Raymond A., and John N. Fujii. *Vectors.* New York: John Wiley & Sons, 1963.

Crowe, Michael J. *A History of Vector Analysis.* Notre Dame, IN: University of Notre Dame Press, 1967.

Tallack, J.C. *Introduction to Vector Analysis.* Cambridge, UK: Cambridge University Press, 1970.

Periodicals

Slauterbeck, James. "Gender differences among sagittal plane knee kinematic and ground reaction force characteristics during a rapid sprint and cut maneuver." *Research Quarterly for Exercise and Sport.* Vol. 75, No. 1 (2004): 31–38.

Web sites

Olive, Jenny. "Working With Vectors." September 2003. <http://www.netcomuk.co.uk/~jenolive/homevec.html> (March 1, 2005).

Roal, Jim. "Automobile physics." AllFordMustangs.com. July 2003. <http://www.allfordmustangs.com/Detailed/29.shtml> (March 7, 2005).

"Vector Math for 3D Computer Graphics." Central Connecticut State University, Computer Science Department. July 2003. <http://chortle.ccsu.ctstateu.edu/VectorLessons/vectorIndex.html> (March 1, 2005).

Volume

Overview

An object's volume describes the amount of space it contains. Calculations and measurements of volume are used in medicine, architecture, science, construction, and business. Gasses and liquids such as propane, gasoline, and water are sold by volume, as are many groceries and construction materials.

Fundamental Mathematical Concepts and Terms

UNITS OF VOLUME

Volume is measured in units based on length: cubic feet, cubic meters, cubic miles, and so on. A cubic meter, for instance, is the amount of volume inside a box 1 meter (m) tall, 1 m wide, and 1 m deep. Such a box is a 1-meter cube, so this much volume is said to be one "cubic" meter. An object doesn't have to be a cube to contain a cubic meter: one cubic meter is also the space inside a sphere 1.24 meters across.

Cubic units are written by using exponent notation: that is, 1 cubic meter is written "1 m^3." This is why raising any number to the third power—that is, multiplying it by itself three times, as in $2^3 = 2 \times 2 \times 2$—is called "cubing" the number.

VOLUME OF A BOX

There are standard formulas for calculating the volumes of simple shapes. The simplest and most commonly used of these is the formula for the volume of a box. (By "box," we mean a solid with rectangular sides whose edges meet at right angles—what the language of geometry also calls a "cuboid," "right prism," or "rectangular parallelepiped.") To find the volume of a box, first measure the lengths of its edges. If the box is L centimeters (cm) long, W cm wide, and H cm high, then its volume, V, is given by the formula $V = L \text{ cm} \times W \text{ cm} \times H \text{ cm}$. This can be written more shortly as $V = LWH \text{ cm}^3$.

The units of length used do not make any difference to the formula for volume: inches or feet will do just as well as centimeters. For example, a room that is 20 feet (ft) long, 10 ft wide, and 12 feet high has volume $V = 20 \text{ ft} \times 10 \text{ ft} \times 12 \text{ ft} = 2{,}400 \text{ ft}^3$ (cubic feet).

VOLUMES OF COMMON SOLIDS

There are standard formulas for finding the volumes of other simple solids, too. Figure 1 shows some of these formulas.

Solid Shape	Dimensions	Formula for Volume
box	Length L, width W, height H	$V = LWH$
cube	Length = width= height = L	$V = L^3$
sphere	radius R	$V = \frac{4}{3}\pi R^3$
cylinder	radius R, height H	$V = \pi R^2 H$
cone	Base radius R, height H	$V = \frac{1}{3}\pi R^2 H$
pyramid	base area A, height H	$V = \frac{1}{3}AH$
torus (doughnut)	distance from center of torus to center of tube D, radius of tube R	$V = 2DR^2$

Figure 1: Standard formula to calculate volume.

In all these formulas, three measures of length are multiplied—not added. This means that whenever an object is made larger without changing its shape, its volume increases faster than its size as measured using a ruler or tape measure. For example, a sphere 4 m across (a sphere with a radius of 2 m) has a volume of $V = 4/3\ \pi 2^3 = 33.5\ m^3$, whereas a sphere that is twice as wide (radius of 4 m) has a volume of $V = 4/3\ \pi\ 4^3 = 268.1\ m^3$. Doubling the radius does not double the volume, but makes it 8 times larger. In general, since the radius is cubed in calculating the volume, we say that a sphere's volume "increases in proportion to" or "goes as" the cube of its radius. This is true for objects of all shapes, not just spheres: Increasing the size of an object without changing its shape makes its volume grow in proportion to the cube (third power) of the size increase.

The formula for an object's volume can be compared to the formula for its area. The area of a sphere of radius R, for example, is $A = 4\ \pi\ R^2$. The radius appears only as a squared term (R^2) in this formula, whereas in the volume formula it appears as a cubed term (R^3). Dividing the volume formula by the area formula yields an interesting and useful result:

$$\frac{V}{A} = \frac{\frac{4}{3}\pi R^3}{4\pi R^2}$$

Crossing out terms that are the same on the top and bottom of the fraction, we have

$$\frac{V}{A} = \frac{1}{3}R$$

which, if we multiply both sides by A, becomes

$$V = \frac{AR}{3}$$

This means that when we increase the radius R of a sphere, area and volume both increase, but volume increases by the increased area times $R/3$. Volume increases faster than area. This fact has important consequences for real-world objects. For example, how easily an animal can cool itself depends on its surface area, because its surface is the only place it can give heat away to the air; but how much heat an animal produces depends on its volume, because all the cubic inches of flesh it contains must burn calories to stay alive. Therefore, the larger an animal gets (while keeping the same shape), the fewer square inches of heat-radiating skin it has per pound: its volume increases faster than its area. A large animal in a cold climate should, therefore, have an easier time staying warm. And in fact, animals in the far North tend to be bigger than their close relatives farther south. Polar bears, for example, are the world's largest bears. They have evolved to large size because it is easier for them to stay warm. On the other hand, a large animal in a hot climate has a harder time staying cool. This is why elephants have big ears: the ears have tremendous surface area, and help the elephant stay cool.

Volume can be described in terms of an amount of the space an object assumes, such as water in a bucket. ROYALTY-FREE/CORBIS.

A Brief History of Discovery and Development

Weights, lengths, areas, and volumes were the earliest measurements made by humankind. Not only are they easier to measure than other physical quantities, like velocity and temperature, but they have an immediate money value. Measuring lengths, builders can build more complex structures, such as temples; measuring area, landowners can know how much land, exactly, they are buying and selling; measuring volume, traders can tell how much grain a basket holds, or how much water a cistern (holding tank) holds. Therefore it is no surprise to find that the Egyptians, Sumerians, Greeks, and ancient Chinese all knew the concept of volume and knew many of the standard equations for calculating it. In 250 B.C. (over 2,200 years ago), the Greek mathematician Archimedes wrote down formulas for the volume of a sphere and cylinder. In approximately 100 B.C., the Chinese had formulas for the volumes of cubes, cuboids, prisms, spheres, cylinders, and other shapes (using, like the Greeks, approximate values for π ranging from rough to excellent).

Such formulas are useful but do not give any way of exactly calculating the volume of a shape whose surface is not described by flat planes or by circles (as are the curved sides of a cylinder, or the surface of a sphere). New progress in the calculation of volumes had to wait almost 2,000 years, until the invention of the branch of mathematics known as calculus in the 1600s. One of the two basic mathematical operations of calculus is called "integration." Integration, as it was first invented, allowed mathematicians to exactly calculate the area under any mathematically defined curve or any part of such a curve; it was soon discovered, however, that integration was not restricted to flat surfaces and areas. It could be generalized to three dimensions—that is, to ordinary space. It had now become possible to calculate exactly the volumes of complexly-shaped objects, as long as their surfaces could be described by mathematical equations.

The next great revolution in volume calculation came with computers. Since computers can add many numbers very quickly, they have made it possible to calculate areas and volumes for complex shapes even when the shapes cannot be described by nice, neat mathematical equations. Today, the calculation of volumes of simple shapes is still routine in many fields, but the use of calculus and computers for complex shapes such as airplane wings and the human brain is increasingly common.

Real-life Applications

PRICING

Volume is closely related to density, which is how much a given volume of a substance weighs. For instance, the density of gold is 19.3 grams per cubic centimeter, that is, one cubic centimeter of gold weighs 19.3 grams, which is 19.3 times as much as one cubic centimeter of water. Silver, platinum, and other metals all have different densities. This fact is used by some jewelry makers to decide how much to charge for their jewelry.

Different metals not only have different densities, they have different costs: at a 2005 price of about $850 per ounce, for example, platinum cost about twice as much as gold. So when a jewelry maker uses a blend of gold and platinum in a piece of jewelry, they need to know exactly how much of each they have used in order to know how much to charge for the piece. Now, a blend of two metals (called an alloy) has a density that is somewhere between the densities of the two original metals. Therefore, determining the average density of a piece (say, a ring) will tell a manufacturer how much gold and platinum it contains, regardless of how complicated the piece is. Volume and weight together are used to determine density. The finished piece is suspended in water by a thread. Any object submerged in water experiences an upward force that depends only on the volume of water the object displaces. Therefore, by weighing the piece of jewelry as it hangs in water, and comparing that weight to its weight out of water, the jeweler can measure exactly what weight of water it displaces. Since the density of water is known (1 gram per cubic centimeter), this water weight tells the jeweler the exact volume of the piece. Finally, knowing both the volume of the piece and its weight, the jeweler can calculate its density by the equation density

= weight/volume. The jewelry maker's wholesale price will be determined partly by this calculation, and so will the retail price in the store.

MEDICAL APPLICATIONS

In medicine, volume measurements are used to characterize brain damage, lung function, sexual maturity, anemia, body fat percentage, and many other aspects of health. A few of these uses of volume are described below.

Brain Damage from Alcohol Using modern medical imaging technologies such as magnetic resonance imaging (MRI), doctors can take three-dimensional digital pictures of organs inside the body, including the brain. Computers can then measure the volumes of different parts of the brain from these digital pictures, using geometry and calculus to calculate volumes from raw image data.

MRI volume studies show that many parts of the brain shrink over time in people who are addicted to alcohol. The frontal lobes—the wrinkled part of the brain surface that is just behind the forehead—are strongly affected. It is this part of the brain that we use for reasoning, making judgments, and problem solving. But other parts of the brain shrink, too, including structures involved in memory and muscular coordination. Alcoholics who stop drinking may regain some of the lost brain volume, but not all. MRI studies also show that male and female alcoholics lose the same amount of brain volume, even though women alcoholics tend to drink much less. Doctors conclude from this that women are probably even more vulnerable to brain damage from alcohol than are men.

Diagnosing Disease Almost half of Americans alive today who live to be more than 85 years old will suffer eventually from Alzheimer's disease. Alzheimer's disease is a loss of brain function. In its early stages, its victims sometimes have trouble remembering the names for common objects, or how they got somewhere, or where they parked their car; in its late stages, they may become incurably angry or distressed, forget their own names, and forget who other people are. Doctors are trying understand the causes of Alzheimer's disease and develop treatments for it. All agree that preventing the brain damage of Alzheimer's—starting treatment in the early stages—is likely to be much more effective than trying to treat the late stages. But how can Alzheimer's be detected before it is already damaging the mental powers of the victim?

Recent research has shown that the part of the brain called the hippocampus, which is a small area of the brain located in the temporal lobe (just below the ear), is the first part of the brain to be damaged by Alzheimer's. The hippocampus helps the brain store memories, which is why forgetting is one of Alzheimer's first symptoms. But instead of waiting for memory to fail badly, doctors can measure the volume of the hippocampus using MRI. A shrinking hippocampus can be observed at least 4 years before Alzheimer's disease is bad enough to diagnose from memory loss alone.

Pollution's Effects on Teenagers Polychlorinated aromatic hydrocarbons (PCAHs) are a type of toxic chemical that is produced by bleaching paper to make it white, improper garbage incineration, and the manufacture of pesticides (bug-killing chemicals). These chemicals, which are present almost everywhere today, get into the human body when we eat and drink. In 2002 scientists in Belgium studied the effects of PCAHs on the sexual maturation of boys and girls living in a polluted suburb. They compared how early boys and girls in the polluted suburb went through puberty (grew to sexual maturity) compared to children in cleaner areas. They found that high levels of PCAH-related chemicals in the blood significantly increased the chances of both boys and girls of having delayed sexual maturity. Once again, volume measurements proved useful in assessing health. The researchers estimated the volume of the testicles as a way of measuring sexual maturity in boys, while they assessed sexual maturity in girls by noting breast development. This study, and others, show that some pollutants can injure human health and development even in very low concentrations. Testicular volume measurements are also used in diagnosing infertility in men.

Body Fat Doctors speak of "body composition" to refer to how much of a person's body consists of fat, muscle, and bone, and where the fat and muscle are located on the body. Measuring body composition is important to monitoring the effects of diet and exercise programs and tracking the progress of some diseases. Volume measurement is used to measure some aspects of body composition. For example, the overall density of the body can be used estimate what percentage of the body consists of fat. Measuring body density requires the measurement of the body's weight—which can be done easily, using a scale—and two volumes.

The first volume needed is the volume of the body as a whole. Since the body is not made of simple shapes like cubes and cylinders, its volume cannot be found by taking a few measurements and using standard geometric formulas. Instead, its volume must be measured by submerging it in water. The body's overall volume can then be found by measuring how much the water level rises or, alternatively, by weighing the body while it is underwater to see how much water it has displaced. (Underwater weighing is the same method used to measure the density of jewelry containing mixed metals, as described earlier in this article.) The body's overall volume is equal to the water displaced.

However, doctors want to know the weight of the solid part of the body; the air in the lungs does not count. And even when a person has pushed all the air they can out of their lungs, there is still some left, the "residual lung volume." Residual lung volume must therefore also be measured, as well as overall body volume. This is done using special machines that measure how much gas remains in the lungs when the person exhales. The body's true, solid volume is approximately calculated by subtracting the residual lung volume from the body's water displacement volume.

Dividing the body's weight by its true, non-air volume gives its density. This is used to estimate body fat percentage by a standard mathematical formula.

BUILDING AND ARCHITECTURE

Many building materials are purchased by area or volume. Area-purchased materials include flooring, siding, roofing, wallpaper, and paint. Volume-purchased materials include concrete for pouring foundations and other structures, sand or crushed rock, and grout (a kind of thin cement used to fill up masonry joints). All these materials are ordered by units of the cubic yard. (One cubic yard equals about .765 cubic meters.) In practice, simple volume formulas for boxes and cylinders are used to calculate how many cubic yards of cement must be ordered to build simple structures like housing foundations. A simple foundation, shaped like a box without a top, can be broken into three slab-shaped boxes, namely the four walls and the floor. Multiplying the length by the width by the thickness of each of these slabs gives a volume: the sum of these volumes is the cubic yardage that the cement truck must deliver. For concrete columns, the formula for the volume of a cylinder is used. For complex structures with curving shapes, a computer uses calculus-based methods to calculate volumes based on digital blueprints for the structure.

The same principle is used in designing machine parts. It is necessary to know the volume of a machine part while it is still just a drawing in order to know what its weight will be: its weight must be known to calculate how much it will weigh, and (if it is a moving part) how much force it will exert on other parts when it moves. For parts that are not too complicated in shape, the volume of the piece is calculated as a sum of volumes of simple elements: box, cylinder, cone, and the like. Computers take over when it is necessary to calculate the volumes of pieces with strange or curvy shapes.

COMPRESSION RATIOS IN ENGINES

Internal combustion engines are engines that burn mixtures of fuel and air inside cylinders. Almost all engines that drive cars and trucks are of this type. In an internal combustion engine, the source of power is the cylinder: a round, hollow shaft sealed at one end and with a plug of metal (the piston) that can slide back and forth inside the shaft. When the piston is withdrawn as far as it will go, the cylinder contains the maximum volume of air that it can hold: when the piston is pushed in as far as it will go, the cylinder contains the minimum volume of air. To generate power, the cylinder is filled with air at its maximum volume. Then the piston is pushed along the cylinder to compress the air. This makes the air hotter, according to the well-known Ideal Gas Law of basic physics—just how hot depends on how small the minimum volume is. Fuel is squirted into the small, hot volume of air inside the cylinder. The mixture of fuel and air is then ignited (either by sheer heat of compression, as in a diesel engine, or by a spark plug, as in a regular engine) and the expanding gas from the miniature explosion pushes the piston back out of the cylinder. The ratio of the cylinder's largest volume to its smallest is the "compression ratio" of the engine: a typical compression ratio would be about 10 to 1. Engines with high compression ratios tend to burn hotter, and therefore more efficiently. They are also more powerful. Unfortunately, there is a dilemma: burning very hot (high compression ratio) allows the nitrogen in air to combine with the oxygen, forming the pollutant nitrogen oxide; burning relatively cool (low compression ratio) allows the carbon in the fuel to combine only partly with the oxygen in the air, forming the pollutant carbon monoxide (rather than the non-poisonous greenhouse gas carbon dioxide).

GLOWING BUBBLES: SONOLUMINESCENCE

When small atoms come together to make a single heavier atom, energy is released. This process is called "fusion" because in it, two atoms fuse into one. All stars, including the Sun, get their energy from fusion. Some nuclear weapons are also based on fusion. But fusion is difficult to control on Earth, because atoms only fuse under extreme heat. If fusion could be controlled, rather than exploding as a bomb, it could be used to generate electricity. Many billions of dollars have been spent on trying to figure out how to make atoms trapped inside magnetic fields fuse—so far without success.

Yet there is a new possibility. Some reputable scientists claim that they can produce fusion using nothing more expensive or exotic than a jar full of room-temperature liquid bombarded by sound waves. This claim—which has not yet been tested by other researchers—is related to the effect called "sonoluminescence," which means "sound-light." Sonoluminescence depends on changes in volume of bubbles in liquid. Under certain conditions, tiny bubbles form and disappear in any liquid that is squeezed and stretched by strong sound waves; when the bubbles collapse, they can emit flashes of light. This happens as follows: Pummeled by high-frequency sound waves, a bubble forms and expands. When the bubble collapses, its radius decreases very rapidly as its surface moves inward at several times the speed of sound. Because the volume of a sphere is proportional to the cube (third power) of its radius, when a bubble's radius decreases to 1/10 of its starting value, its volume decreases to $(1/10)^3 = 1/1{,}000$ of its starting value. (These are typical figures for the collapse of a sonoluminescence bubble.) This decrease in volume squeezes the gas inside the bubble, and, according to laws of physics, when a gas is squeezed its temperature goes up. Also, the compression happens very quickly—too quickly for much heat to escape from the bubble. Therefore, the bubble's rapid shrinkage causes a fast rise in temperature inside the bubble. The temperature has been shown to rise to tens of thousands of degrees, and may reach over two hundred thousand degrees. Such heat rivals that at the heart of the Sun and makes the gas in the bubble glow. It may also do something else: in 2002 scientists at Oak Ridge National Laboratory claimed to have detected neutrons flying out of a beaker of fluid in which sonoluminescence was occurring. Neutrons would be a sign that fusion was occurring. If it is, then there is a close resemblance between bubble fusion and the diesel engines found in trucks: both devices work by rapidly decreasing the volume of a gas in order to heat it to the point where energy is released. In a diesel engine, the energy is released by a chemical reaction. In a fusion bubble, it would be released by a nuclear reaction.

As of 2005, the reality of bubble fusion had been neither proved nor disproved. If it is proved, it might eventually mean that producing electricity from fusion could be done more cheaply than scientists had ever before dreamed. Describing changes in bubble volume mathematically is basic to all attempts to understand and control sonoluminescence and bubble fusion.

SEA LEVEL CHANGES

One of the potential threats to human well-being from possible global climate change is the rising of sea levels. The International Panel on Climate Change predicts that ocean levels will rise by 3.5 inches to 34.5 inches (about 9 to 88 centimeters) by the year 2100, with a best guess of 1.6 ft (about 50 centimeters) with the ocean continuing to rise. Hundreds of millions of people live near sea level worldwide, and their homes might be flooded or at greater risk from flooding during storms. Also, many small island nations might be completely flooded.

Sea level rises when the volume of water in the ocean increases. There are two ways in which a warmer Earth causes the volume of water in the ocean to increase. First, there is the melting of ice. Ice exists on Earth mostly in the form of glaciers perched on mountain ranges and the ice caps at the north and south poles. Second, there is the volume increase of water as it gets warmer. Like most substances, water expands as it gets warmer: a cubic centimeter of seawater gains about .00021 cubic centimeters of volume if it is made 1 degree Centigrade warmer. Therefore, the oceans get bigger just by getting warmer. In fact, the International Panel on Climate Change predicts that most of the sea-level rise that will occur in this century will be caused by water expansion, rather than by ice melting and increasing the mass of the sea. Calculations of the volume of water that will be added to the ocean by melting glaciers and icecaps and by thermal expansion are at the heart of predicting the effects of global warming on sea levels.

WHY THERMOMETERS WORK

The fact that liquids expand as they get warmer (until they start to boil) is used to measure temperature in old-fashioned mercury or colored-alcohol thermometers. Geometry is used to amplify or multiply the expansion effect: a thin cylinder connected attached to a sphere (the "bulb"). The bulb is full of liquid. If the radius of the thermometer bulb is r_B, then its volume (V_B, for "volume, bulb") is given by the standard volume formula for a sphere as

$$V_B = \frac{4}{3}\pi r_B^3$$

If the cylinder's radius is r_C, then the volume of liquid in the cylinder (V_C, for "volume, cylinder") is given by the standard volume formula for a cylinder as $V_C = \pi r_C^2 H$, where H is the height of the fluid in the cylinder. We read the temperature from a thermometer of this type by reading H from marks on the cylinder.

There is room in the cylinder for more liquid, but there is no room in the sphere, which is full. If the thermometer contains a liquid that has a "volume thermal coefficient" of $\alpha = .0001$, a cubic centimeter of the liquid will gain .0001 cubic centimeters of volume if it is warmed by 1 degree Centigrade. Say that the thermometer starts out with no fluid in the cylinder and the bulb perfectly full. Then the temperature of the thermometer goes up by 1°C. This causes the volume of the fluid in the bulb, V_B before it is warmed, to increase by .0001 V_B. But this extra volume has nowhere to go in the bulb, which is full, so it goes up the cylinder. The amount of fluid in the cylinder is then $V_C = \pi r_C^2 H = .0001 V_B$. If we divide both sides of this equation by πr_C^2, we find that

$$H = \frac{.001 V_B}{\pi r_C^2}$$

Because V_B is on top of the fraction, making it bigger makes H bigger. That is, the bigger the bulb, the bigger the change in the height of the fluid in the cylinder when the temperature goes up. Since r_C is on the bottom of the fraction, making it smaller also makes H bigger. That is, the narrower the cylinder, the bigger the change in the height of the fluid in the cylinder when the temperature goes up. This is why thermometers have very narrow cylinders attached to fat bulbs—so it is easy to see how far the fluid goes up or down the cylinder when the temperature changes.

MISLEADING GRAPHICS

Many newspapers and magazines think that statistics are dull, and so they have the people who work in their graphics departments make them more visually appealing. For example, to illustrate money inflation (how a Euro or a dollar buys less every year), they will show you a picture of shrinking bill—a big bill, then a smaller bill below it, and a smaller below that, and so forth. Or, to illustrate the increasing price of oil, they will show you a picture of a row of oil barrels, each bigger than the last.

Such pictures can create a very false impression, because it is usually the lengths of the dollar bills or the oil barrels (or whatever the object is), not their areas or volumes, that matches the statistic the art is trying to communicate. So, to show the price of oil going up by 10%, a publication will often show two barrels, one 10% taller and wider than the other. But the equation for the volume of a barrel, which is a cylinder, is $V = \pi r^2 H$, where r is the radius of the barrel and H is its height. Increasing r or H by 10% is the same as multiplying it by 1.1, so increasing the dimensions of the barrel by 10% shows us a barrel whose volume is $V_{\text{bigger}} = \pi (1.1\ r)^2 (1.1)H$. If we multiply out the factors of 1.1, we find that $V_{\text{bigger}} = 1.331 V$—that is, the volume of the larger barrel in the picture, the amount of oil it would contain, is not 10% larger but 33.1% larger. Because volume increases by the cube of the change in size, the larger the size change, the more misleading the picture.

Look carefully at any illustration that shows growing or shrinking two-dimensional or three-dimensional objects to illustrate one-dimensional data (plain old numbers that are getting larger or smaller). Does the artwork exaggerate?

SWIMMING POOL MAINTENANCE

Everyone who owns a swimming pool knows that they have to add chemicals to keep the water healthy for swimming. It's not enough to just dump in a bucket or two of aluminum sulfate or calcium hypochlorite, though—the dose has to be proportioned to the volume of water in the pool.

Some pools have simple, box-like shapes: their volume can be calculated using the standard formula for the volume of a box, volume equals length times width times height. A standard formula can also be used for a circular pool with a flat volume, which is simply a cylinder of water. Many pools have more complex shapes, though, and even a rectangular pool often has a deep end and a shallow end. The deep and shallow ends may be flat, with a step between, or the bottom of the pool may slope. Some pools are elliptical (shaped like a stretched circle), and an elliptical pool may also have a sloping bottom.

To calculate the correct chemical dose for a swimming pool, it is necessary, then, to take some measurements. A pool with a complex shape has to be divided into sections with simpler shapes, and the volumes of the separate pieces calculated and added up. More complex formulas are needed for, say, the volume of an elliptical pool with a sloping bottom; calculus is needed to find these formulas. Fortunately for the owners of complexly shaped pools, volume-calculation computer software exists that will calculate a pool's volume given the basic measurements of its shape. For an elliptical pool with a sloping bottom, you would need to measure the length of the pool, the width of the pool, the maximum depth, and the minimum depth.

BIOMETRIC MEASUREMENTS

On average, men's brains tend to be larger than women's, occupying more volume and weighing more. Before the invention of modern medical imaging machines like CAT (computerized axial tomography) scanners, brain volumes were measured by measuring the volumes of men's and women's skulls after they were dead. Beads, seeds, or ball bearings were poured into the empty skull to see how much the skull would hold, then they were weighed. More beads, seeds, or bearings meant more brain volume. Today, brain volume can be measured in living people using computer software that uses three-dimensional medical scans of the brain to count how many cubic centimeters of volume the brain occupies.

But the fact that men, on average, have slightly larger brains (about 10% larger) does not mean that men are smarter than women. To begin with, a bigger brain does not mean a more intelligent mind, and there is great individual variation among people of both sexes. Some famous scholars have been found, after death, to have brains only half the size of other scholars. People of famous intelligence, like Einstein, usually do not have larger-than-average brain volume. Second, about half of the average size difference is accounted for by the fact that men tend to be larger than women. Brain size goes, on average, with body size: taller, more muscular men tend to have larger brains than smaller, less muscular men. Elephants and whales have larger brains several times larger than those of human beings, but are not more intelligent. To some extent, therefore, men have larger brains only because their bodies are larger, too.

In the nineteenth and early twentieth centuries, brain-volume measurements were used to justify laws that allowed only men to vote and hold some other legal rights. This is a classic case of accurate measurements being interpreted in a completely misleading way.

RUNOFF

Runoff is water from rain or melting snow that runs off the ground into streams and rivers instead of soaking into the ground. Scientists and engineers who study flood control, sewage management, generating electricity from rivers, shipping goods on rivers, or recreation on rivers make determinations of water volume to estimate supply. To make an educated guess, they initially estimate the volume of water that will be added by snowmelt and rainfall during a given period of time. This indicates how much water will arrive, and when and how fast, in various rivers or lakes.

Hydrogeologists and weather scientists use complex mathematical equations, satellite data, soil-test data, and computer programs to predict runoff volumes. Some of the factors that they must take into account include rain amount, intensity, duration, and location; soil type and wetness; snowpack depth and location; temperature and sunshine; time of year; ground slope; and the type and health of the vegetation covering the ground. All this information goes into a mathematical model of the stream, lake, or reservoir basin into which the water is draining. Given the exact shape of the basin receiving the water, water volume can be translated into water depth. In some places, water can be drained from reservoirs to make room for the volume of water that has been forecast to flow from higher ground, thus preventing floods.

Where to Learn More

Books

Tufte, Edward R. *The Visual Display of Quantitative Information.* Cheshire, CT: Graphics Press, 2001.

Web sites

"Causes of Sea Level Rise." Columbia University, 2005. <http://www.columbia.edu/~epg40/elissa/webpages/Causes_of_Sea_Level_Rise.html> (April 4, 2005).

"Making a River Forecast." US National Weather Service, Sep. 21, 2004. <http://www.srh.noaa.gov/wgrfc/resources/making_forecast.html> (April 6, 2005).

"Volume." Mathworld. 2005. <http://mathworld.wolfram.com/Volume.html> (April 4, 2005).

Word Problems

Overview

The ability to communicate and the development of language have paralleled the progression in society of mathematical and scientific developments. Humans think and imagine in language and pictures, so it is hardly surprising that much of mathematics deals with the translation from words to expressions. The word translate can be used because many people view math as a language in its own right. After all, it has its own rules of grammar and layout. It should also be perfectly logical.

It is often observed that a good mathematician is one who can translate complicated real-life situations into logical mathematical sentences that can then be solved.

Fundamental Mathematical Concepts and Terms

There are two distinct types of word problems, both relevant to today's world. First, there is the statement believed to true. Mathematics can often be used to establish the validity of the statement. This proposition is often called a hypothesis. Often a written statement can be proven to hold true without exception. These ideas branch out into a large mathematical area called proof. There are many different ways of proving things. These proofs can often have tremendous impact on the real world because people can the use these ideas completely and confidently.

Second, there is the word problem, to which the solution happens to involve mathematics. Mathematical modeling is considered to be the process of turning real-life problems into the more abstract and rigorous language of mathematics. It generally involves assumptions and simplifications required to express the complex situation as one that can be solved.

These solutions are then compared to the actual readings or observations. Alterations are then made to the model to try to achieve a more realistic solution. These alterations are often referred to as refinements. This process of solving, comparing, and refining is called the modeling process. It is used to solve many of the problems in the real world. It is used because it is often impossible to exactly model the frequently immeasurable possibilities in real life. Simplifications often lead to a realistic and useable model.

Diagrams are also used to simplify situations. The key elements can be marked and these are then used within the model. One of the key facts that should be considered is that a diagram will help simplify even the most complex of problems.

A Brief History of Discovery and Development

It is frequently the case that the person involved as a manager behind a job will have the ideas but not the mathematical ability to solve the problem. It is for this reason that mathematics, whether through mathematicians, engineers, scientists, or statisticians, is thus employed.

Possibly one of the early cases of such an idea was the building of ancient monuments some of which, it is now believed, tell time and measure the passing of seasons. The most famous example includes the building of the pyramids. The pharaohs, wanting to express their might and wealth, commanded the building of these tombs without the slightest idea of the mathematics behind them. It was the engineers who set to work, translating the request into achievable, long-lasting designs.

As the years have progressed, so the requests and subsequent designs have become and more detailed and complicated. War, however terrible, has forced great strides in our technologies. Requests for fighting machines have driven much of the mathematics behind flight, engines, and electronics. Progress in trade and finance has also forced people into solving problems involving money. Though these calculations generally use the four basic operators, (add, subtract, divide, and multiply), the ability to translate between statements and calculations is a highly sought after skill. The more complex finance has become, so the complexity of problems met in the real world has increased.

Perhaps the biggest driving force is the current emphasis towards efficiency. It is increasingly the case that the best solutions, often referred to as optimal solutions, are required. Today, only the very best will do.

Real-life Applications

TEACHERS

Teachers spend most of their time trying to construct real-life problems. It is widely believed that understanding the mathematics behind actual problems assists in grasping the more theoretical, fundamental, and abstract ideas that underpin mathematics. It also makes the subject more accessible, relevant, and interesting. Indeed, it is the application to real life that has driven many of the advancements in mathematics. The more abstract side of mathematics is a beautiful area, and application to the real world provides a stepping-stone into this complex and remarkable subject.

COMPUTER PROGRAMMING

Computers are built with an underlying logic behind them. This logic is used to then program software or games. The computer designer will have ideas about how to make the interface look and how to program the operating software to allow for a suitable user-friendly environment.

SOFTWARE DESIGN

The design of software goes through various processes. First, the creative department will come up with ideas for a suitable game. This will often be deduced through market research. The department will then pass on ideas to the programmers, who will translate the creative ideas into programming code. Programming code is an example of the use of mathematics. It follows a logical structure and obeys the many structures underlying mathematics.

CREATIVE DESIGN

The artistic idea behind animation, computer graphics, or a storyline will often be verbal. This then has to be turned into motion through the work of computer designers. Highly competent mathematicians will program these packages. The concepts behind three dimensions, perspective, etc. have to be converted into machine code. These are effectively strings of mathematical statements. They will use vast arrays (data storage) that are then manipulated.

INSURANCE

Insurance involves almost exclusively real-life situations. A client will provide a list of items that need to be insured against loss, and the insurance company will then try to offer an attractive premium that the client will be willing to pay to insure his items. The evaluating of such premiums can be a highly complex task. The people involved, who are often referred to as actuaries, need to simplify all the variables involved and work out the various probabilities. Not only do they want to encourage the client to pay the premium, they must also ensure that, on average, the company will not lose vast sums of money in event of a claim.

Actuaries evaluate what is often referred to as the expected monetary value of the situation. This is simply the expected financial outcome of a given financial situation.

They will often draw a simple tree diagram, upon which expected occurrences are labeled. They can then work out from this the best possible premium for the situation.

This allows solutions to such questions as, What is the best premium? How much should be charged? It also allows the consumer to evaluate the best deal being offered. Everyone, at some point in life, will be faced with the prospect of buying insurance. Every first-time driver will be expected to pay a premium that is much greater that experienced drivers.

CRYPTOGRAPHY

Cryptography is the ability to send encoded data that, in theory, will be unreadable without a key. Authorities need to be able to control and often intercept messages and then read them. In modern times, where terrorism is often referred to as a significant threat, it is essential to be able to understand what such groups are saying. By its very nature, cryptography deals problems involving words.

There are many different ways of coding data, yet an awareness of the different possibilities means that, with powerful computers, a piece of writing can be unscrambled in many different ways until the correct key is found. The ability to decode information can hinge on knowledge of the actual language used. However a coding is applied, the frequency of certain letters within the language can be used to try to decode simple situations. During World War II, decoding was often found to be difficult due to the placing of random letters into specific sections of the text, but the decoders generally prevailed.

MEDICINE AND CURES

Research in medicine is frequently concerned with questioning the benefits of drugs as well as assessing their possible side effects. It is an extremely difficult area to research, because people's lives are so heavily mixed into the equation. It is impossible to test all drugs on all people and record which ones work while recording the visible effects on the patients. So, how does a question such as "Does smoking cause cancer?" actually get solved mathematically?

These are questions involving causality. Namely, does smoking actually cause cancer? It is often the case that, even though there appears to be a direct link, it is either a fluke or a third variable is causing the apparent situation. To determine this, strict statistical tests need to be carried out using a control group, made up of people that have no link to the drug in question. Another group is then selected, who are given just the drug. These people would have to be selected randomly to reduce the chance of a third variable. The outcomes can then be compared and inferences drawn.

HYPOTHESIS TESTING

This is an important area of mathematics. It is equivalent to a court case, in which a party is only found guilty if the evidence is of sufficient nature. For instance, it is believed that playing computer games has caused a decrease in the number of people reading books. To prove this, the situation is set up extremely systematically. A null hypothesis is defined. This often states a simple belief that there has been no change: computers have not caused a decrease in literacy.

An alternative hypothesis is defined. This would be a statement indicating that there has been a change. In this case, computers have decreased literacy. A statement is then made indicating how much evidence is required to decide on the alternative hypothesis. This is called the significance. The statistician would then pick a random sample of people relevant to the survey. These would have to be drawn from the whole population. The statistician would then take a survey on reading and computer habits and compare this to data from the past. If the change (presuming a change) were to be sufficient, it would be stated that there existed enough evidence for the alternative hypothesis.

In hypothesis testing it is essential to define the significance before the test, otherwise the conclusion may be compromised.

ARCHAEOLOGY

Archaeology uses many mathematical ideas to analyze many different aspects, from dating individual objects to how the landscape has changed. These facts are then pieced together to provide an overall picture to help in understanding the past.

ENGINEERING

The conversion of ideas into safe and workable designs involves a lot of detailed mathematics. For instance, how does water arrive through the tap? The many different stages in the process would be separated and each part solved progressively. The whole system involves forces, which allow the water to flow around the system. This in turn puts pressure on the system; hence it needs to be strong enough and yet cheap enough to run. A single error in calculation along the way and the whole process would have to be thought through again at much

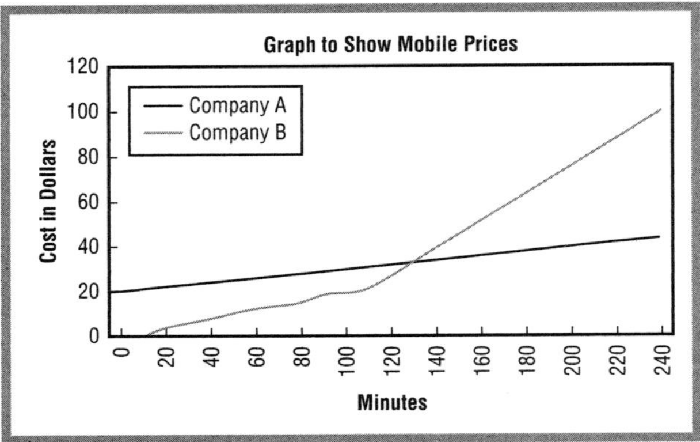

Figure 1.

expense. The sewerage and water system beneath any major city is a great engineering and mathematical feat.

COMPARISONS

Statements are often made concerning views on sports persons or other famous figures such as pop stars. Frequent allusions are made to the best ever sportsman or the most successful singer. Mathematics is used to solve such problems using the concept of averages. There are three main types of averages: mean, median, and mode, each having an exact meaning.

For example, a teacher has stated that Sam is better at math than David. This is because Sam averages 70, while David averages 65. Sam's scores were 40, 70, and 80; David's scores were 65, 65, and 100. It is perhaps immediately apparent that David has the better scores overall. When solving problems involving averages, it is also useful to indicate how spread out the data is. This indicates how consistent someone or the object in question is.

PERCENTAGES

Everyday, the consumer is confronted by billboards offering massive savings and bargain prices in an attempt by retailers to tempt the customer in. The customer must see a way around any potential pitfalls. For instance, if a store suggests that 40% of their competitors are worse than they are, the clever consumer would logically deduce that 60% are as good or better!

EXCHANGE RATES

The difference in currency from one country to the next can cause many problems for consumers. There is also a variation from one day to the next. Some currency exchange companies may charge an extra amount; this is referred to as commission. Being aware of these facts allows the consumer to correctly evaluate the relative amount they are spending while abroad. They need to ask themselves, "Which is the more expensive: a coat costing $10 or one costing 15 euros?" The concept of ratios can be used to solve this particular problem: If that day's ratio is $1 to 1.2 euros, then $10 = 12 euros. Hence, the $10 coat is the better deal. Obviously, it pays to be aware of exchange rates.

PHONE COMPANIES

It can be difficult choosing the best company to use for a mobile phone. They all offer different rates and different incentives. A graph is a good way to compare different phone options. It may save money in the long run. For example, company A has a fixed charge of $20, and charges $1 for every 10 minutes; company B has no fixed charge, but charges $1 for every five minutes for the first two hours and then $3 every 5 minutes thereafter. Figure 1 shows a comparison graph. If the consumer uses the phone for less than 130 minutes a month, then option A is the better deal; otherwise company B offers the better deal.

TRAVEL AND RACING

Before setting out on a trip, it is important to assess travel times. To work out how long a 100 kilometer journey would take, one could make an approximation of 80 km/hour, which would therefore make the trip take 1 1/4 hours.

Another example is a man taking part in a rally. The overall length is 120 kilometers. He completes the first 60 kilometers in 1 hour and twelve minutes. To win the prize he needs to average over 100 kilometers an hour for the whole race. It would be impossible, because even if he travels at phenomenal speeds, he still wouldn't get his average speed above 100 kilometers an hour. In fact, even assuming he could arrive at the finishing post instantaneously, he still would only match the target, not beat it.

PROPORTION AND INVERSE PROPORTION

Many problems in real life have simple proportional laws and so are easy to solve. If 10 people on average can produce a factory output of 1,000 units, then 20 people on average should be able to produce 2,000 units. This deduction is called direct proportion. Unfortunately, it is not always that simple; careful reasoning is required before stating what could be the wrong solution.Suppose it takes 10 people 10 hours to do a job. How long would it take two people? The answer is not two hours! There are less people and so the job should take longer. This particular case is an example of inverse proportion. It can be worked out using the unitary method: 10 people: 10 hours; 1 person: 100 hours; 2 people: 50 hours.

Even though proportion appears easy, when it is applied to other real-life problems it can get much more complex. For example, a company is producing boxes for storing model cars. The boxes are 2 cm by 2 cm by 2 cm. For a special edition, they want to create a box with a volume that is twice as big. What should the length of the sides be? The apparently obvious, yet incorrect, answer is for the sides to be 4 cm long. But the 2 cm sides give a volume of 8 cm^3, while the 4 cm sides give a volume of 64 cm^3. Much too big! By doubling the sides, the volume becomes 8 times as big. This is called cubic proportion.

If solving a problem that involves proportion, it should be determined whether it is direct proportion or not. It is also a good idea to always check answers afterwards.

ECOLOGY

A problem facing ecologists at the moment is the saving of endangered animals. Statements are frequently made concerning those dwindling in stock, and radical solutions are suggested. Yet, it is essential that the solutions be explored before any action is taken.

To model situations encountered in ecology, mathematical equations are set up that are indicative of the way the population changes as time progresses. These can be referred to as differential equations. These indicate how a population continually changes from second to second. This can be a bad model for species that breed at specific times. Such a population will have very distinct, regular changes.

The type of equation used to solve these situations can be known as difference equations. This would be used to illustrate changes over discrete periods of time. A list of equations, often referred to as a series of equations, is produced. These equations would each correspond to a different variable within the ecosystem in question. These are then solved, often using computers, to suggest the outcomes if different methods are used. If an equation is solved using computers, it is often referred to as an analytical solution.

A simple example to consider is that of rabbits and foxes. The ecologist will consider that the more rabbits there are, the quicker they will breed and hence the population will increase. If there are more rabbits, there is more for the foxes to eat, and so the foxes thrive and their population increases. Conversely, more rabbits are eaten, so their population decreases. Each of these lines could be represented by an equation and these could be used as indications of how the populations will develop.

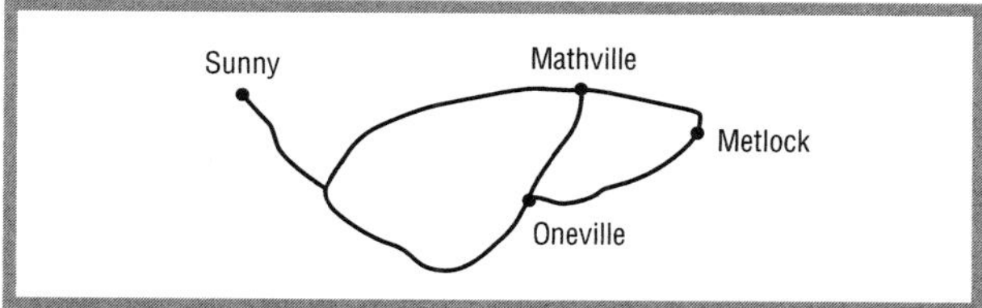

Figure 2.

TRANSLATION

As the commercial possibilities expand, and more and more cultures mix and work together, the ability to communicate is becoming increasing essential. Yet it is virtually impossible for a human translator to be present at all times to assist between different languages. It is for this reason, as well as cost consideration, that the concept of computerized translation is very appealing. Yet the ability to turn a random phrase in English into Spanish is difficult, if it is to be done efficiently. The simplest solution would be to have all conceivable phrases stored somewhere for each language, and to then link them. This is often called a one-to-one (functional) solution.

Careful consideration should, however, reveal the limitations of such an idea. The number of possible sentences in a language is unimaginably vast. The aim is therefore to program the computer with a sense of grammar and language structure. When a sentence is typed in, the computer recognizes whether words are verbs, nouns, or prepositions, converts these into the required counterpart, and then applies the correct grammar. This in itself is a remarkably complex task. Computers are still poor translators. However, the continual development of computers is allowing advances in such areas.

NAVIGATION

Strictly speaking, for many transportation companies, navigation is concerned with getting from point A to point B in the shortest time and cheapest way possible. A company will set out with the sole objective of finding this route. Finding the shortest distance is a large discipline of mathematics and often goes under the overall umbrella of decision mathematics.

To solve this problem, the company would make a map indicating all the possible routes and their respective costs. Figure 2 is an example of such a simplification.

Paradox

A paradox is a statement that seems to contradict expected reasoning. There are many famous paradoxes within mathematics and they often lead to exploration into new areas to try to evaluate why they occur. For example, the Sorites paradox. Sorites is Greek for heap and describes a set of thinking problems. At what point does a pile of sand denote a mound of sand? One grain clearly isn't a mound; add one more grain to this, and little difference has been made. By this definition, adding one grain each time still means there is no mound. At what point is a mound achieved? Conversely, if there is a mound and a grain of sand is removed, there is still presumably a mound. Keep removing one grain, and when is there no mound? Is it just the limitations of language that cause the apparent paradox?

Another paradox, originally expressed in ancient Greek, is well-known. A man fires an arrow at a moving target, albeit one that is slower than the arrow. Unfortunately, the arrow never hits the target. This is because by the time the arrow would have caught up with the target, this object has moved that much further on. So the arrow needs to travel a bit further, but by this time the target has once again moved. And so the argument persists. This entire argument has now been resolved and indeed is linked to a whole area of mathematics often referred to as convergence and divergence in sequences. These are extremely important areas in number theory.

Cost is a generic term used to denote the area of consideration. This could be time, or distance, or cost, or even gasoline consumption. An algorithm is then used to solve this problem. There are many methods available; the main one used is called Dijkstra's algorithm. Any electronic route-finder on cars will probably apply this method. A more complete algorithm used is called Floyd's. This is a repeated version of Dijkstra and finds the shortest distance between all points on a map.

The maps used are always simplified versions of the real-life situation. They will never resemble visually the actual physical situation. These maps are referred to as graphs, the roads are often called arcs, and the places where roads diverge or converge are called nodes, or vertices. This leads to a large area of real-life mathematics called graph theory.

GRAPH THEORY

Graph theory is often used to solve real-life problems, often those expressed in words that appear complex on the face of it. For example, the problem is to find the most efficient way to build a car using the minimal number of people, while completing the task within a prescribed time. The way to solve this problem is to identify the tasks and to construct a precedence diagram for the situation. The diagram merely indicates the order in which certain tasks need to be performed. It is obvious, for instance, that the engine cannot be placed in the car before the car itself has been built.

The situation thus described would then be solved using a method often called critical path analysis. Diagrams to show number of workers can also be drawn, which show how many people are required at any one time and would be used during the hiring process and to plan wages. These concepts are important to learn when considering a career in management and business.

LINEAR PROGRAMMING

Linear programming is used to solve such problems as how to maximize profit and minimize costs. The situation is simplified into a series of simple equations, and these are solved to present the optimal, or best, solution. For example, a company wants to produce two items of candy. Candy A will sell for \$1.50; Candy B will sell for \$2. The company wants to produce at most 1,000,000 candy bars altogether. Due to demand, it wants to make at least twice as much of A as of B. The ratio of the secret ingredient X in the two candy bars is 2:5. The company has 7,000,000 parts of ingredient X. How much of each should they produce to maximize profit?

The problem is solved as follows: They let x = the amount of Candy A made, and y = amount of Candy B made. Then, they want to maximize $1.50x + 2y$, since this denotes profit, subject to: $x + y < 1{,}000{,}000$ (total number of bars less than one million); $x > 2y$ i.e. $x - 2y > 0$ (twice as many of A as of B); $2x + 5y < 7{,}000{,}000$ (Total amount of ingredient X is less than 7,000,000 parts).

These equations can then be solved to find the optimal solution. They can be expressed graphically, using x- and y-coordinates to represent amount of candy A and candy B. These equations are linear because the coefficient of both x and y is 1. It is best solved using a computer. A method that is most efficient is called the simplex method. A computer is able to use the algorithm quickly and give the optimal solution in virtually no time at all.

TRAVELING SALESPERSON

Most companies need to travel either to market their product or to make deliveries. It is essential that this be done as efficiently as possible. Often a delivery will do a circular trip, calling at all required places. To save gas, the shortest route is found, though this may be in terms of time, or gas, or cost, or a combination of many factors. This requires graph theory to find a solution. Nodes are drawn to represent the places required and arcs are used to represent possible journeys.

There is no easy way to find an optimal solution. For extremely large routes, even a computer would take years to reach an optimal solution. For this reason, a trial and improvement technique is used. This is an important concept in mathematics. Estimates for worst-case and best-case scenarios are found. A logical search (often referred to as an inductive process) must take place. Gradually, improvements are made, until the company is satisfied with the solution. They may stumble upon a better solution later. The company that achieves the better solution will be the one that survives.

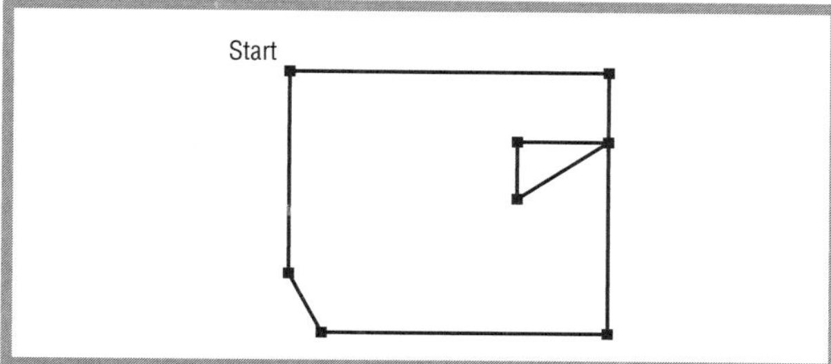

Figure 3.

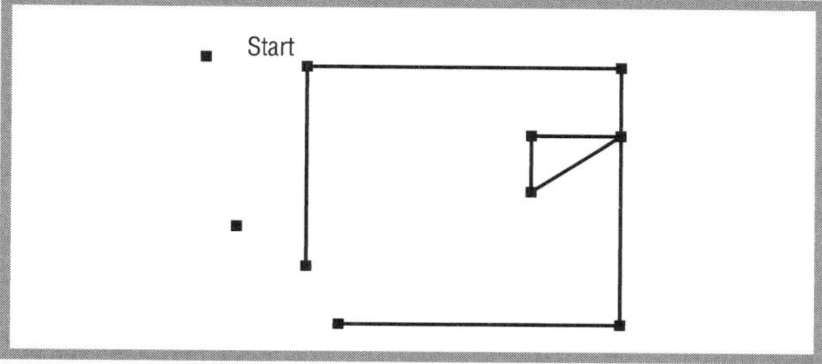

Figure 4.

POSTMAN

A mailman who needs to walk down all streets in a particular precinct will want to take the shortest route possible, and avoid repetitions, if possible. Consider Figures 3 and 4.

In Figure 3, all of the roads (arcs) are complete. However, Figure 4 has one of the roads (arcs) removed. Even though there are fewer "roads" to go down, the actual solution takes longer to perform. It is actually the case that a good solution exists if all nodes have an even number of roads/arcs leading out of them. If there is a node with an odd number of roads coming out of them, then the problem becomes more complex.

To solve the problem, a consideration is taken of the odd nodes. As a reminder, this means the nodes with an odd number of roads coming out of them. The shortest arcs between such nodes are then doubled up. This is equivalent to walking up and down the road twice. It is like meeting a dead-end and the postman has to double back.

There are many different jobs where such analysis is required. Many bulk delivery firms will use such ideas. It can also be used for hypothetical problems such as where the arcs represent tasks and where all the tasks need to be performed, though not in any particular order.

ROTA AND TIMETABLES

One of the more complex aspects of any business is that of staffing levels and evaluating when staff should work. Many food outlets require shift patterns to be established, and the average high school will have many hundreds of teachers that need to be organized. A careful, logical approach is required to meet the demands.

SHORTEST LINKS TO ESTABLISH ELECTRICITY TO A WHOLE TOWN

What is the most efficient way to connect a whole town to a main electricity supply? Clearly, the most efficient solution would be the one using the smallest length of cable. There are two established techniques for solving this problem.

Drawing a graph is required to solve this problem. Nodes are used to represent houses, and arcs are used to represent all the possible connections available. The graph will be a complete graph. This is because all the different possibilities will be considered. One of the two following efficient methods will solve the problem.

In the Kruskal's algorithm, all the different possible cable lines are ranked from shortest (best) to longest (worst). Then cable is progressively added in until all the houses are connected. In the Prim's algorithm, it is the houses that are progressively joined by lengths of cable. Starting with the house that is closest each time, all houses are joined together.

RANKING TEST SCORES

Ranking a long list of numbers occurs often in real life. This seems like a trivial task until there is a list of

Connecting Four Towns

Consider four towns, each located at a vertex. A rail network is required to connect these four towns. Which of the following two solutions, Option A or Option B, is the optimal solution?

It turns out that Option B is the better solution. Indeed, by formulating mathematical expressions for the railway tracks, calculus can be used to evaluate what the length of the horizontal section must be for the smallest route. This will depend on the exact distances between each of the towns.

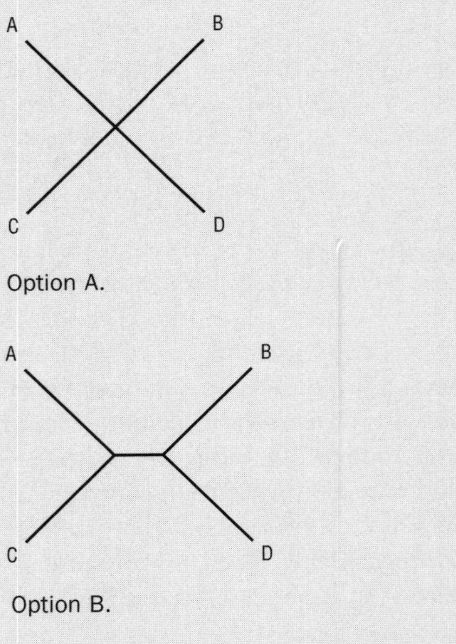

Option A.

Option B.

substantial size. Suddenly, a logical method is required. There are many different methods used, all going under the name of sorting algorithms. They all have different advantages and disadvantages. These algorithms may be programmed into software to allow computers to do the hard work. A computer needs an explicit set of instructions if it is to complete a task. The programmer must consider the amount of coding required to get the sort function to work.

SEARCHING IN AN INDEX

With a lot of information, it can be difficult to find one precise piece. It is for this reason that a dictionary is ordered sequentially. In another example, a student may have a large amount of school notes, each page numbered and in order, and the student needs to find a specific page to study for an exam. The method to use is called binary search.

This method requires a numbered list. This would be the case in most examples of filing. A good starting point would be the halfway point in the list. The student can look through the upper half first, then the lower half, until the specific page is found. This is much quicker method than randomly looking at pages. Obviously, a computer would be much quicker!

EFFICIENT PACKING AND ORGANIZATION

To pack the most objects in a given space requires careful mathematics. One method is extremely good at these packing situations. The rule is to order the objects first, from largest to smallest, and then pack them in that particular order.

SEEDING IN TOURNAMENTS

One of the prerequisites for many sporting events is that the best players don't meet each other until the later stages of the game. To accomplish this, players are allocated seeds, or rankings based upon their past and current performance. The players are then often pooled into different groups and the fixtures are arranged initially within groups. This will ensure that seed 1 and seed 2 will not meet until later in the tournament.

ARCHITECTURE

Buildings must be designed by taking several factors into consideration. It is to resolve the myriad issues that architectural design is so important. Architects are workers with a fully functional knowledge of the mathematics behind construction.

Objects of such magnitude as buildings must be constructed of materials that support the extreme forces exerted on them. The tensile strength of a material involves how much it can be stretched without deforming. The compressive strength corresponds to the ability to withstand compressive forces. (It would be disastrous if the walls of a building began to shrink!)

The shape and structure of the building is also important. Certain configurations are recognized as having a much greater stability. Often, geometry will be used to ensure that angles of adjoining structures maximize the strength required.

COOKING INSTRUCTIONS

Many meals require precise instructions, depending on oven type and power. It is then up to the consumer to evaluate the cooking time for the product. Many pieces of meat have times prescribed according to mass. For example, a chicken may require 30 minutes cooking, plus an extra 30 minutes per 500 grams. It is obviously important to be able to understand such instructions.

RECIPES

Recipes are real-life examples of word problems. They provide exact quantities to make a meal for a specific number of people. It is then up to the individual to adjust the ratio accordingly. This is an example of direct proportion. It is an essential skill for those involved in mass catering or indeed in any production to be able to scale up required ingredients to satisfy variable orders.

LOTTERIES AND GAMBLING

Many millions of people gamble every day. They are often enticed by vocabulary, such as even chance or good chance, without really knowing what the phrases mean. The odds in horse racing always start as a ratio; it is up to the betters to understand the relative merits of the odds and make a judgment accordingly.

BANKS, INTEREST RATES, AND INTRODUCTORY RATES

The modern banking market is extremely competitive. One of the main concerns when establishing a savings account is that of interest. Each bank may offer a slightly different level, and some offer initial rates that soon change.

There are two different types of interest. The main type is called compound interest. This is normally paid yearly and is evaluated from the amount currently in the account. The second type is simple interest. This is a fixed amount. It is often worked out by looking at the initial amount deposited into the account.

An example would be look at savings account A, which has an initial deposit of $1,000 that offered a yearly interest rate of $100 fixed; savings account B offered 8% yearly. The progression of account A would be 1,000, 1,100, 1,200, 1,300, 1,400, 1,500, 1,600, 1,700; the progression of account B would be 1,000, 1,080, 1,166, 1,259, 1,360, 1,469, 1,586, 1,713. Clearly, option B is relatively slower to start off with. However, after seven years the amount in account B overtakes that in account A. It is always important to look in detail at a mathematical situation and not just take a short-sighted view of the problem.

FINANCE

A company will often lay down objectives for the forthcoming year. These will be in the form of a business plan that describes the growth desired and what expenditures can be used, among other factors. It is often up to consultants to suggest ideas for how such objectives can be achieved. Economics can be modeled through a range of equations and economic principles are often applied to the stock market and growth of countries and cities. A consultant would be able to use the initial data and work out the best way the resources can be used to ensure the company achieves good results.

The study of economics is highly mathematical. There are many accepted models used within the business world.

DISEASE CONTROL

Many scientists currently monitor disease and try to evaluate likely outbreaks. The World Health Organization (WHO) may be interested in the likelihood of an outbreak of malaria in a certain part of Africa. Mathematical models are constructed, using data available, to evaluate possibilities. These models will frequently involve past data, as well as expected data. Understanding the probabilities of recurrences and the likelihood of location would be a useful tool in combating the many serious diseases.

GEOLOGY

Geology is the study of the physical Earth, and most aspects would be considered relevant to the real world. As of 2005, due to the Asian tsunami disaster occurring in 2004, an awareness of the forces of nature is at the forefront of people's consciousness. The question that many officials may ask is "Will this happen again?" or "When would such an occurrence happen?" or "How would a tsunami affect us if it occurred closer to our country?"

The mathematician would work out the many different possibilities that could occur. Perhaps by studying the effects of the recent disaster more information will be accessible and further developments made. Yet to do this, it would be broken down into the following key areas, such as where could such an event occur, how unstable is the area, how deep are the oceans, and what effect would this have?

The mathematician would then be able to apply models to each of these situations and produce a logical answer giving the range of expected possibilities. The study of dynamics, especially in fluids such as the oceans, is a vast area of applied mathematics. Many famous mathematicians (for example, Euler) spend years of their life studying such issues.

SURVEYING

When building on a new site, a company would first of all be expected to analyze the area to ensure no dangers are around. Yet to solve this, consideration would have to be taken into what safe actually means within the context, and compare it to the construction being built. The situation would be simplified into key areas, including what sort of weight can the land tolerate and what effect on the environment would the project have? Such questions would be explored mathematically through a consideration of the weight of the engineering project and the stability of the surface.

STORE ASSISTANTS

Store assistants are constantly faced with word problems that may need immediate response. A customer may ask how much a group of items would cost and the assistant may not have a calculator at hand. The sales assistant must be able to give an immediate response.

STOCK KEEPING

Store managers must work out how much stock to order. If too much is ordered, it may be wasted; yet if too little is ordered, customers will be dissatisfied. Managers develop their own techniques for solving such questions, however much of what they do will depend upon instinct and experience. Many real word problems require experience to be solved. This can be paralleled in pure mathematics. A good store manager will analyze sales of the same period for previous years. They will evaluate averages and use these figures to determine the amount that will be required. They may also produce graphs to show how the average amount is changing. These are referred to as moving average problems. For examples, average sales may have gone up by $10, then $20, then $30; consequently, a fair estimate may be made that the next increase will be $40. The manager then uses this figure when deciding how many units to order. Once again, the problem is solved through converting the real-life situation into exact mathematical figures. These allow for simple conclusions that can be backed up with fact.

ACCOUNTS AND VAT

Deciphering monetary information often requires a mathematical answer.VAT is a tax paid on items that are not essential and is required by law within the European Union. Any U.S. company selling into the EU has to, by law, charge VAT at the required level.

If an item's basic cost is known, then VAT is easy to work out. The tax is the required percentage of the total cost. For example, a coat exported to the United Kingdom cost $85.11 before VAT was added. If the U.K. VAT is 17.5% then the cost of the coat (rounded in dollars) becomes $\$85.11 + \$85.11 \times (17.5/100) = \$100.00$. The person is able to claim the VAT tax of $14.89 back from the U.K. government if the coat is essential for his employment.

BEARINGS AND DIRECTIONS OF TRAVEL

The shortest route between two points on a flat surface is the straight line connecting the two points. However, how is motion achieved in that straight line? This is a question that transport companies, especially nautical-related transport, need to consider all the time because other factors are continuously trying to influence the motion of the vessel. There will be currents and wind trying to steer the vessel off course. The ship would therefore have to steer a course that compensates for these extra factors. These problems can be solved using bearings and trigonometry. Today, of course, sensors will detect the forces present and computers will be able to adjust the steering as required.

QUALITY CONTROL

It is important for companies to monitor output to ensure that goods meet standards. The authorities often define these standards, and not meeting them could lead to heavy fines and/or closure. For example, the criteria are that only 5% of products are below a required size and the company produces one million of these items a day. How do they monitor their output?

A system is often used called systematic sampling. Every one hundredth item produced is checked against the required criteria. The company will then keep a running total of items failing or passing the test. As long as a sufficient number is above the required standard, the company will keep producing. The authorities will normally publish guidelines, and the company uses those.

Sampling is used to solve a wide range of such problems. In different situations, different sampling techniques are used. Samples are used because it is often impossible to test or analyze every single item in a population.

WHAT IS THE AVERAGE HEIGHT IN A NEIGHBORHOOD?

Manufacturers of items ask this sort of question all the time when the size of people, for example, has direct relevance on production. It would be a bad business decision to produce small clothes if the population happened to be a tall one. Yet, how would a company evaluate the average height?

The company would first identify the target market. This is important if their line of production happens to be jackets for women. They would then need to pick a random sample, which reduces the potential for bias. Often the company will do a form of quota sampling. This is a method to ensure that people of all ages are picked. A quota is a group. The company will identify all the relevant groups and pick out a random people from each. The formula used to find the number of people in a random sample or quota group is normally the square root of the entire targeted population.

OPINION POLLS

Opinion polls are used to answer such questions as "Who is the most popular politician?" Politicians can use them as propaganda, in both a positive and negative way. Opinion polls, however, are often biased. Mathematically speaking, opinion polls are not necessarily considered to be sound. They frequently target only a select group in a population and thus lead to often conflicting and contradictory evidence.

WEATHER

Forecasts are used and needed across many spheres in many different occupations. It is not possible to say what will happen; instead forecasters deal with what is most likely to happen. The reason weather cannot be predicted with much accuracy is due to a mathematical idea called chaos theory. Basically, there are so many interactions happening at both the macroscopic and microscopic level that any slight perturbation in any of these interactions could seriously affect the weather's outcome. Many sporting events and agricultural areas rely exclusively on forecasts to plan their daily tasks.

Riddles

A riddle is a written or verbal statement that requires exact logic to solve. The answer should be unique and make exact sense; otherwise, it is insolvable. Riddles parallel a lot of work done in mathematics in real life. They require sentences to be simplified into understandable ideas. Solutions can then be posed, until the correct solution is acquired. The solution of a riddle mimics the modeling method in mathematics.

To solve a riddle, one must consider the set of solutions that solve each sentence. The solution that overlaps all parts of the riddle is the final solution. Consider the following challenging riddle:It is better than God and more evil than the devil.Rich people want it, poor people need it.You die if you eat it.

What is the riddle's solution? (The answer is "nothing.")

The fundamental concepts behind weather forecasting are the understanding of the interactions in the atmosphere and the modeling of this using mathematics. Powerful computers are today used to predict the likely outcome, churning out vast output of data. The art of predicting weather is often referred to as meteorology. It is certainly not an exact science. To try to get a realistic answer to the problem of weather forecasting, the super computers produce different outputs with a slightly different starting point (a forced perturbation). The average can then be taken. These small perturbations often lead to dramatic changes in the output. There is frequently a dramatic divergence in solutions, especially when one begins to predict more than just three or four days in the future.

THROWING A BALL

How one throws a ball to maximize the distance achieved is of particular relevance within the sporting world. The answer is solved through a series of assumptions. If it is assumed that the ball is thrown approximately from ground level and that the only force acting on the ball is gravity, the solution is that the angle should be 45°. It is clear why the angle affects the solution. If the ball is thrown vertically upwards, it will cover no distance, but if it is thrown horizontally, it will fall quickly to the ground. This model can then be improved and different solutions will be thus arrived. However, this gives the mathematician a starting point from which to develop a theory.

MEASURING THE HEIGHT OF WELL

The problem when constructing a working well for a village in Africa is that there is a chasm already present. There is a simple way to approximate its depth. If a stone is dropped down the well, the time taken to reach the bottom can be measured. A distinct sound would be heard as it hits the water. The depth of the well can be approximated using the formula: $d = 4.9 \times t^2$.

DECORATING

When setting out on a renovation project, one of the first questions will be a consideration of the materials required. To minimize the cost of decoration it would be advisable to use careful mathematics to evaluate the quantity of material required. A professional decorator will not want to mix a required hue only to find that there is not enough to finish the whole room.

These types of problems can be easily solved through a consideration of area. Rooms are generally regular. A simple calculation involving width and height would give the amount of wall space involved. The materials should have indications on the labels informing the consumer how much area they will cover. It is then a simple case of using proportion to evaluate the amount of material needed.

DOES GLOBAL WARMING EXIST?

There are many different arguments on either side of the debate of global warming. Mathematics provides a way of looking at such issues and problems in a non-emotive way, allowing for careful and logical reasoning. It is, however, easy to manipulate many ideas involved and the issue must be studied free from influence either political or otherwise. This underpins the mathematics behind independent surveys. It is a tool. Like all tools it can be used flexibly in ways that are not obvious to the layman.

DOES MMR (MEASLES, MUMPS, RUBELLA) IMMUNIZATION CAUSE AUTISM?

There is a reported link between immunization and subsequent disease. Mathematics, especially statistical ideology, is used to test the likelihood of such a link existing. Unfortunately, the mathematics is often lost beneath emotion and ideology until the evidence itself is discounted or stated to be invalid. This is the main reason why statistical tests used to investigate links need to be done as rigorously as possible. There will always be an element of doubt in the conclusions reached. The reduction of this doubt will lead to more convincing arguments, and so results can be displayed and credible conclusions reached. Recent research does not establish a link between MMR immunization and autism.

Potential Applications

The existence of word problems and their necessity within society will never cease. Language will continue to develop and so will the mathematical thirst to solve and to explain. The ability to solve such problems and the skills to explain in simple terms will always be considered an essential skill in all areas of employment.

As time passes, mathematical models will become more and more sophisticated and the advent of more powerful computing will allow more accurate solutions. More and more advanced questions about the universe and the inherent mathematics that underpins it will continue to be pursued. Who knows how far the solutions will take us?

Where to Learn More

Books

Parramore, K., J. Stephens, G. Rigby, and C. Compton. *MEI Decision and Discrete Mathematics* London: Hodder Arnold H&S, 2004.

Porkess, R., et al. *MEI Statistics 2* London: Hodder Arnold H&S, 2005.

Web sites

Value Added Tax. Online Resources. <http://www.vat.com/faq.html> (March 1, 2005).

Zero-Sum Games

Overview

A zero-sum game is a game in which whatever is lost by one player is gained by the other player or players. The study of zero-sum games is the foundation of game theory, which is a branch of mathematics devoted to decision-making in games.

In mathematics, all situations in which there are two or more parties—people, companies, teams, or nations—making decisions that affect some measurable outcome are "games." The decisions made by a game player make up that player's "strategy." The goal of game theory is to calculate the best strategy for a given game. Zero-sum games are a special part of game theory that can be applied in law, military strategy, biology, and economics.

Games are not necessarily played for fun. They can be deadly serious. Chess, cards, and football are considered "games" in game theory, but so are business and war. Not all the pastimes we call "games" are games in the game-theory sense. The children's card game called War is an example of a game that is not a game (mathematically speaking). In War, the players repeatedly match cards, one from each player, and the player with the higher card takes the pair. They continue until one player holds all the cards. Which player ends up with all the cards depends only on how the cards have been shuffled and dealt. No decisions are made by either player, so there is no way to choose a strategy. The winner is decided by pure chance.

True games can, however, involve an element of chance. In football, for instance, a player can slip on wet turf, make a freak catch, or get confused and throw the ball the wrong way. Sometimes the winning team is even decided by such an event. But football coaches still plan strategies, and strategy does make a difference.

Fundamental Mathematical Concepts and Terms

In a zero-sum game, the players compete for shares of something that is in limited supply. One player's loss is the other player's gain: if your slice of pie is bigger, mine must be smaller.

The term "zero-sum" refers to the numbers that are assigned to different game endings. If winning a game of chess is assigned a value of $+1$, then losing a game has the value -1 and the sum of the loser's score and the winner's score for every game is $1 - 1 = 0$, "zero sum." When there is a draw, both players get 0 points and the game remains zero-sum because $0 + 0 = 0$.

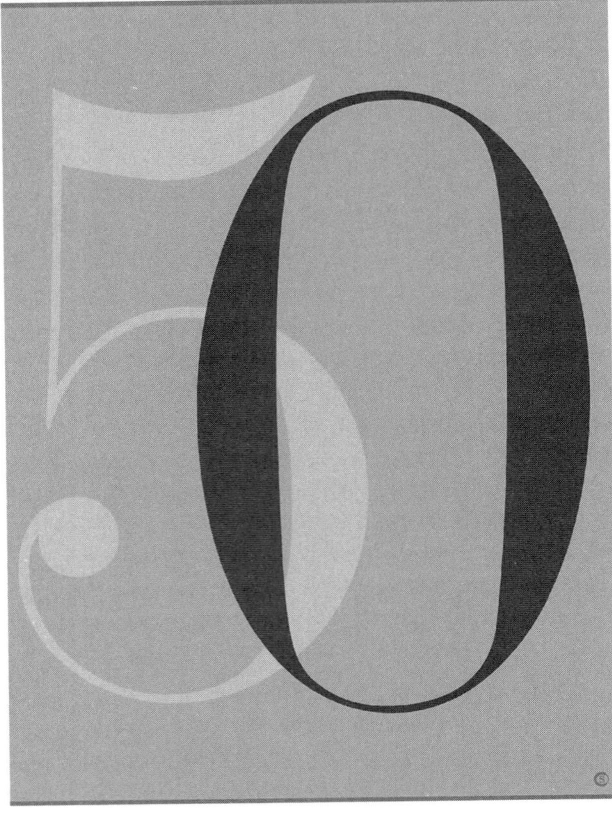

In zero sum games, winners entail losers. STEVE COLLIER; COLLIER STUDIO/CORBIS.

Two-player zero-sum games are also called strictly competitive games. Games may also have more than two players, as in poker or Monopoly. When three or more players play a zero-sum game, some players may team up or collaborate against the others, so multi-player zero-sum games are not "strictly competitive."

The theory of zero-sum games is the starting point for the theory of all other games, which can be lumped under the term "non-zero-sum games." Non-zero-sum games are games which are not played for fixed stakes.

The most famous non-zero-sum game is the Prisoner's Dilemma, first proposed by Merrill Flood and Melvin Dresher at the Rand Corporation in 1950. In this situation, there are two prisoners who have committed a serious crime. The police put each prisoner in a separate cell and try to get them to confess by telling each prisoner (falsely) that the other prisoner has already confessed, and that if they will also confess, they will get a reduced sentence. But, the police add, if the prisoner does not confess, they will get a heavy sentence.

If both prisoners confess, they will both get reduced sentences. If only one confesses, then the one that confesses will get a reduced sentence and the other will get a heavy sentence. If neither confesses, then both will be freed. Obviously, it would be best for both prisoners if they refused to confess. Yet, it can be shown by game theory that the most mathematically "rational" thing for each prisoner to do is to confess. This is a "dilemma" or no-win situation because the best strategy is to confess and take a reduced sentence rather than to refuse to confess, because each prisoner cannot guarantee what the other will do. Though not confessing might result in no sentence at all, a heavy sentence could result for a prisoner who does not confess when the other does. The guessing game played by the two prisoners is a non-zero-sum game because both prisoners might win (go free) or lose (get sentences) at the same time: there is not a fixed number of years of imprisonment that must be divided between the prisoners.

Real-life Applications

GAMBLING

Competitive gambling for money is usually a zero-sum game because the money won by one player must be lost by another. There is a fixed amount of money, and rolling dice or dealing cards cannot destroy it or create any more. Zero-sum game theory can therefore be used to find the best possible strategies for such games. This applies to games in which there is an element of choice or strategy, such as poker. In fact, the game of poker was what inspired Hungarian-born American mathematician John Von Neumann (1903–1957) to invent modern game theory, which he did starting with his 1928 article, "Theory of Parlor Games."

However, not all gambling games are "games" in the game-theory sense. Playing a slot machine is not a game, for example, because it is a matter of pure chance, all the player does is pull the handle or push the button. Game theory has nothing to say about activities like slots, roulette, dice, or lotteries because they allow no choices to the player and therefore no strategy. The only choice the player has is to play or not play. Mathematics can deal with games of pure chance, but this is done using probability theory, not game theory. Probability theory is used in game theory to deal with games that mix strategy with chance.

EXPERIMENTAL GAMING

Psychologists have used game theory to study how human beings make real-world decisions. They do this by asking volunteers to play a game. The psychologists use game theory to calculate the best or optimal strategy for the game and compare the behavior of the volunteers to the results of game theory. Psychologists have studied behavior in both zero-sum and non-zero-sum situations. They have often found that people do not behave in the way that game theory says is most "rational."

This does not necessarily mean, however, that people act foolishly. People may simply disagree with the mathematical definition of rationality. For example, if people are offered an (imaginary) choice of $1,000 in cash or a black box that has a 50% chance of containing either nothing or $10,000, they usually take the cash. Mathematics, however, says that the player's most "rational" choice is to maximize their expected or average winnings by choosing the black box. If the game were played many times over, a player who always chose the box would make more money on average (about $5 thousand) than a player who always took the $1 thousand. In this sense it is more "rational" to take the box.

But there is something artificial about saying that the behavior of a player who takes the cash is not rational. Why should a person take a 50% chance of getting nothing when they could get money without risk? This desire to avoid drastic risk is an example of what game theorists call "risk aversion." People usually prefer a strategy that protects them from disaster to a strategy that offers them big potential winnings but exposes them to possible disaster.

CURRENCY, FUTURES, AND STOCK MARKETS

Currency and futures trading are zero-sum games. Currency trading is a form of money investment in which speculators buy up one kind of money—dollars, pounds, euros, yen, or other—and then sell it again, trying to make a profit. For example, if 1 US dollar can buy 1.01 euros in Germany, and 1.01 euros can buy 1.02 yen in Japan, and 1.02 yen can buy 1.03 dollars in the U.S., then an investor can make $.03 by taking $1, buying a euro with it, buying a yen with the euro, and buying a dollar with the yen. This would be a way of getting something for nothing, except that for every penny made in the currency-trading market somebody loses a penny in the currency-trading market. The market does not generate new wealth: like a poker game, it only moves money around. Currency trading is therefore a zero-sum game. In addition, such trading as outlined above does not take into account fees that brokers charge to make transactions.

In futures trading, speculators gamble on whether unprocessed commodities like grain, beef, or oil will be worth more, less, or the same in the near future. Since a loss for the seller of the commodity is a gain for the buyer of the commodity and vice versa, the futures market is also a form of a zero-sum game. The commodity markets allow producers to fix sale prices ahead of delivery and therefore manage their risk of losing money.

There is debate about whether the stock market is a zero-sum game, but most economists agree that it is not. In the stock market, investors buy shares of ownership in companies. For instance, buying a single share might make you the owner of one millionth of the ABC Corporation. These shares can be bought and sold. As long as the value of the companies being owned remains fixed, buying and selling stock in them is a zero-sum game; however, the companies are real-world enterprises that may decrease or increase in value. Demand for a product might increase or decrease, or a vital resource (like oil) might run out, a company might go out of business, or a new technology might be developed that increases productivity and makes more real wealth. Any of these events changes the amount of wealth that the stock-market game is being played for.

WAR

War as such is not a zero-sum game. In almost any case, if both sides helped each other instead of fighting, they would be better off than if they fought. And, if the war is destructive enough, both sides, even the "winner," may end up worse off than before.

However, particular battles are often zero-sum games. The military forces fighting a battle are trying to destroy each other's resources—to kill soldiers and to destroy weapons, vehicles, and supplies. A loss for one side is a gain for the other, which is the primary feature of zero-sum games. Military strategists do in fact study battle strategy in terms of zero-sum games as well as in terms of more complex, non-zero-sum game theory.

Where to Learn More

Books

Colman, Andrew M. *Game Theory and Its Applications in the Social and Biological Sciences.* New York: Routledge, 1999.

Davis, Morton D. *Game Theory: A Nontechnical Introduction.* New York: Basic Books, 1970.

Straffin, Philip D. *Game Theory and Strategy.* Washington, DC: Mathematical Association of America, 1993.

Glossary

80/20 rule: A general statement summing up the tendency for a few items to consume a disproportionate share of resources, such as cases in which 20% of a store's customers lodge 80% of the total complaints.

Acceleration: A change of velocity (either in magnitude or direction).

Actuary: A mathematical expert who evaluates the statistical likelihood of various insurable events for underwriting purposes.

Algebra: A collection of rules: rules for translating words into the symbolic notation of mathematics, rules for formulating mathematical statements using symbolic notation, and rules for rewriting mathematical statements in a manner that leaves their truth unchanged.

Algorithm: A set of mathematical steps used as a group to solve a problem.

Analogue: A continuously variable medium, for use as a method of storing, processing, or transmitting information.

Analytic geometry: A branch of mathematics that uses algebraic equations to describe the size and position of geometric figures on a coordinate system. Developed during the seventeenth century, it is also known as Cartesian geometry or coordinate geometry. The use of a coordinate system to relate geometric points to real numbers is the central idea of analytic geometry. By defining each point with a unique set of real numbers, geometric figures such as lines, circles, and conics can be described with algebraic equations. Analytic geometry has found important applications in science and industry alike.

Angle: A geometric figure formed by two lines diverging from a common point or two planes diverging from a common line often measured in degrees.

Area: The measurement of a surface bounded by a set of curves as measured in square units.

Arithmetic: The study of the basic mathematical operations performed on numbers.

Array: A rectangular arrangement of numerical data in rows and columns, as in a matrix.

Average: A numeral that expresses a set of numbers as a single quantity. It is the sum of the numbers divided by the number of numbers in the set.

Axis: Lines labeled with numbers that are used to locate a coordinate.

Balance: An amount left over, such as the portion of a credit card bill that remains unpaid and is carried over until the following billing period.

Bankruptcy: A legal declaration that one's debts are larger than one's assets; in common language, when one is unable to pay his bills and seeks relief from the legal system.

Bicentric perspective: Perspective illustrated from two separate viewing points.

Binary code: A string of zeros and ones used to represent most information in computers.

Bit: The smallest unit of storage in computers. A bit stores binary values.

Boolean algebra: The algebra of logic. Named after English mathematician George Boole, who was the first to apply algebraic techniques to logical methodology. Boole showed that logical propositions and their connectives could be expressed in the language of set theory.

Bouncing a check: The result of writing a check without adequate funds in the checking account, in which the bank declines to pay the check. Fees and penalties are normally imposed on the check writer.

Byte: A byte is a group of eight bits.

Calculator: A tool for performing mathematical operations on numbers.

Calculus: A branch of mathematics that deals with the way that relationships between certain sets (or functions) are affected by tiny changes in one of their variables.

Cartesian coordinate: A coordinate system where the axes are at 90 degrees to each other, with the *x* axis along the horizontal.

Centric perspective: Perspective illustrated from a single viewing point.

Chi-square test: The most commonly used method for comparing frequencies or proportions. It is a statistical test used to determine if observed data deviate from those expected under a particular hypothesis. The chi-square test is also referred to as a test of a measure of fit or "goodness of fit" between data. Typically, the hypothesis tested is whether or not two samples are different enough in a particular characteristic to be considered members of different populations. Chi-square analysis belongs to the family of univariate analysis, i.e., those tests that evaluate the possible effect of one variable (often called the independent variable) upon an outcome (often called the dependent variable).

Chord: A straight line connecting any two points on a curve.

Coefficient: A coefficient is any part of a term, except the whole, where term means an adding of an algebraic expression (taking addition to include subtraction as is usually done in algebra). Most commonly, however, the word coefficient refers to what is, strictly speaking, the numerical coefficient. Thus, the numerical coefficients of the expression $5xy^2$ $3x + 2y$ are considered to be 5, -3, and $+2$. In many formulas, especially in statistics, certain numbers are considered coefficients, such as correlation coefficients.

Combinatorics: The study of combining objects by various rules to create new arrangements of objects. The objects can be anything from points and numbers to apples and oranges. Combinatorics, like algebra, numerical analysis and topology, is an important branch of mathematics. Examples of combinatorial questions are whether we can make a certain arrangement, how many arrangements can be made, and what is the best arrangement for a set of objects. Combinatorics can be grouped into two categories: enumeration, which is the study of counting and arranging objects; and graph theory, or the study of graphs. Combinatorics makes important contributions to fields such as computer science, operations research, probability theory, and cryptology.

Common denominator: A common denominator for a set of fractions is simply the same (common) lower symbol (denominator). In practice the common denominator is chosen to be a number that is divisible by all of the denominators in an addition or subtraction problem. Thus for the fractions 2/3, 1/10, and 7/15, a common denominator is 30. Other common denominators are 60, 90, etc. The smallest of the common denominators is 30 and so it is called the least common denominator.

Complex numbers: Complex numbers are so called because they are made up of two parts which cannot be combined. Even though the parts are joined by a plus sign, the addition cannot be performed. The expression must be left as an indicated sum.

Concentration: The ratio of one substance mixed into another substance.

Congruent: Two triangles are congruent if they are alike in every geometric respect except, perhaps, one. That one possible exception is in the triangle's "handedness." There are only six parts of a triangle that can be seen and measured: the three angles and the three sides. The six features of a triangle are all involved with congruence.

Conic section: The plane curve formed by the intersection of a plane and a right-circular, two-napped cone.

Constant: A value that does not change.

Convenience sampling: Sampling done based on the easy availability of the elements.

Coordinate: A set of two or more numbers or letters used to locate a point in space. For example, in two dimensions a coordinate is written as (x,y).

Cross-section: The two-dimensional figure outlined by slicing a three-dimensional object.

Cubed root: The relation of the volume of a cube to one of its edges.

Cubic equation: A cubic equation is one of the forms of $ax^3+bx^2+cx+d = 0$ where a,b,c, and d are real numbers.

Curve: A curved or straight geometric element generated by a moving point that has extension only along the one-dimensional path of the point.

Data point: A point in a graph or other display that depicts a specific value given by a function or calculation.

Decimal: Relating to the base power of ten.

Decimal fraction: A numeral that uses the numeration system, based on ten, to represent fractional numbers. For example, a decimal fraction for 2 and 1/4 is 2.25.

Decimal number system: A base-10 number system that requires ten different digits to represent numbers (0 through 9) where the value of a number is defined by its place (a place value system where a "1" could be valued at "one," "ten," "one hundred," "one thousand," etc.).

Decryption: The process of using a mathematical algorithm to return an encrypted message to its original form.

Degree: The word "degree" as used in algebra refers to a property of polynomials. The degree of a polynomial in one variable (a monomial), such as $5x^3$, is the exponent, 3, of the variable. The degree of a monomial involving more than one variable, such as $3x^2y$, is the sum of the exponents; in this case, $2 + 1 = 3$.

Dependent variable: What is being modeled; the output that results from a function or calculation.

Derivative: The limiting value of the ratio expressing a change in a particular function that corresponds to a change in its independent variable. Also, the instantaneous rate of change or the slope of the line tangent to a graph of a function at a given point.

Differentiate: The process of determining the derivative or differential of a particular function.

Digital: Of or relating to data in the form of numerical digits.

Dimension: The number of unique directions it is possible for a point to move in space. The world is normally thought of as having three dimensions. Flat surfaces have two dimensional and more advanced physical concepts that require the use of more than three dimensions.

Distributive property: The distributive property states that the multiplication "distributes" over addition. Thus $a \times (b + c) = a \times b + a \times c$ and $(b + c) \times a = b \times a + c \times a$ for all real or complex numbers a, b, and c.

Dividend: A mathematical term for the beginning value in a division equation, literally the quantity to be divided. Also a financial term referring to company earnings which are to be distributed to, or divided among, the firm's owners.

Divisibility: The ability to divide a number by another number without leaving a remainder.

Domain: The domain of a relation is the set that contains all the first elements, x, from the ordered pairs (x,y) that make up the relation. In mathematics, a relation is defined as a set of ordered pairs (x,y) for which each y depends on x in a predetermined way. If x represents an element from the set X, and y represents an element from the set Y, the Cartesian product of X and Y is the set of all possible ordered pairs (x,y) that can be formed.

Encryption: Using a mathematical algorithm to code a message or make it unintelligible.

Enumeration: The study of counting and arranging objects.

Equation: A mathematical statement involving an equal sign.

Equivalent fractions: Two fractions are equivalent if they stand for the same number (that is, if they are equal). The fractions 1/2 and 2/4 are equivalent.

Estimation: A process that arrives at an answer that approximates the correct answer.

Exponent: Also referred to as a power, a symbol written above and to the right of a quantity to indicate how many times the quantity is multiplied by itself.

Exponential growth: A growth process in which a number grows proportional to its size. Examples include viruses, animal populations, and compound interest paid on bank deposits. The rate of growth is proportional to the size of the sample or population (i.e., a relation between the size of the dependent variable and rate of growth).

Fibonacci numbers: The numbers in the series, 1, 1, 2, 3, 5, 8, 13, 21, 34, 55, 89, 144 . . . , which are formed by adding the two previous numbers together.

Formula: A general fact, rule, or principle expressed using mathematical symbols.

Fractal: A self-similar shape that is repeated over and over to form a complex shape.

Fraction: The quotient of two quantities, such as 1/4.

Frequency: Number of times that a repeated event occurs in a given time period, typically within one second.

Function: A mathematical relationship between two sets of real numbers. These sets of numbers are related to each other by a rule that assigns each value from one set to exactly one value in the other set. The standard notation for a function $y = f(x)$, developed in the eighteenth century, is read "y equals f of x." Other representations of functions include graphs and tables. Functions are classified by the types of rules which govern their relationships.

Gambling: A popular form of entertainment in which players select one of several possible outcomes and wager money on that outcome.

Game theory: A branch of mathematics concerned with the analysis of conflict situations. It involves determining a strategy for a given situation and the costs or benefits realized by using the strategy. First developed in the early twentieth century, it was originally applied to parlor games such as bridge, chess, and poker. Now, game theory is applied to a wide range of subjects such as economics, behavioral sciences, sociology, military science, and political science.

Geometry: A fundamental branch of mathematics that deals with the measurement, properties, and relationships of points, lines, angles, surfaces, and solids.

Golden ratio: The number 1.61538 that is found in many places in nature.

Greatest common divisor: The largest number that is a divisor of two numbers.

Hypotenuse: The longest leg of a right triangle, located opposite the right angle.

Improper fraction: A fraction whose value is greater than or equal to 1.

Independent variable: Input data to a function. The input data used to develop a model where the outcomes or results are determined by function and/or calculation.

Inequality: A statement about the relative order of members of a set. For instance, if S is the set of positive integers, and the symbol $<$ is taken to mean less than, then the statement $5 < 6$ (read "5 is less than 6") is a true statement about the relative order of 5 and 6 within the set of positive integers.

Infinity The term infinity conveys the mathematical concept of large without bound, and is given the symbol ∞.

Inflation: A steady rise in prices, leading to reduced buying power for a given amount of currency.

Input: What is used to develop a model, the independent variables.

Integer: The positive and negative whole numbers. $-4, -3, -2, -1, 0, 1, 2, \ldots$ The name "integer" comes directly from the Latin word for "whole." The set of integers can be generated from the set of natural numbers by adding zero and the negatives of the natural numbers. To do this, one defines zero to be a number which, added to any number, equals the same number.

Integral: A quantity expressible in terms of integers (the positive and negative whole numbers). Also, a quantity representing a limiting process in which the domain of a function is divided into small units.

Integral calculus: A branch of mathematics used for purposes such as calculating such values as volumes displaced, distances traveled, or areas under a curve.

Interest: Money paid for a loan, or for the privilege of using another's money.

Irrational number: A number that cannot be expressed as a fraction, that is, it cannot be written as the quotient of two whole numbers. As a decimal, an irrational number is shown by an infinitely long non-repeating sequence of numbers. Examples of irrational numbers are pi (the ratio of circumference to diameter of a circle), e (base of the natural logarithms).

Iteration: Iteration consists of repeating an operation of a value obtained by the same operation. It is often used in making successive approximations, each one more accurate than the one that preceded it. One begins with an approximate solution and substitutes it into an appropriate formula to obtain a better approximation. This approximation is subsequently substituted into the same formula to arrive at a still better approximation, and so on.

Key: A number or set of numbers used for encryption or decryption of a message.

Knot theory: A branch of mathematics that studies the way that knots are formed.

Least-terms fraction: A fraction whose numerator and denominator do not have any factors in common. The fraction 2/3 is a least-terms fraction; the fraction 8/16 is not.

Line: A straight geometric element generated by a moving point that has extension only along the one-dimensional path of the point.

Linear algebra: Includes the topics of vector algebra, matrix algebra, and the theory of vector spaces. Linear algebra originated as the study of linear equations, including the solution of simultaneous linear equations. An equation is linear if no variable in it is multiplied by itself or any other variable. Thus, the equation $3x + 2y + z = 0$ is a linear equation in three variables.

Linear equation: An equation on which the left-hand side is made up of a sum of terms (each of which consists of a constant multiplying a variable), and the right-hand side which consists of a constant. For example, $2x_0 + 3x_1 = 4$.

Linear programming: A method of optimizing an outcome (e.g., profit) defined by a linear equation but constrained by a number of linear inequalities. The inequalities are recast as linear equation and the resulting system is solved using matrix algebra.

Logarithm: The power to which a base number, usually 10, has to be raised to in order to produce a specific number.

Logic: The study of the rules which underlie plausible reasoning in mathematics, science, law, and other disciplines.

Long odds: Poor odds, or odds which suggest an event is highly unlikely to occur.

Lottery: A contest in which entries are sold and a winner is randomly selected from the entries to receive a prize.

Mathematics: The systematic study of relationships in the physical world and relationships between symbols which need not pertain to the real world. In relation to the world, mathematics is the language of science. It operates within the laws and constraints of science as it examines physical phenomena.

Matrix: A rectangular array of variables or numbers, often shown with square brackets enclosing the array. Here "rectangular" means composed of columns of equal length, not two-dimensional. A matrix equation can represent a system of linear equations.

Median: A measure of central tendency, like an average. It is a way of describing a group of items or characteristics instead of mentioning all of them. If the items are arranged in ascending order of magnitude, the median is the value of the middle item.

Metric system: The metric system of measurement is an internationally agreed-upon set of units for expressing the amounts of various quantities such as length, mass, time, temperature, and so on.

Mode: A set of numbers is the number that occurs most frequently. There may be more than one mode. In the set (1,4,5,7), all four numbers are modes. But in the set (1,4,4,6), 4 is the only mode. The mode is one of the measures of central tendency, the others being the mean and the median.

Model: A system of theoretical ideas, information, and inferences presented as a mathematical description of an entity or characteristic.

Modulus: An operator that divides a number by another number and returns the remainder.

Mortgage: A loan made for the purpose of purchasing a house or other real property.

Nash equilibrium: A set of strategies, named after John Nash, that results in the maximum benefit of each player.

Natural numbers: The ordinary numbers, 1, 2, 3, . . . with which we count. Sometimes they are called "the counting numbers."

Negative numbers: Numbers that have a value less than zero.

Nth term: The phrase 'nth term' is used to describe any term in a sequence. The n refers to its ordered place in the sequence.

Number theory: Number theory is the study of natural, or counting numbers, including prime numbers.

Odds: A shorthand method for expressing probabilities of particular events. The probability of one particular event occurring out of six possible events would be 1 in 6, also expressed as 1:6 or in fractional form as 1/6.

Operation: A method of combining the members of a set so the result is also a member of the set. Addition, subtraction, multiplication, and division of real numbers are everyday examples of mathematical operations.

Orthogonals: In art, the diagonal lines that run from the edges of the composition to the vanishing point.

Output: Output data to a function. The output data, the dependent variable(s), that define a model.

Parabola: The open curve formed by the intersection of a plane and a right circular cone. It occurs when the plane is parallel to one of the generatrices of the cone.

Parallel: Two or more lines (or planes) are said to be parallel if they lie in the same plane (or space) and have no point in common, no matter how far they are extended.

Parallelogram: A plane figure of four sides whose opposite sides are parallel. A rhombus is a parallelogram with all four sides of equal length; a rectangle is a parallelogram whose adjacent sides are perpendicular; and a square is a parallelogram whose adjacent sides are both perpendicular and equal in length.

Percent: From Latin for *per centum* meaning per hundred, a special type of ratio in which the second value is used to represent the amount present with respect to the whole. Expressed as a percentage, the ratio times 100 (e.g., 78/100 = .78 and so .78 × 100 = 78%).

Perfect number: A number that is equal to the sum of its divisors.

Permutations: All of the potential choices or outcomes available from any given point.

Pi: The ratio of the circumference of a circle to the diameter: $\pi = C/d$ where C is the circumference and d is the diameter. This fact was known to the ancient Egyptians who used π for the number 22/7 (3.14159) which is accurate enough for most applications.

Pixel: Short for "picture unit," a pixel is the smallest unit of a computer graphic or image. It is also represented as a binary number.

Player: In game theory, a decision maker.

Plays: In game theory, choices that can be made.

Point: A geometric element defined only by an ordered set of coordinates.

Polar angle: The angle between the line drawn from a point to the center of a circle and the *x* axis. The angle is taken by rotating counterclockwise from the *x* axis.

Polar coordinate: A two-dimensional coordinate system that is based on circular symmetry. It has two coordinates, the radius and the polar angle.

Polar-coordinate system: One of the several systems for addressing points in the plane is the polar-coordinate system. In this system a point P is identified with an ordered pair (r,θ) where r is a distance and θ an angle.

Positive numbers: Commonly defined as numbers greater than zero, the numbers to the right of zero on the number line. Zero is not a positive number. The opposite, or additive inverse, of a positive number is a negative number. Negative numbers are always preceded by a negative sign (−), while positive numbers are only preceded by a positive sign (+) when it is required to avoid confusion.

Powers: The number of times that a base is to be multiplied by itself.

Prime factorization: The process of finding all the divisors of a number that are prime numbers.

Prime number: Any number greater than 1 that can only be divided by 1 and itself.

Probability: The likelihood that a particular event will occur within a specified period of time. A branch of mathematics used to predict future events.

Probability distribution: The expected pattern of random occurrences in nature.

Probability theory: A branch of mathematics concerned with determining the long run frequency or chance that a given event will occur. This chance is determined by dividing the number of selected events by the number of total events possible.

Program: A sequence of instructions, written in a mathematical language, that accomplish a certain task.

Proper fraction: A fraction whose value is less than 1.

Proportion: Two quantities with equal ratios.

Public key system: A cryptographic algorithm that uses one key for encryption and a second key for decryption.

Pythagorean theorem: A theorem of geometry, often attributed to Pythagoras of Samos (Greece) in the sixth century B.C., states the sides a, b, and c of a right triangle satisfy the relation $c^2=a^2+b^2$ where c is the length of the hypotenuse of the triangle and a and b are the lengths of the other two sides.

Quadrilateral: A polygon with four sides. Special cases of a quadrilateral are: (1) A trapezium—A quadrilateral with no pairs of opposite sides parallel; (2) A trapezoid—A quadrilateral with one pair of sides parallel; (3) A parallelogram—A quadrilateral with two pairs of sides parallel; (4) A rectangle—A parallelogram with all angles right angles; (5) A square.

Radius: The distance from the center of a circle to its perimeter.

Rate: A comparison of the change in one quantity, such as distance, temperature, weight, or time, to the change in a second quantity of this type. The comparison is often shown as a formula, a ratio, or a fraction, dividing the change in the first quantity by the change in the second quantity. When the changes being compared occur over a measurable period of time, their ratio determines an average rate of change.

Ratio: The ratio of a to b is a way to convey the idea of relative magnitude of two amounts. Thus if the number a is always twice the number b, we can say that the ratio of a to b is "2 to 1." This ratio is sometimes written 2:1. Today, however, it is more common to write a ratio as a fraction, in this case 2/1.

Rational number: A number that can be expressed as the ratio of two integers such as 3/4 (the ration of 3 to 4) or −5:10 (the ration of −5 to 10).

Real number: Any number which can be represented by a point on a number line. The numbers 3.5, −.003, 2/3, etc. are all real numbers.

Reciprocal: The reciprocal of a number is 1 divided by the number. Thus the reciprocal of 3 is 1/3. If a number a is the reciprocal of the number b, then b is the reciprocal of a. The product of a number and its reciprocal is 1.

Reconcile: To make two accounts match; specifically, the process of making one's personal records match the latest records issued by a bank or financial institution.

Rectangle: A quadrilateral whose angles are all right angles. The opposite sides of a rectangle are parallel and equal in length. Any side can be chosen as the base and the altitude is the length of a perpendicular line segment between the base and the opposite side. A diagonal is either of the line segments joining opposite vertices.

Reflection: The operation of moving all the points to an equal distance, on the opposite side of a line of reflection.

Register: A record of spending, such as a check register, which is used to track checks written for later reconciliation.

Root: The solutions of a polynomial equation, of which the square and cube root are special cases.

Rotation: The operation of moving all the points of an object through a fixed angle around a fixed point.

Scale: The ratio of the size of an object to the size of its representation.

Scientific notation: A shorthand way to write very large or very small numbers.

Segment: A portion truncated from a geometric figure by one or more points, lines, or planes; the finite part of a line bounded by two points in the line.

Set: A collection of elements.

Simple random sampling: A sampling method that provides every element equal chance of being selected.

Statistics: Branch of mathematics devoted to the collection, compilation, display, and interpretation of numerical data. In general, the field can be divided into two major subgroups, descriptive statistics and inferential statistics. The former subject deals primarily with the accumulation and presentation of numerical data, while the latter focuses on predictions.

Stockholder: The partial owner of a public corporation, whose ownership is contained in one or more shares of stock. Also called a shareholder.

Stratified sampling: In this type of random sampling, elements are grouped together before sampling.

Symmetric key system: A cryptographic algorithm that uses the same key for encryption and decryption.

Symmetry: An object that is left unchanged by an operation has a symmetry.

Symmetry, or balance: A design is symmetrical if its two opposite sides divided by a line in the center are identical, or nearly identical.

System of equations: A group of equations that all involve the same variables.

Systematic sampling: In this type of sampling, there are intervals between each selection for sampling.

Term: A number, variable, or product of numbers and variables, separated in an equation by the signs of addition and equality.

Translation: The operation of moving each point a fixed distance in the same direction.

Trigonometry: A branch of applied mathematics concerned with the relationship between angles and their sides and the calculations based on them. First developed as a branch of geometry focusing on triangles during the third century B.C, trigonometry was used extensively for astronomical measurements. The major trigonometric functions, including sine, cosine, and tangent, were first defined as ratios of sides in a right triangle.

Unit fraction: A fraction with 1 in the numerator.

Vanishing point: In art, the place on the horizon toward which all other lines converge; a focus point.

Variable: A symbol representing a quantity that may assume any value within a predefined range.

Vector: A quantity consisting of magnitude and direction, usually represented by an arrow whose length represents the magnitude and whose orientation in space represents the direction.

Volume: The amount of space occupied by a three-dimensional object as measured in cubic units.

Whole number: Any positive number, including zero, with no fraction or decimal.

Zero: The absence of a quality (normalizing) or absence of quantity (numerical). It can also be a reference point, such as 0° on a temperature scale. In a mathematical system, zero is the additive identity. It is a number that can be added to any given number to yield a sum equal to the given number. Symbolically, it is a number 0, such that $a + 0 = a$ for any number a.

Zero-sum game: An outcome of a game where players choices have produced neither a win or a draw for all of the players.

Field of Application Index

An *italicized* page number indicates a photo, figure (f), illustration or table (t).

A

B

C

E

G

H

M

N

P

S

T

General Index

A **boldface** page number indicates the main essay for a topic. An *italicized* page number indicates a photo, figure (f), illustration, or table (t).

B

C

E

F

G

H

O

P

Q

R

S

T